25.00

Laboratory Text and Study Guide

Essentials of
Anatomy & Physiology

DATE DUE

Laboratory Text and Study Guide

Michael G. Wood, M.S.
Del Mar College, Texas

Essentials of
Anatomy & Physiology

Prentice Hall, Upper Saddle River, New Jersey 07458

Safety Notification

Some of the exercises in this laboratory text involve the use of chemicals, heat, glass, sharp objects, and body tissues and fluids. If handled or used improperly, these materials may be hazardous. Safety precautions are outlined in Exercise One of the text and should be followed by students at all times. Your school may have additional rules and regulations pertaining to laboratory safety. Always ask your laboratory instructor if you are unsure of a procedure.

Production Editor: *Naomi Sysak*
Supplement Acquisitions Editor: *Linda Schreiber*
Executive Editor: *David K. Brake*
Manufacturing Buyer: *Ben Smith*

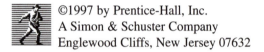
©1997 by Prentice-Hall, Inc.
A Simon & Schuster Company
Englewood Cliffs, New Jersey 07632

Printed in the United States of America

10 9 8 7 6 5 4 3 2 1

ISBN 0-13-255670-7

Prentice-Hall International (UK) Limited, *London*
Prentice-Hall of Australia Pty. Limited, *Sydney*
Prentice-Hall Canada Inc., *Toronto*
Prentice-Hall Hispanoamericana, S.A., *Mexico*
Prentice-Hall of India Private Limited, *New Delhi*
Prentice-Hall of Japan, Inc., *Tokyo*
Simon & Schuster Asia Pte. Ltd., *Singapore*
Editora Prentice-Hall do Brasil, Ltda., *Rio de Janeiro*

This laboratory text is dedicated to my wonderful family—my wife Laurie, and daughters Abi and Beth. Each has made sacrifices during this extended writing project, and I deeply thank them for their encouragement and devotion.

Contents

Preface

This text laboratory text and study guide is designed for students enrolled in an introductory anatomy and physiology course. Although this text directly accompanies the first edition of *Essentials of Anatomy and Physiology* by Frederic Martini, it also integrates with other introductory text books. The text accurately presents background material and laboratory procedures for each exercise. Self reference provide students relevancy and perspective of the material. The laboratory text is designed around the following elements:

ORGANIZATION

The laboratory text is arranged into twenty exercises that correspond to most essentials text books. Longer exercises are segmented into parts with unique student objectives, a terminology guide, materials list, and laboratory activities. The modular approach enables faculty to easily assign specific sections of an exercise. Dividing the exercises into self-contained segments assists students in planning and helps to gauge their progress.

STUDENT ACTIVITIES

The text and study guide encourages students to become more involved with their learning through labeling figures, recording observations, enhancing language skills, and completing laboratory results. Completion of all activities will help students arrive at a better understanding of human form and function.

STUDENT OBJECTIVES

Each exercise includes objectives to identify major themes students should master during this exercise. Each part of longer exercises has specific student objectives. The laboratory reports at the end of each exercise test for mastery of each objective with a variety of labeling and written activities.

WORD POWER

A major challenge to students of anatomy and physiology is the technical terminology of science. To foster language skills, each exercise and section has a list of relevant prefixes, suffixes, and root words. The manual's emphasis on writing bolsters proficiency with the terminology.

MATERIALS

A list of required laboratory equipment, models, prepared microscope slides, and other materials is included at the start of each exercise and section. Faculty will find the lists useful in preparing the laboratory. Students can use the lists to determine the materials necessary to accomplish the laboratory activities.

LABORATORY ACTIVITIES

Laboratory observations and written exercises follow each section of specific information. These activities provide concise instructions and procedures to guide students through the exercise. Reference to laboratory models, microscopic materials, and text figures clearly outlines the assignments required to master the student objectives.

ART PROGRAM

Illustrations and photographs have been carefully selected to parallel the descriptive text. Figures include reference icons, accurate labeling, and a macro to micro approach to guide students from familiar to detailed structures. Many figures incorporate written labeling assignments.

LABORATORY REPORTS

Comprehensive laboratory reports are located at the conclusion of each exercise. The reports conform to the sectional approach of the exercises and they include a variety of tasks to reinforce the laboratory objectives. Instructors may utilize the reports for out-of-class assignments.

STUDY GUIDE

Many sections are designed to augment lecture concepts through directed review questions and labeling activities. Students enrolled in a lecture-based course may use the text as a study guide to enhance their studies of anatomy and physiology.

Acknowledgments

I thank David K. Brake, Executive Editor, for introducing my writing to Prentice Hall; Linda Schreiber, Supplement Acquisitions Editor, for managing the enormous details of the project; and Naomi Sysak, Production Editor, for her superb design and layout.

This *Laboratory Text and Study Guide* is derived from a full version manual currently in development. Anil Rao from Metropolitan State College of Denver, and Joe Wheeler at North Lake College, made manuscript contributions to the full version from which this manuscript originated. Laboratory reports include materials from Charles Seiger's *Study Guide*, a supplement to *Fundamentals of Anatomy and Physiology* by Fredric Martini. Gerardo Cobarruvias from Del Mar College provided many outstanding photographs.

Many reviewers devoted hours proofreading the manuscript drafts. Their insightful suggestions helped to compose a better laboratory text.

Reviewers for the essentials text include

Karen Jones, *Pitt Community College*

Frank Schwartz, *Cuyahogo Community College*

Caryl Tickner, *Stark Technical College*

Reviewers for the full version manuscript were

Linda Banta, *Sierra College*

Barton Bergquist, *University of Northern Iowa*

William M. Clark, *Kingwood College*

Kim Cooper, *Arizona State University*

David Evans, *Penn College*

Ruby Fogg, *New Hampshire Technical Institute*

Angie Huxley, *Pima Community College*

Martha Newsome, *Tomball College*

Karen Roy, *Los Angeles Valley College*

Sandra Steingraber, *Columbia College*

Frank Veselovsky, *South Puget Sound Community College*

Any errors or omissions in this first edition work are my responsibility and are not a reflection of the editorial and review team. Comments from faculty and students are welcomed and may be directed to me at the address below. I will consider each suggestion in the preparation of subsequent editions.

Michael G. Wood
c/o Prentice Hall
1208 E. Broadway, Suite 200
Tempe, AZ 85282

1 LABORATORY SAFETY

OBJECTIVES

On completion of this exercise, you should be able to:

- Locate all safety equipment in the laboratory.
- Show how to safely handle glassware, including insertion and removal of glass rods into stoppers.
- Demonstrate how to safely clean up and dispose of broken glass.
- Demonstrate how to safely plug and unplug electrical devices.
- Explain how to protect yourself from and dispose of body fluids.
- Explain how to mix solutions and measure chemicals safely.
- Describe the potential dangers of each lab instrument.
- Discuss the disposal techniques for glass, chemicals, body fluids, and other hazardous materials.

Experiments and exercises in the anatomy and physiology laboratory are, by design, safe. Some of the hazards are identical to risks found in your home such as broken glass and the chance of electrical shock. The major hazards can be grouped into the following categories: electrical, chemical, body fluid, preservatives, and instrumentation. The following is a discussion of hazards each category poses and a listing of safety guidelines students should follow to prevent injury to themselves and others while in the lab. Proper disposal of biological and chemical wastes ensures these contaminants are not released into your local environment.

A. Laboratory Safety Rules

The following guidelines are necessary to ensure the laboratory is a safe environment for students and faculty alike.

1. No unauthorized persons are allowed in the laboratory. Only students enrolled in the course and faculty are to enter the laboratory.
2. No unauthorized experiments or deviations of experiments in this manual or class handouts are allowed without the faculty's consent.
3. No smoking, eating, gum chewing or drinking is allowed in the laboratory.
4. Always wash your hands before and after each laboratory exercise involving chemicals, preserved materials, or body fluids and after cleaning up spills.
5. Shoes must always be worn while in the laboratory.
6. Be alert to unsafe conditions and actions of other individuals in the laboratory. Call attention to those activities. Someone else's accident can be as dangerous to you as an accident you may have.
7. Glass rods called pipettes are commonly used to measure and transfer solutions. Never fill a pipette by drawing on it with your mouth. Always use a pipette bulb.
8. Immediately report all spills and injuries to the laboratory faculty.

9. You must inform the laboratory faculty of any medical condition that may limit your activities in laboratory.

B. Location of Safety Equipment

Write the location of each piece of safety equipment as your instructor explains how and when to use each item.

first aid kit _____.

nearest telephone _____.

fire exits _____.

fire extinguisher _____.

eye wash station _____.

chemical spill kit _____.

fan switches _____.

biohazard container _____.

C. Glassware

Glassware is perhaps the most dangerous item in the laboratory. Broken glass has the sharpest edges known, and it must be cleaned up and disposed of safely. Other glassware-related accidents occur when a glass rod or tube is broken while attempting to insert it into a cork or stopper.

BROKEN GLASS

- Broken glass should be swept up immediately. Never use your hands to pick up broken glass. A whisk broom and a dust pan should be used to sweep the area of all glass shards.
- All broken glass and other sharp objects should be packaged in a box that will contain the small fragments. Tape the box closed, and write "**BROKEN GLASS INSIDE**" in large letters across the box. This will alert custodians and other waste collectors of the hazard inside. Your lab instructor will arrange for the disposal of the sealed box.

INSERTING GLASS INTO A STOPPER

- Never force a glass rod, tube, or pipette into a cork or rubber stopper. Use a lubricant such as glycerine or soapy water to ease the glass through the stopper.
- Always push on a glass tube or rod near the stopper you are trying to insert the glass through. This reduces the length of glass between the stopper and your hand and greatly reduces the chance of breaking the rod and jamming glass into your hand.

D. Electrical Equipment

The electrical hazards in the laboratory are similar to the hazards in your house. A few commonsense guidelines will almost eliminate the risk of electrical shock.

- Unravel electrical cords completely before plugging the cord into an electrical outlet. Electrical cords are often wrapped tightly around the base of microscopes, and users often unravel just enough cord to plug the microscope in. Moving the focusing mechanism of the microscope may pinch the electrical cord and shock the user. Inspect the cord for frays and the plug for secure connections.
- Do not force an electrical plug into an outlet. If the plug does not easily fit into the outlet, inform your laboratory instructor.

- Unplug all electrical cords by pulling on the plug, not by tugging on the cord. Many electrical accidents occur by users pulling the cord out of the plug, leaving the plug in the outlet.
- Never plug or unplug an electrical device if the area is wet.

E. Body Fluids

Three body fluids are most frequently used in the study of anatomy of physiology: saliva, urine, and blood. Because body fluids can harbor infectious organisms, safe handling and disposal procedures must be followed to prevent infecting yourself and others.

- Work only with your body fluids. It is beyond the scope of this course to collect and experiment on body fluids from another individual.
- Never allow a body fluid to touch your unprotected skin. Always wear gloves and safety glasses when working with body fluids—even when using your own fluids.
- Always assume that a body fluid can infect you with a disease. This extreme measure prepares you for working in a clinical setting where you may be responsible for handling body fluids from the general population.
- Clean all body fluid spills with a 10% bleach solution or a commercially prepared disinfectant labeled for this purpose. Always wear gloves during the clean up and dispose of contaminated towels and rags in a biohazard container.

F. Chemical Hazards

Most chemicals used in laboratories are safe. Following a few simple guidelines will protect you from chemical hazards.

- Read the label of the chemical you are going to work with. Be aware of chemicals that may irritate your skin or stain your clothing. Chemical containers are usually labeled to show contents and potential hazards related to the contents. Most laboratories or chemical stockroom managers keep copies of technical chemical specifications called Material Safety Data Sheets (M.S.D.S.). These publications from the chemical manufacturer detail the proper usage of the chemical and known adverse effects the chemical may cause. All individuals have a federal right to inspect these documents. Ask your laboratory instructor for more information about the M.S.D.S.
- Never touch a chemical with your unprotected hands. Wear gloves and safety glasses when weighing and measuring chemicals and during all experimental procedures involving chemicals.
- Always use an appropriate spoon or spatula for retrieving dry chemical samples from a large storage container. Do not shake chemicals out of the jar. This may result in dumping the entire container of chemical onto yourself and your workstation.
- If pipetting is required, always use a pipette bulb and never pipette by mouth. Your instructor will explain how to use the particular type of pipette bulb available in the laboratory.
- When pouring acids and other solutions in a large container, always pour the approximate amount required into a smaller beaker first. Use this smaller container to fill your glassware with the solution. Attempting to pour from the large storage container may result in spills that come in contact with skin and clothing.
- Do not return unused portions of a chemical to its original container. This will prevent a possible contamination of the chemicals in the original container. Dispose of the excess chemical as directed by your laboratory instructor. Do not pour unused or used chemicals down the sink unless directed by your laboratory instructor.
- When mixing solutions, always add a chemical to water, never add water to the chemical. By diluting chemicals in water, you reduce the chance of a strong chemical reaction from occurring.

G. Instrumentation

You will use a variety of scientific instrumentation in the laboratory. Safety guidelines for specific instruments are included in the appropriate exercise. This discussion is for instruments most frequently used in laboratory exercises.

MICROSCOPES

The microscope is the main instrument you will use in the study of anatomy. Exercise 3 of this manual is devoted to use and care of the microscope.

DISSECTION TOOLS

Working with sharp blades and points always presents the possibility of a cutting injury. Remember to always cut away from yourself and not to force a blade through a tissue. Use small knife strokes for increased blade control rather than large cutting motions with the blade. Always use a sharp blade and dispose of used blades in a specially designated "Sharps" container. Carefully wash and dry all instruments on completion of each dissection.

Special care is necessary while changing disposable scalpel blades. Your laboratory instructor may demonstrate the proper technique for blade replacement. Wash the used blade before you attempt to remove it from the handle. Examine the handle and blade, and decide how the blade fits onto the handle. Do not force the blade during the removal process. If you experience difficulty while changing blades, do not hesitate to ask your instructor for assistance. Discard the used blade in the appropriate container for sharp objects.

WATER BATHS

Water baths are used to incubate laboratory samples at a specific temperature. Potential hazards involving water baths include electrical shock due to contact with water, and burn-related injuries from touching hot surfaces or from spilling hot solutions. **Electrical hazards** are minimized by following the safety rules concerning plugging and unplugging electrical devices. **Burns** are reduced by using tongs to immerse or remove samples from a water bath. Point all glassware with samples away from yourself and others. If the sample boils, it could splatter out and burn your skin. Use racks in the water bath to support all glassware and place hot samples removed from a water bath on a cooling rack. Monitor the temperature and water level of all water baths. Excessively high temperatures increase the chance of burns and will usually ruin an experiment. With boiling water baths, add water frequently and do not allow all the water to evaporate.

MICROCENTRIFUGE

The microcentrifuge is used in blood and urine analyses. The instrument spins at speeds exceeding thousands of revolutions per minute. Although the moving parts are housed in a protective casing, it is important to keep all loose hair, clothing, and jewelry away from the instrument. Never open the safety lid while the centrifuge is on or spinning. *Do not* attempt to stop a spinning centrifuge with your hand. The instrument has an internal braking mechanism that will quickly stop the centrifuge for you.

H. Handling and Disposal of Preserved Specimens

Most animal and tissue specimens used in the laboratory have been treated with chemicals to prevent decay. These preservatives are irritants and should not contact your skin or your mucous membranes (linings of the eyes, nose, and mouth, and urinary, digestive, and reproductive openings). The following guidelines will protect you from these hazards:

- If you are pregnant, limit your exposure to all preservatives. Discuss the laboratory exercise with your lab instructor. Perhaps you can observe rather than perform the dissection.
- Always wear gloves and safety glasses when working with preserved material.
- Your laboratory may be equipped with exhaust fans to ventilate preservative fumes during dissection exercises. Do not hesitate to ask your lab instructor to turn on the fans if the odor of the preservatives becomes bothersome.
- Many preservatives are toxic and require special handling techniques. Drain the preservative out of the specimen before the dissection or observation. Pour the fluid into the specimen's storage container or into a container provided by your lab instructor. Do not pour the preservative down the drain.
- Promptly wipe up all spills and clean your work area when you have completed the dissection. Keep your gloves on during the clean up, and dispose of gloves and paper towels in the proper biohazard container.
- Do not dispose of biological tissues or specimens in the general trash. All preserved materials must be placed in a biohazard container for proper disposal. Specimens should be wrapped in large plastic bags that are filled with an absorbent material such as kitty litter. All sealed bags are placed in a large cardboard box for pickup by a hazardous waste company, that will incinerate the specimens.
- To dispose of preservatives, such as formalin, a central storage container is maintained until collected by a waste management company. Formalin solutions contain formaldehyde, a cancer-causing agent, and must be handled by trained individuals. Under no circumstances should preservatives be discarded down the drain.

I.   Disposal of Chemical and Biological Wastes

To safeguard the environment and individuals employed in waste collection, it is important to dispose of all potentially hazardous wastes in specially designed containers. State and federal guidelines detail the storage and handling procedures for chemical and biological wastes. Your lab instructor will manage the wastes produced in this course.

CHEMICAL WASTE

Most chemicals used in undergraduate laboratories are harmless and may be diluted in water and disposed of by pouring down the drain. Your lab instructor will show during each lab that chemicals may be discarded in this manner. Other chemicals should be stored in a waste container until a waste disposal company is notified to collect the waste.

BODY FLUIDS

Objects contaminated with body fluids are considered a high-risk biohazard and must be disposed of properly. Special "BioHazard" containers will be available during labs that use body fluids. For example, during the lab on blood typing, all used blood lances, slides, alcohol wipes, gloves, and paper towels will be disposed of in a clearly marked biohazard container. A special biohazard "Sharps" container may be provided to dispose of all sharp objects such as glass, needles, and lancets.

When full, biohazard containers and their contents are sterilized by autoclaving at a high temperature and pressure to kill all pathogens associated with the body fluids. The biohazard container is then disposed of according to state and federal regulations (usually incinerated by a waste disposal company).

Name _____   Date _____

Section _____   Number _____

LABORATORY SAFETY
Exercise 1 Laboratory Report

A. Discussion

1. Discuss how to protect yourself from body fluids such as saliva and blood.

2. Why should you consider a body fluid capable of infecting you with a disease?

3. Describe how to dispose of materials contaminated with body fluids.

4. Explain how to safely plug and unplug an electrical device.

5. Discuss how to protect yourself from preservatives used on biological specimens.

6. Why are special biohazard containers used for biological wastes?

7. Explain how to clean up broken glass.

8. List the location of the following safety items in the laboratory:
   first aid kit _____.
   nearest telephone _____.
   fire exits _____.
   fire extinguisher _____.
   eye wash station _____.
   chemical spill kit _____.
   fan switches _____.
   biohazard container _____.

9. Your instructor informs you that a chemical is not dangerous. How may you dispose of the chemical?

10. What precautions should you take while using a centrifuge?

11. How are preservatives such as formalin correctly discarded?

12. Discuss how to safely measure and mix chemicals.

2 INTRODUCTION TO THE HUMAN BODY

WORD POWER

superior (super—over, above)
inferior (infer—low, underneath)
anterior (anter—in front of)
posterior (post—behind)
dorsum (dors—the back)
ventral (vent—underside, belly)
cephalic (cepha—the head)

caudal (caud—the tail)
medial (medi—the middle)
lateral (later—the side)
proximal (proxim—near)
distal (dist—distance)
transverse (trans—across, through)

MATERIALS

torso model
sectioned objects

Knowledge of what lies beneath the skin and how it works has been slowly amassed over a span of nearly 3,000 years. Obviously, any logical practice of medicine depends on an accurate knowledge of human anatomy. Yet people have not always realized this. Through most of human history, corpses have been viewed with superstitious awe and dread. Observations of anatomy by dissection were illegal, and medicine therefore remained an elusive practice that often harmed rather than cured the unfortunate patient. Yet, despite superstitions and prohibitions, there have been, throughout the ages, scientists who desperately wanted to know the human body as it really is, rather than how it was imagined to be.

Study Tip

To benefit more from your laboratory studies, prepare for each laboratory meeting. Before class, read the appropriate chapter(s) in this manual, and complete the labeling of as many figures as possible. Then, relate the laboratory exercises with the theory concepts in the lecture textbook. Approaching the laboratory in this manner will maximize your hands-on time with laboratory materials.

PART ONE   INTRODUCTION TO ANATOMY AND PHYSIOLOGY

OBJECTIVES

On completion of this part of the exercise, you should be able to:

- Define anatomy and physiology and discuss the specializations of each.
- Describe each level of organization of the body.
- Describe anatomical position and its importance in anatomical studies.
- Use directional terminology to describe the relations of your surface anatomy.
- Describe and identify the major planes and sections of the body.
- Locate all abdominopelvic quadrants and regions on laboratory models.

- Locate the organs of each organ system on laboratory models.
- State the basic functions of each organ system.

A. Organization of the Body

Anatomy is the study of body structures. Early anatomists described the body's **gross anatomy**, the large parts such as muscles and bones. As knowledge of the body advanced and scientific tools enabled more detailed observations, the field of anatomy began to diversify. For example, **microanatomy** is the study of microscopic structures. **Cytology** is the study of cells, and **histology** is the study of tissues.

Physiology is the study of organ function. As in anatomy, a physiologist may investigate organ functions from a molecular, cellular, histological, or more complex level. Physiology may be regarded as the work that cells must do to keep the body stable and operating efficiently. **Homeostasis** is the maintenance of a relatively steady internal environment through physiological work. Stress, inadequate diet, and diseases disrupt the normal physiological processes and may lead to serious health problems or death of the individual.

Anatomists and physiologists study the body from many different levels of structural and functional detail. These **levels of organization** are reflected in the fields of specialization in anatomy and physiology. Figure 2.1 illustrates the levels of organization of the cardiovascular system. Each higher level increases in structural and functional complexity, progressing from chemicals, to cells, tissues, organs, and organ systems that function to maintain the organism.

The smallest level of organization is the **chemical and molecular level**. Atoms, such as carbon and hydrogen, bond together and form molecules. Molecules are organized into cellular structures, called organelles, which have distinct shapes and functions. The organelles collectively compose the next level of organization, the **cellular level**. The cellular level is the fundamental level of biological organization because it is cells, not molecules, that are alive. Living organisms can reproduce and grow, to move and to control cellular chemical reactions to obtain useful energy and process wastes. Different types of cells working together constitute the **tissue level**. Although tissues lack a specific shape, they are distinguishable by the variety of cell types. Tissues combine at the **organ level**. Each organ has a distinct three-dimensional shape and a broader range of functions than individual cells or tissues. The **organ system level** includes all the organs that work together to accomplish a common goal. All organ systems make up the individual or the **organism level**.

Laboratory Activity   LEVELS OF ORGANIZATION _____

MATERIALS

A variety of objects to represent each level of organization (charts, models, cell and tissue specimens, and so on).

PROCEDURES

Classify each object set on display by your laboratory instructor to the level of organization it represents. Write your answers in the space provided.

- Chemical and molecular level_____
- Cellular level _____
- Tissue level _____
- Organ level_____
- Organ system level_____
- Organism level _____

●Figure 2.1
Levels of Organization

Interacting atoms form molecules that combine in the protein fibers of heart muscle cells. These cells interlock, creating heart muscle tissue that constitutes most of the walls of a three-dimensional organ, the heart. The heart is one component of the cardiovascular system, which also includes the blood and blood vessels. All of the organ systems combine to create an organism, a living human being.

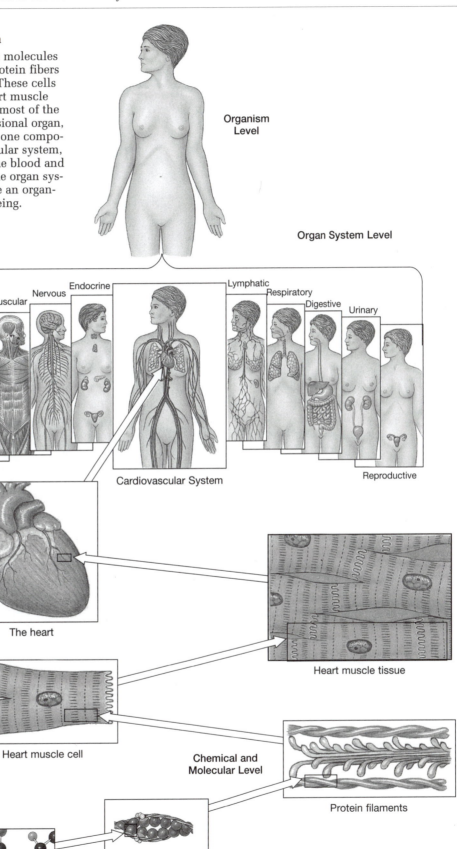

Organism Level

Organ System Level

Skeletal   Muscular   Nervous   Endocrine   Lymphatic   Respiratory   Digestive   Urinary

Integumentary   Cardiovascular System   Reproductive

The heart

Heart muscle tissue

Heart muscle cell

Chemical and Molecular Level

Protein filaments

Atoms in combination

B. Anatomical Position

The human body can bend and stretch in a variety of directions. Although this flexibility allows us to move and manipulate our environment, it can cause ambiguity when describing and comparing structures. For example, what is the correct relation between the wrist and the elbow? If your arm is raised above your head, you might reply that the wrist is above the elbow. With your arms at your sides, you would respond that the wrist is below the elbow. Each response appears correct, yet which is the proper anatomical relation?

To avoid confusion, the body is always referred to in a universal position called **anatomical position**. In anatomical position, the individual is standing erect with the feet pointed forward, eyes straight ahead, and the palms of the hands facing forward with the arms at the sides (see Figure 2.5). Stand up and assume the anatomical position. The entire front of your body should be oriented forward. Notice that the position is not the natural posture of the body. The arms and hands have rotated forward to bring the forearm and palms to the front. While lying on the back in anatomical position, the individual is said to be **supine**; when lying face down, **prone**.

When observing a person in anatomical position, his or her right is on your left, and his or her left is on your right. For example, to shake someone's hand, your right hand must cross over to meet his or her right hand. The positioning of right and left is important to remember when viewing structures, such as the heart, which differs structurally between right and left sides.

C. Directional Terminology

Imagine attempting to give someone directions if you could not use terms like "north" and "south" or "left" and "right." These words have a unique meaning and guide the traveler to the correct location. Describing the body also requires specific terminology. Words such as "near," "close to," "around," and "on top of" are too vague for anatomical descriptions. To prevent misunderstandings, concise terms are used to describe the location and associations of anatomy. These terms have their roots in Greek and Latin languages. Figure 2.2 and Table 2.1 summarize the most frequently used directional terms. Notice that most of the directional terms may be grouped into opposing pairs or antonyms.

Students often use directional terms interchangeably; however, there is usually a specific term for the desired description. For example, "superior" and "proximal" both describe the upper region of arm and leg bones. If you are discussing the point of attachment of a bone, the term *proximal* is the most descriptive. When comparing the location of a bone to an inferior bone, the term *superior* is used.

In four-legged animals such as cats, anatomical position is standing with all four limbs on the ground. The trunk of the body is now horizontal to the ground rather than vertical as in humans; therefore, the meanings of directional terms change. "Superior" now refers to the back or **dorsal** surface, whereas "inferior" relates to the belly or **ventral** surface. **Cephalic** means toward the front or anterior, and **caudal** refers to posterior structures.

Laboratory Activity   DIRECTIONAL TERMINOLOGY_____

MATERIALS

torso models, charts, anatomical models

PROCEDURES

1. Review each directional term presented in Figure 2.2 and Table 2.1.
2. Use the available laboratory models and your body to practice using directional terms while comparing anatomy. The Laboratory Report at the end of this exercise may be used as a guide for comparisons.

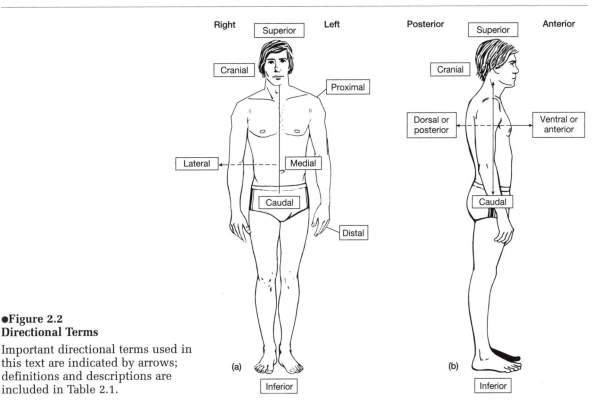

●Figure 2.2
Directional Terms

Important directional terms used in
this text are indicated by arrows;
definitions and descriptions are
included in Table 2.1.

D. Planes and Sections

To observe the body's internal organization, it must be cut or **sectioned**. All structures,
such as the body trunk, knee, arm, and eyeball can be sectioned. The imaginary line of
a section is called a **plane**. The orientation of the plane on the body will determine the
shape and appearance of the exposed internal anatomy. Imagine cutting a soda straw
crosswise (transversely) and another straw lengthwise (sagittally). The transverse section would produce a circle, and the sagittal section would produce a long U shape.

Three major types of sections are employed in the study of anatomy, two vertical
sections and one transverse section (see Figure 2.3). **Transverse** sections are perpendicular to the vertical orientation of the body, the body axis. These sections are often
called cross sections because they go across the body axis. Superior and inferior structures are divided by transverse sections.

Vertical sections are parallel to the axis of the body and include sagittal and frontal
sections. **Sagittal** sections divide parts into right and left portions. A **midsagittal** section
divides structures into equal left and right parts. **Parasagittal** sections are sagittal sections
lateral or medial to the midline. A **frontal** or coronal section separates anterior and
posterior structures.

Laboratory Activity   PLANES AND SECTIONS _____

MATERIALS

Sectioned anatomical models

PROCEDURES

1. Review each plane and section in Figure 2.3.
2. Identify the sections on models and other material presented by your lab instructor.
3. Section several common objects along their sagittal and transverse planes.
   Compare the resulting appearance of each sectioned object.

TABLE 2.1	**Directional Terms (see Figure 2.2)**	
Term	*Region or Reference*	*Example*
Anterior	The front; before	The navel is on the *anterior (ventral)* surface of the trunk.
Ventral	The belly side (equivalent to anterior when referring to human body)	
Posterior	The back; behind	The shoulder blade is located posterior (*dorsal*) to the rib cage.
Dorsal	The back (equivalent to posterior when referring to human body)	The *dorsal* body cavity encloses the brain and spinal cord.
Cranial or cephalic	The head	The *cranial*, or *cephalic*, border of the pelvis is superior to the thigh.
Superior	Above; at a higher level (in human body, toward the head)	
Caudal	The tail (coccyx in humans)	The hips are *caudal* to the waist.
Inferior	Below; at a lower level	The knees are *inferior* to the hips.
Medial	Toward the body's longitudinal axis	The *medial* surfaces of the thighs may be in contact; moving medially from the arm across the chest surface brings you to the sternum.
Lateral	Away from the body's longitudinal axis	The thigh articulates with the *lateral* surface of the pelvis; moving laterally from the nose brings you to the eyes.
Proximal	Toward an attached base	The thigh is *proximal* to the foot; moving proximally from the wrist brings you to the elbow.
Distal	Away from an attached base	The fingers are *distal* to the wrist; moving distally from the elbow brings you to the wrist.
Superficial	At, near, or relatively close to the body surface	The skin is *superficial* to underlying structures.
Deep	Farther from the body surface	The bone of the thigh is *deep* to the surrounding skeletal muscles.

E. Regional Terminology

Approaching the body from a regional perspective often simplifies the study of anatomy. Many internal structures are named after the overlying surface structures. For example, the knee is called the popliteal region, and the major artery in the knee is the popliteal artery. Surface features are also used as anatomical landmarks to locate internal structures. Figure 2.4 presents the major regions of the body. Use the following list of regional terms to label Figure 2.4:

HEAD, NECK, AND TRUNK

- **Cephalon**—the head; organized into the cranium and the face
- **Cranium**—the upper rounded part of the skull that contains most of the brain
- **Face**—the front of the skull containing the eyes, nose, mouth, and ears
- **Cervical**—the neck region
- **Trunk**—the chest and abdominal regions; the neck, arms, and legs attach to the trunk
- **Thorax**—the upper portion of the trunk or chest
- **Abdomen**—the trunk below the chest
- **Dorsum**—the back or posterior surface of the trunk
- **Flank**—the side of the trunk below the ribs
- **Loin**—the lower back below the ribs
- **Gluteal**—the buttocks

●Figure 2.3
Planes of Section

The three primary planes of section are indicated here. Table 2.1 defines and describes them.

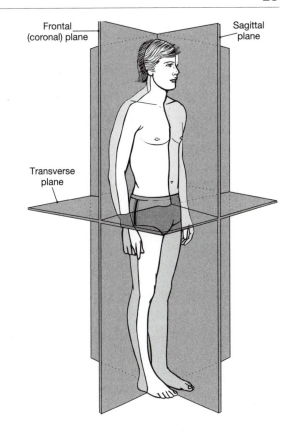

THE SHOULDER AND ARM

- **Scapula**—the shoulder
- **Brachium**—the arm
- **Axilla**—the armpit
- **Antebrachium**—the forearm
- **Antecubitis**—the elbow
- **Carpus**—the wrist
- **Palm**—the anterior surface of the hand
- **Digits**—the fingers and toes

THE PELVIS AND LEG

- **Pelvis**—the hips
- **Thigh**—the upper leg
- **Popliteal**—the knee
- **Calf**—the lower leg
- **Tarsus**—the ankle
- **Plantus**—the sole of the foot

The position of internal organs is simplified by partitioning the trunk into four equal **quadrants**. Observe in Figure 2.5a the midsagittal and transverse planes used to delineate the quadrants. Quadrants are used to describe the position of internal organs. The stomach, for example, is mostly located in the left upper quadrant.

For more detailed descriptions, the abdominal surface is divided into nine **abdomino-pelvic** regions, shown in Figure 2.5b. Four planes are used to define the regions, two vertical and two transverse planes arranged in the familiar "tic-tac-toe" pattern. The vertical planes, called the right and left **lateral planes**, are just medial to the nipples.

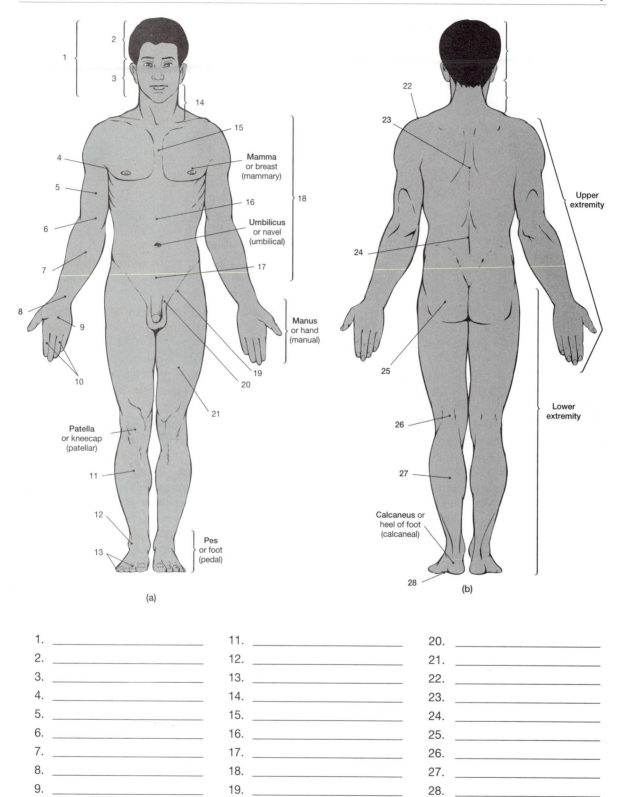

(a)

(b)

●Figure 2.4
Regional Terminology

1. _____
2. _____
3. _____
4. _____
5. _____
6. _____
7. _____
8. _____
9. _____
10. _____

11. _____
12. _____
13. _____
14. _____
15. _____
16. _____
17. _____
18. _____
19. _____

20. _____
21. _____
22. _____
23. _____
24. _____
25. _____
26. _____
27. _____
28. _____

They divide the trunk into three nearly equal vertical regions. A pair of **transverse** planes cross the vertical planes and separate the trunk horizontally.

The nine abdominopelvic regions are as follows. The **umbilical** region surrounds the umbilicus or navel. Lateral to this region are the right and left **lumbar** regions. Above the umbilical region is the epigastric region containing the stomach and much of the liver. The right and left **hypochondriac** regions are lateral to the **epigastric** region. Inferior to the umbilical region is the **hypogastric** or **pubis** region. The right and left **iliac** regions border the hypogastric region laterally.

Laboratory Activity   REGIONAL REFERENCES _____

MATERIALS

torso models
charts

PROCEDURES

1. Label and review the regional terminology in Figure 2.4.

2. Use a lab model or yourself and identify the regional anatomy as presented in Figure 2.4.

●**Figure 2.5**
Abdominopelvic Quadrants and Regions

(a) Abdominopelvic quadrants divide the area into four sections. These terms, or their abbreviations, are most often used in clinical discussions.
(b) More precise regional descriptions are provided by reference to the appropriate abdominopelvic region.
(c) Quadrants or regions are useful because there is a known relationship between superficial anatomical landmarks and underlying organs.

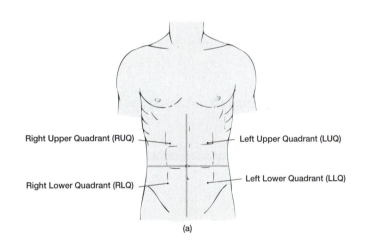

Right Upper Quadrant (RUQ)

Left Upper Quadrant (LUQ)

Left Lower Quadrant (LLQ)

Right Lower Quadrant (RLQ)

(a)

1. _____
2. _____
3. _____
4. _____
5. _____
6. _____
7. _____
8. _____
9. _____

(b)

Liver
Gallbladder
Stomach
Large intestine
Small intestine
Appendix
Spleen

(c)

3. Identify each abdominopelvic quadrant and region in Figures 2.5a and 2.5b and on laboratory models.

F. **Introduction to Organ Systems**

The human body comprises 11 organ systems, each responsible for a specific function. Most anatomy and physiology courses are designed to progress through the lower levels of organization first and then examine each organ system. Because organ systems work together to maintain the organism, it is important that you have an understanding of the function of each organ system. Examine Table 2.2, an introduction to organ systems, and learn the major organs and the basic function of each organ system.

Laboratory Activity   IDENTIFICATION OF ORGAN SYSTEMS _____

MATERIALS

torso model with internal organs
charts

PROCEDURES

1. Locate the principal organs of each organ system on lab models.
2. On your body, identify the general location of as many organs as possible.

PART TWO   BODY CAVITIES AND MEMBRANES

WORD POWER	**MATERIALS**
pericardium (peri—around) pleural (pleura—lung)	torso model anatomical charts articulated skeleton heart model, lung model intestinal model

OBJECTIVES

On completion of this part of the exercise, you should be able to:

- Describe and identify the location of the dorsal cavity and its divisions.
- Describe and identify the location of the ventral cavity and its divisions.
- Describe and identify the serous membranes of the body.

Rub your hand rapidly against a desk top for 10 to 15 seconds. What did you feel—heat, friction? Consider what would happen to your hand if you rubbed it like this for an entire day. Now put your hand over your heart and feel it move in your chest. Your heart beats tens of thousands of times daily yet does not suffer from friction and abrasion. Why not? In this laboratory section, you will study the galleries where your organs hang, the internal body cavities, and the surrounding protective membranes.

Body cavities are internal spaces that house organs. These spaces are not, however, empty chambers in the body. Each cavity surrounds an organ as a small area between the body wall and the organ. The walls of a body cavity support and protect the internal anatomy. In addition, most internal organs are lined with special double-layered **serous**

TABLE 2.2	An Introduction to Organ Systems
Organ System	*Major Functions*
Integumentary system	Protection from environmental hazards, temperature control
Skeletal system	Support, protection of soft tissues, mineral storage, blood formation
Muscular system	Locomotion, support, heat production
Nervous system	Directing immediate responses to stimuli, usually by coordinating the activities of other organ systems
Endocrine system	Directing long-term changes in the activities of other organ systems
Cardiovascular system	Internal transportation of cells and dissolved materials, including nutrients, wastes, and gases
Lymphatic system	Defense against infection and disease
Respiratory system	Delivery of air to sites where gas exchange can occur between the air and circulating blood
Digstive system	Processing of food and absorption of nutrients, minerals, vitamins, and water
Urinary system	Elimination of excess water, salts, and waste products
Reproductive system	Production of sex cells and hormones

membranes that isolate the organ from surrounding structures and reduce friction and abrasion on the organ surface.

A. Body Cavities

The body has two major body cavities, the dorsal cavity and the ventral cavity, which contain internal organs. The dorsal cavity houses the brain and spinal cord in fluid-filled spaces protected by bone. The ventral cavity contains the organs of the chest, abdomen, and pelvis.

THE DORSAL CAVITY

The **dorsal body cavity** contains the central nervous system, the brain and spinal cord. It is subdivided into the **cranial** and the **spinal cavities**, shown in Figure 2.6. The cranial cavity is the space inside the skull where the brain is found. This cavity is formed by bones of the skull that protect the delicate brain. The spinal cavity is a canal that passes through the spine or vertebral column. The spinal cord occurs in the spinal cavity. The vertebrae have holes called vertebral foramina that collectively form the walls of the spinal cavity. The cranial and spinal cavities are interconnected at the base of the skull where the spinal cord joins the brain.

THE VENTRAL BODY CAVITY

The ventral body cavity is the entire space of the body trunk anterior to the vertebral column and posterior to the sternum (breast bone) and the abdominal muscle wall. Using Figure 2.6 as a guide, trace the outline of the ventral body cavity on the anterior surface of your body. This large cavity is divided into two major cavities, the **thoracic cavity** and the **abdominopelvic cavity**. These cavities, in turn, are further subdivided into the specific cavities that surround individual organs. The thoracic cavity, or chest cavity, contains the heart, lungs, trachea, larynx, esophagus, thymus gland, and many large blood vessels. Walls of the thoracic cavity are muscle and bone. The **pericardial cavity** surrounds the heart. Each lung is contained within a **pleural cavity**. The middle of the thoracic cavity contains a space called the **mediastinum** which contains the heart, large vessels of the heart, the thymus gland, and the trachea and esophagus. The lungs are located outside the mediastinum.

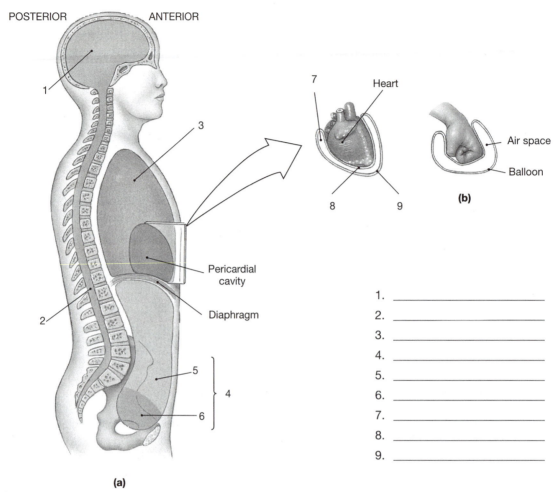

POSTERIOR   ANTERIOR

7   Heart

Air space

Balloon

(b)

8   9

Pericardial
cavity

Diaphragm

1. _____
2. _____
3. _____
4. _____
5. _____
6. _____
7. _____
8. _____
9. _____

(a)

●**Figure 2.6**
Body Cavities

(a) The dorsal body cavity is bounded by the bones of the skull and vertebral column.
The muscular diagram divides the ventral cavity into a superior thoracic (chest) cavity
and an inferior abdominopelvic cavity. The pericardial cavity is located inside the chest
cavity. **(b)** The heart is suspended within the pericardial cavity like a fist pushed into a
balloon. The attachment site, corresponding to the wrist of the hand in the model, lies
at the connection between the heart and major blood vessels.

The **abdominopelvic cavity** is separated from the thoracic cavity by a dome-shaped
muscle, the diaphragm. Locate the approximate position of the diaphragm on your
anterior surface and in the various views of Figure 2.6. The cavity is the space between
the diaphragm and the floor of the pelvis. This cavity is subdivided into the abdominal
cavity and the pelvic cavity. The **abdominal cavity** contains most of the digestive system
organs such as the liver, gallbladder, stomach, pancreas, kidneys, and small and large
intestine. The **pelvic cavity** is the small cavity enclosed by the pelvic girdle of the hips.
This cavity contains internal reproductive organs, parts of the large intestine, the rectum,
and urinary bladder.

Laboratory Activity   BODY CAVITIES_____

MATERIALS

torso model   articulated skeleton
laboratory models   charts

PROCEDURES

1. Label the various body cavities in Figure 2.6.
2. Locate the subdivisions of the dorsal body cavity on your body and on the available laboratory models.
3. Locate each body cavity on models, charts, and skeleton. With your finger, trace the location of each ventral body cavity on the anterior of your body trunk.
4. Identify the organ(s) of each body cavity on the laboratory models.

B. Serous Membranes

The heart, lungs, and intestines are encased in specialized **serous membranes**, double membranes comprising two layers with a lubricating fluid between. Directly attached to the exposed surface of an internal organ is the **visceral layer** (VIS-er-al; organ in body cavity) of the serous membrane. The **parietal layer** (pa-RI-e-tal; wall) is superficial to the visceral layer and lines the wall of the body cavity. Between these layers is the organ's cavity. This minute space contains a slippery lubricant called **serous fluid**.

Figure 2.6a highlights the anatomy of the serous membrane of the heart, the **pericardium**. It is composed of an outer **parietal pericardium**, a fibrous sac that anchors the heart in position and prevents overexpansion of the heart. The **visceral pericardium** is attached to the surface of the heart. The **pericardial cavity** is between the serous layers. Imagine pushing your fist into a water balloon; your fist is the heart and the balloon is the serous membrane. The balloon immediately surrounding your hand represents the visceral pericardium, and the outer layer of the balloon is the parietal pericardium. The water-filled space is the pericardial cavity with water representing serous fluid.

Each lung is isolated in a pleural cavity. The **parietal pleura** lines the thoracic wall, and the **visceral pleura** is attached to the surface of the lung. Because each lung is contained inside a separate pleural cavity, injury and subsequent collapse of one lung will not cause the deflation of the opposite lung.

Most of the digestive organs are encased in the **peritoneum** (pe-ri-to-NE-um), the serous membrane of the abdomen. The **parietal peritoneum** has many folds that wrap around and attach the abdominal organs to the posterior abdominal wall. The **visceral peritoneum** lines the organ surface. The **peritoneal cavity** is between the parietal and visceral peritoneal layers. The peritoneum has many blood vessels, lymphatic vessels, and nerves to support the digestive organs. The kidneys are **retroperitoneal** and are located outside the peritoneal cavity.

Laboratory Activity SEROUS MEMBRANES_____

MATERIALS

toso model
anatomical models
charts

PROCEDURES

1. Label the layers of each serous membrane in Figure 2.6.
2. Identify the pericardium, pleura, and peritoneum on models and charts.
3. Identify the organ(s) of each body cavity on the lab models.
4. In Figure 2.6, for each serous membrane, highlight the parietal and visceral layers with different colored pencils.

INTRODUCTION TO THE HUMAN BODY
Laboratory Report Exercise 2

A. Fill in the Blanks

1. Your heart is located in a small cavity called the _Pericardia_ that is located in a larger cavity, the _____.

2. Your intestines are surrounded by a double membrane called a _____ membrane.

3. The kidneys are _____ because they are located outside the _____.

4. A plane at the hips separates the abdominal cavity from the _____.

5. The inner membrane layer surrounding the heart is the _____.

6. The brain and spinal cord are contained in the _____.

7. The lubricating substance in body cavities is called _____.

8. The large medial cavity of the chest is called the _____.

9. The muscle that divides the ventral body cavity is the _____.

10. The outer layer of a serous membrane is the _____ layer.

B. Matching

Match each directional term listed on the left with the correct description on the right.

1. _____	anterior	A.	to the side
2. _____	lateral	B.	away from a point of attachment
3. _____	proximal	C.	close to the body surface
4. _____	inferior	D.	the front
5. _____	posterior	E.	away from the body surface
6. _____	medial	F.	above, on top of
7. _____	distal	G.	toward a point of attachment
8. _____	superficial	H.	below, a lower level
9. _____	superior	I.	the back
10. _____	deep	J.	to the middle

C. Drawing

1. Draw a picture of a doughnut sectioned by two perpendicular planes.

2. Sketch the body trunk and the planes that designate the abdominopelvic regions.

D. Discussion Questions

1. Describe the levels of organization in the body.

2. Why is anatomical position important when describing structures?

3. List the nine abdominopelvic regions and the location of each.

4. Compare the study of anatomy with that of physiology.

E. Directional and Regional Terminology

Use the correct directional term to show the relationship between the following structures:

1. The chin is _____ to the nose.
2. The brachium is _____ to your antecubitis.
3. The thumb is _____ to the ring finger.
4. The skin is _____ to the muscles.
5. The trunk is _____ to the pubis.
6. The ring finger is _____ to the little finger.
7. The upper arm is _____ to the elbow.
8. The thigh near the knee is _____ to the upper thigh.
9. The ears are _____ to the eyes.
10. The buttock is _____ to the pubis.

F. Completion

1. Label Figures 2.8, 2.9, and 2.10.

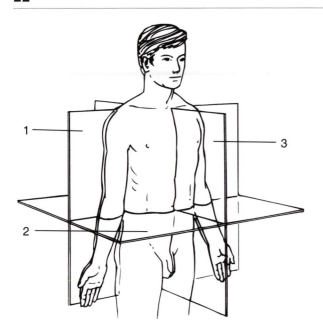

●Figure 2.8
Planes of the Body

1. _____
2. _____
3. _____

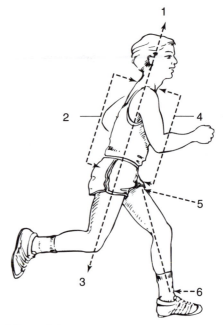

●Figure 2.9
Human Body Orientation and Direction

1. _____
2. _____
3. _____
4. _____
5. _____
6. _____

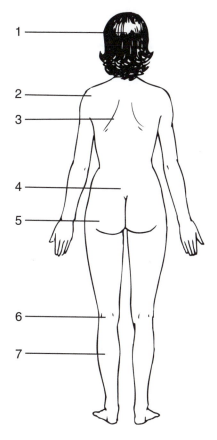

Posterior view
(dorsal)

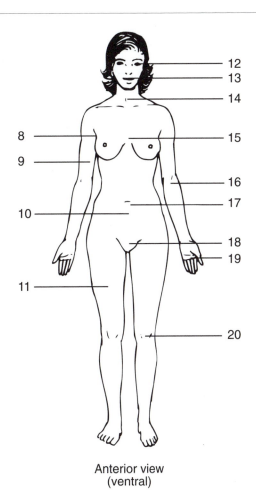

Anterior view
(ventral)

●Figure 2.10
Regional Body References

1. _____

2. _____

3. _____

4. _____

5. _____

6. _____

7. _____

8. _____

9. _____

10. _____

11. _____

12. _____

13. _____

14. _____

15. _____

16. _____

17. _____

18. _____

19. _____

20. _____

3 USE OF THE MICROSCOPE

WORD POWER	MATERIALS
oculus (the eye, ocular lens)	microscope
aper (open, aperture)	clean microscope slides
magni (large, magnification)	microscope coverslips
	dropper bottle of water
	scissor and newspaper
	millimeter graph slides
	practice slides

OBJECTIVES

On completion of this exercise, you should be able to:

- Describe how to properly carry, clean, and store the microscope.
- Identify the parts of the microscope.
- Focus the microscope on a specimen and adjust the magnification.
- Adjust the light source of the microscope.
- Calculate the total magnification for each objective lens.
- Make a wet mount slide.

As a student of anatomy and physiology, you will explore the organization and structure of cells, tissues, and organs. The basic research tool for your observations is the microscope. The instrument is easy to use once you learn its parts and how to adjust them to produce a clear image of a specimen. It is important that you complete each activity in this exercise and that you can effectively use the microscope by the end of the laboratory period.

The microscope uses several lenses to direct a narrow beam of light through a thin specimen mounted on a glass slide. Focusing knobs move the lenses to bring the specimen into **focus** within the round viewing area of the lenses, the **field of view**. The lenses **magnify** objects so that they appear larger than they are in life. **Resolution** is your eyes' ability to distinguish between two points. Within the physical limits of the light microscope, resolution increases along with increases in magnification.

A. Care and Handling of the Microscope

The microscope is a precision scientific instrument, with delicate optical components. Observe the following guidelines as you use the microscope:

1. Carry the microscope with two hands, one hand on the arm and the other hand supporting the base (see Figure 3.1 for parts of the microscope). Do not swing the microscope as you carry it to your lab station. You may bump the microscope or cause a lens to fall out.

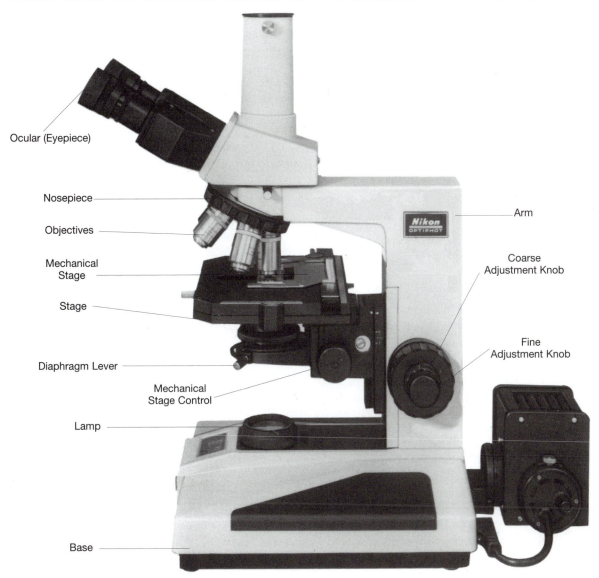

Ocular (Eyepiece)

Nosepiece

Objectives

Mechanical Stage

Stage

Diaphragm Lever

Mechanical Stage Control

Lamp

Base

Arm

Coarse Adjustment Knob

Fine Adjustment Knob

●Figure 3.1
Parts of the Compound Microscope

   2. Completely unravel the electric cord on microscopes with a built-in light source.

   3. To clean the lenses, use only the special lens cleaning fluid and lens paper provided by your lab instructor. Facial tissue paper is made of minute wood chips and fibers that will damage the special optical coating on the lenses.

   4. Store the microscope with the cord wrapped neatly, the low power objective lens in position, and the stage in the uppermost position. Return the microscope to the storage cabinet, or cover it with a dust cover.

B. Parts of the Microscope

Get a microscope from the storage cabinet in the lab, and find each part as it is described. Your laboratory may be equipped with a microscope different from the one shown in Figure 3.1. Your laboratory instructor will discuss the type of microscope you will use.

OCULAR LENS

The ocular lens is the eyepiece where the user places his or her eye(s) to observe the specimen. The magnifying power of most ocular lenses is 10×. This results in an image ten times larger than the actual size of the specimen. **Monocular** microscopes have a single ocular lens; **binocular** microscopes have two ocular lenses, one for each eye. It is necessary to adjust binocular microscopes so that the image from each eyepiece forms a single image. This is easily accomplished by adjusting the distance between the ocular lenses so that it is the same as the distance between your pupils. Move the body tubes apart, and then look in the microscope. If you see two images, slowly move the body tubes closer together until a single **field of view** or circle is seen with both eyes.

NOSEPIECE

The nosepiece is a rotating mechanism at the base of the cylindrical body tube with several objective lenses of different lengths. Turning the nosepiece moves an objective lens into place over the specimen.

OBJECTIVE LENSES

The objective lenses are mounted on the nosepiece. Magnification is determined by the user's choice of objective lens. The longer the objective lens, the greater its magnifying power.

STAGE

The stage is a flat horizontal shelf under the objectives that supports a glass specimen slide. The center of the stage has an **aperture** or hole through which light passes to illuminate the specimen on the slide. A pair of **stage clips** holds the slide steady while you move the slide by hand to view each region on the slide. Some microscopes have a mechanical stage that moves the slide with more precision than is possible manually. The mechanical stage has two adjustment knobs on the side that move a slide over the stage's platform. One knob moves the slide horizontally, and the other moves the slide vertically.

COARSE (GROSS) FOCUS ADJUSTMENT

The coarse focus adjustment is the large dial on the side of the microscope that moves the objective lenses to produce a sharp, clear image. Use the coarse focus knob *only at low magnification* to initially focus a specimen.

FINE FOCUS ADJUSTMENT

The small dial on the side of the microscope is the fine focus adjustment. This knob moves the objective lens for precision focusing after gross focus is achieved. The fine focus knob is used at all magnifications and is the *only focusing knob* used above low magnification.

CONDENSER

The condenser is the small lens under the stage that condenses or narrows the beam of light and directs the light through the slide specimen. A **condenser adjustment knob** moves the condenser vertically. For your studies, the condenser should always be in the high position next to the aperture of the stage.

IRIS DIAPHRAGM

The iris diaphragm is a series of flat metal plates found at the base of the condenser. The plates slide together and create a hole called an **aperture** that regulates the amount of light passing into the condenser. Most microscopes have a small **diaphragm lever** extending from the iris to open or close the aperture to adjust the light for optimal contrast and minimal glare.

LIGHT SOURCE

Most microscopes have a built-in electric light source with a rheostat dial to control the intensity of the light. Microscopes without an electric light source use a mirror to reflect surrounding light into the condenser.

BODY TUBE

The body tube is the cylindrical tube that supports the ocular and extends down to the nosepiece.

ARM

The supportive frame of the microscope is called the arm. It joins the body tube to the base. The microscope is correctly carried with one hand on the arm and the other on the base.

BASE

The base is the broad flat lower support of the microscope.

C. Microscope Basics

Three basic adjustments allow the user to control the appearance and quality of an image under the microscope: (1) illumination controls, (2) objective lens selection, and (3) focusing knobs.

ILLUMINATION CONTROL

Light intensity must be controlled to prevent washing out details of the specimen and reduce strain to the eyes. Light must also be condensed into a narrow beam to pass through a specimen and regulated to produce high contrast and low glare.

On microscopes with an electric light source, the intensity or brightness of light is regulated by a rheostat that controls the output of an electric bulb. Some microscopes use a mirror instead of an electric light source to reflect light toward the specimen. Changing the angle of the mirror varies the light intensity. Regardless of the light source, the light passes through the condenser lens, and narrows into a beam that illuminates the specimen. Associated with the condenser is usually an iris, which regulates the contrast of the light, much as the iris in your eye controls the amount of light striking the lens. As magnification increases, more light is necessary to fully illuminate the specimen.

LENS SELECTION

Magnification is how large an image appears compared to the specimen's natural size. An object viewed at 10× appears ten times larger than life size (1×). Magnification is determined by the objective lens used to view the object. Turning the nosepiece rotates a different objective into place over the specimen and changes the magnification. Figure 3.2 illustrates the optical components of a compound microscope.

A simple lens system, like a hand-held magnifying glass, uses a single lens to magnify an object. Microscopes use a compound lens system with each lens consisting of many pieces of optical glass. The ocular lens, or eyepiece, is usually a 10× lens. Student microscopes usually have three objective lenses: 4×, 10×, and 40× objectives. Magnification of each lens is stamped on the barrel of each objective lens. To calculate

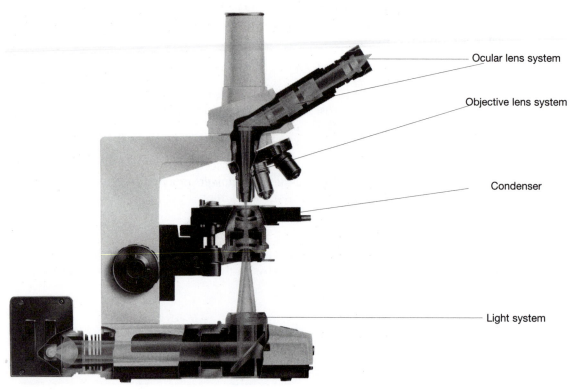

●Figure 3.2
The Optical Components of a Compound Microscope

the total magnification, the magnifications of the ocular lens and the objective lens in use are multiplied. A 10× ocular lens used with a 10× objective lens, for example, produces a total magnification of 100×.

Notice on a microscope that the higher power objectives are longer than the low power lens. As higher power objectives are used, the distance between the specimen and the lens decreases. Therefore, always rotate the low power objective into place before inserting or removing a slide from the stage. This will provide ample working distance between the lens and the stage.

Determine the magnification of the ocular and each objective lens on your lab microscope. Complete Table 3.1 by calculating the total magnification for each combination of ocular and objective lens on your microscope in the lab.

THE FOCUSING KNOBS

The coarse and fine focus knobs are both used to move an objective lens closer to or farther from the slide to achieve a clear image. The coarse focus knob is only used to initially move the low power objective into position to focus on the specimen. The fine

TABLE 3.1	Total Magnification		
Ocular Lens	*Objective Lens*	*Total Magnification*	*Working Distance (estimate in mm)*

focus knob provides a crisp focus and moves the focal point through a specimen at different depths of the slide preparation. Although the specimen on a slide is very thin, it does have layers of cells. Fine focus is the only focus knob used with magnifications above low power. Minimal focusing is needed at higher magnifications because the space between the objective lens and the specimen is very small. Using coarse focus at a high magnification would move the objective lens too much and possibly force the lens into the slide. Further, most microscopes are **parfocal**, and once focus is achieved at low magnification the image will remain in focus as high magnification objectives are used. The fine focus adjustment is therefore the only focusing knob used at magnifications greater than low power.

Depth of field is the portion of a specimen in focus, the layer of the slide currently in focus. This focal depth is greatest at low power and decreases as magnification increases. The reduction in depth of field at high power requires the use of fine focus to move the focal depth and scan the specimen.

Laboratory Activity   USING THE MICROSCOPE

I. Newspaper Print Observation

MATERIALS

microscope   newspaper cut into small pieces
slide and coverslip   dropper bottle of water

Use of the Microscope

1. If the microscope has an internal light source, plug in the microscope and turn the lamp on. Adjust the mirror to reflect light into the condenser if the microscope does not have a built-in light source.
2. To prevent the possibility of breaking a slide, always start with the low power objective. Rotate the nosepiece to swing the low magnification objective lens into position over the aperture. This provides maximum clearance between the objective and the stage for the placement of a slide.
3. Place the slide on the stage of the microscope; use the stage clips or mechanical slide mechanism to secure the slide. Move the slide so that the specimen is over the aperture of the stage.
4. To focus on the specimen, look into the ocular and raise the low power objective lens by slowly turning the coarse focus knob. The image should come into focus. Note that some microscopes focus by moving the stage rather than the objective. To focus a specimen on this type of microscope, move the stage closer to the objective lens by turning the coarse focus knob.
5. Once the image is clear, use the fine focus knob to examine the detailed structure of the specimen.
6. To examine part of the specimen at a higher magnification, move that part of the image over the center of the aperture before changing to a higher magnification objective lens. This will keep the specimen in the smaller field of view at the high magnification.
7. Do not move the focusing knobs before increasing magnification. Most microscopes are parfocal and are designed to stay in focus when a different objective lens is selected. After changing magnification, use the fine focus knob only to adjust the lens.
8. On completion of your observations, reset the microscope to low magnification, remove the slide from the stage, and store the microscope.

PROCEDURES

1. Make a **wet mount slide** of a small piece of the newsprint as follows:
   a. Obtain a slide, coverslip, and small piece of newspaper print.
   b. Place the newsprint on the slide, and add a small drop of water onto it.
   c. Put the coverslip over the newsprint, as shown in Figure 3.3. The coverslip will keep the lenses dry.

2. Select the low magnification objective, and place the slide on the stage. Always view a slide starting at low power. You will see more of the specimen and can quickly select areas on the slide for detailed studies at a higher magnification. What is the total magnification of your microscope at low magnification? _____.

3. Move the slide until the newspaper print is directly over the aperture of the stage. Use the coarse focus knobs to move the objective as close to the specimen as possible without touching the slide.

4. Look into the ocular lens, and slowly turn the coarse focus knob until you see the fibers of the newspaper. Once in focus, adjust the light sources for optimal contrast and resolution.

5. Use the fine focus knob to bring the image into crisp focus. Remember, the microscope you are using costs hundreds of dollars and will produce a clear image if you adjust it correctly. If your specimen is still not in focus, return to the section entitled "Use of the Microscope."

6. Follow each step, and record your observations in the provided spaces.

   a. Once the image is correctly focused, locate the letter "a" or the letter "e." Describe the ink and the paper fibers.

   _____

   b. Move the slide toward you. In which direction did the image of the letter move? _____

   c. Move the slide to the left. In which direction did the image of the letter move? _____

   d. Is the image of the letter oriented in the same direction as it is on the slide?

   _____

II. Depth of Field Observation

To show depth of field, you will examine a slide of colored threads. Notice how the threads are layered and how much of the slide is in focus at each magnification.

MATERIALS

microscope
slide of colored threads
(If colored thread slides are unavailable, students can make a slide using hairs from different individuals.)

PROCEDURES

1. Move the low power objective into position, and place the slide on the stage with the threads over the aperture.

●**Figure 3.3**
Coverslip Placement

Touch the water or stain with the edge of the coverslip, then lower coverslip flat onto slide. Use a paper towel to absorb excess water that has leaked out from under the coverslip.

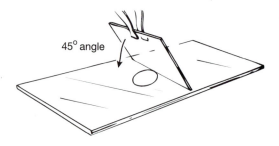

45° angle

2. Use the coarse focus knob to bring the threads into focus. Move the slide to view the area where the threads overlap. Turn the nosepiece on the microscope to select the high power objective.

3. Use the fine focus adjustment and decide which thread is on top, in the middle, and on the bottom. Write your results here.

   Top thread _____.

   Middle thread _____.

   Bottom thread _____ .

4. At which magnification were all the threads in focus?

5. Remove the slide, and return the microscope to the storage area.

III. Additional Practice with the Microscope

MATERIALS

microscope
practice slide
(Your lab instructor will choose a slide of a stained tissue that is visible to the unaided eye.)

PROCEDURES

1. Locate the stained cells on the practice slide.

2. Increase magnification, and examine individual cells.

3. Draw a sketch of a cell in the following space.

4. If fresh tissue is used, dispose of all materials in a biohazard container as described by your instructor. (See exercise 1 in this manual for details on biohazardous materials and their proper disposal.)

USE OF THE MICROSCOPE

Exercise 3 Laboratory Report

A. Discussion Questions

1. Explain what parts of the microscope are used to regulate the intensity and contrast of light. What does each of these parts do?

2. How is magnification controlled in the microscope?

3. Why should you always view a slide at low power first?

4. Briefly explain how to care for the microscope.

5. Describe when you should use coarse focus and when to use fine focus.

B. Matching

Match the part of the microscope listed on the left with the correct description on the right.

1. _____ ocular	A. used for precise focusing
2. _____ aperture	B. lower support of microscope
3. _____ body tube	C. narrows the beam of light
4. _____ mechanical stage	D. hole in stage
5. _____ fine focus	E. used only at low power
6. _____ base	F. has knobs to move slide
7. _____ objective lens	G. special paper for cleaning
8. _____ coarse focus	H. eyepiece
9. _____ condenser	I. holds the ocular lens
10. _____ lens paper	J. lens on nosepiece

4 CELL STRUCTURE AND FUNCTION

Cells were first described in 1665 by a British scientist named Robert Hooke. Hooke examined a thin slice of tree cork with a microscope and observed open spaces in the cork that he called **cells**. Over the next two centuries, scientists examined cells from plants and animals, and formulated the **cell theory**, which states that: (1) all plants and animals are composed of cells, (2) all cells come from preexisting cells, (3) cells are the smallest living units that perform physiological functions, (4) each cell works to maintain itself at the cellular level, and (5) homeostasis is the result of the coordinated activities of all the cells in an organism.

In this exercise, you will examine the anatomy of the cell and how cells divide to produce more cells for growth and repair of the body. Although the body has a variety of cell types, a generalized or composite cell, as shown in Figure 4.1, will be used to discuss cell anatomy. All cells possess an outer boundary, the cell membrane. Cells may also have a nucleus and other internal structures called **organelles**.

PART ONE—CELL STRUCTURE

WORD POWER	MATERIALS
cell (small room)	cell models
organelle (tiny organ)	mitosis models
nucleus (small nut)	microscope
cytokinesis (kinesis—movement)	prepared slides
	Ascaris (round worm)
	whitefish blastula

OBJECTIVES

On completion of this part of the exercise, you should be able to:

- Identify organelles of the cell on charts, models, and other lab material.
- State a function of each organelle.
- Discuss the cell's life cycle, including the stages of interphase and mitosis.
- Identify the stages of mitosis using the round worm (*Ascaris*) slide.

A. The Cell Membrane

The **cell membrane** is the physical boundary of the cell, separating the **extracellular fluid** surrounding the cell from the **cytosol**, the intracellular liquid. The cell membrane regulates the movement of ions, molecules, and other substances into and out of the cell. It is the point of contact with the extracellular fluid of the body. Hormones, enzymes, and other regulatory molecules bind to specific receptor proteins in the cell membrane. Muscle and nerve cell membranes are excitable membranes and produce electrical currents called action potentials. Figure 4.2 illustrates the structure of the cell membrane.

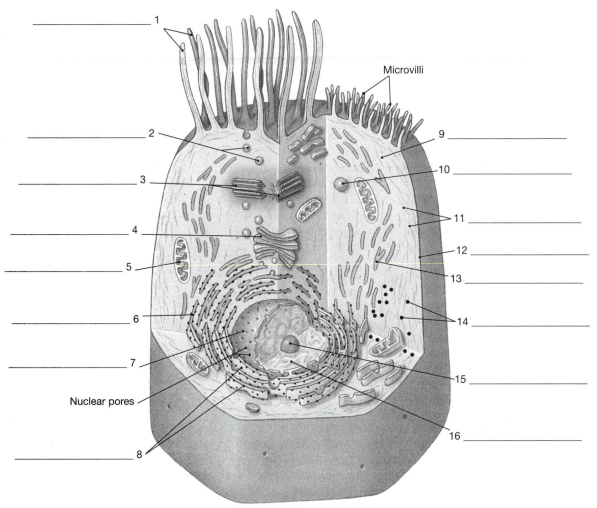

1

Microvilli

2

3

4

5

6

7

Nuclear pores

8

9

10

11

12

13

14

15

16

●Figure 4.1
The Anatomy of a Composite Cell

The cell membrane is composed of a **phospholipid bilayer**, a double layer of phospholipid molecules. Floating like icebergs in the phospholipid bilayer are a variety of **integral proteins**. Some of these proteins have **channels** that regulate the passage of specific ions through the membrane. Loosely attached to the external and internal surfaces of the membrane are **peripheral proteins**.

B. Organelles

Within the cell are **organelles**, the functional pieces of biological machinery. As represented in Figure 4.1, each organelle has a distinct anatomical organization and is specialized for a specific function. Organelles are suspended in the cytosol. The organelles and cytosol together compose the **cytoplasm** of the cell.

Organelles are grouped into two broad classes: membranous organelles and nonmembranous organelles. **Membranous** organelles are enclosed in a lipid membrane that isolates the organelle from the cytosol. The nucleus, endoplasmic retiula, Golgi apparatus, vesicles, and mitochondria are examples of membranous organelles. **Nonmembranous** organelles lack an outer membrane and are in direct exposure to the cytosol. Ribosomes, centrioles, the cytoskeleton, cilia, and flagella are nonmembranous organelles.

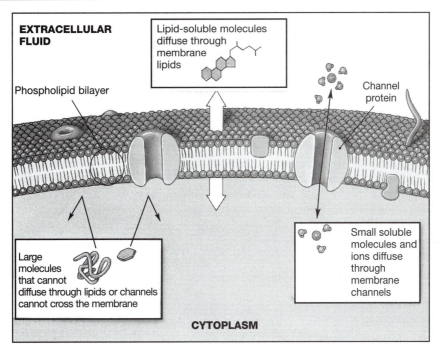

●Figure 4.2
The Cell Membrane
Components of the cell membrane and diffusion of materials through the cell membrane.

Study Tip

Some students have difficulty integrating concepts of anatomy and physiology. Anatomy becomes a memorization process and physiology a confusing hurdle. Because form (anatomy) and function (physiology) are intertwined, it is helpful to link a function to each anatomical structure. While you identify organelles on cell models, consider the function of each organelle.

The **nucleus** controls all activities of the cell. Protein synthesis, gene action, cell division, and metabolic rate are all regulated by the nucleus. The nucleus usually appears as a dark stained structure in the cell. The stained material is **chromatin**, uncoiled chromosomes consisting of DNA and protein molecules. The **nuclear enve-lope** surrounds the nuclear material and contains pores through which instructions from the nucleus pass into the cytosol. A darker stained **nucleolus** is the site for the synthesis of RNA molecules that combine with proteins to form ribosomes.

Ribosomes orchestrate protein synthesis. Instructions on how to make each protein are copied from DNA into RNA molecules that carry the instructions to a ribosome. Ribosomes comprise two subunits that clamp together around an RNA molecule during protein synthesis. Ribosomes occur **free** in the cytoplasm or **attached** to a membranous system, the endoplasmic reticulum.

Surrounding the nucleus is a complex membrane system, the **endoplasmic reticulum (ER)**. Two types of ER occur: **rough ER** with attached ribosomes on the surface of the membrane and **smooth ER** that lacks ribosomes and appears smooth. Generally, the endoplasmic reticulum functions in synthesis of organic molecules, transport of materials within the cell, and storage of molecules, such as calcium ions in muscle cells. Materials in the ER may pass into the Golgi apparatus for eventual transport out of the cell.

The **Golgi apparatus** is a series of flattened saccules adjoining the endoplasmic reticulum. The ER can pass protein molecules to the Golgi apparatus for modification and secretion. Cell products such as mucus are synthesized, packaged, and secreted by the Golgi apparatus. Small **secretory vesicles** pinch off the Golgi saccules, fuse with the cell membrane, and rupture to release their contents. The empty vesicles are added to the plasma membrane for renewal and repair. This transport out of the cell is called **exocytosis**.

One type of vesicle produced by the Golgi apparatus is the **lysosome**. Lysosomes are filled with powerful enzymes and function in removal of damaged organelles or microbes. White blood cells, for example, ingest bacteria by surrounding them with cell membrane extensions. Once surrounded, the membrane pinches off to form a vesicle that is moved inside the cell. Lysosomes fuse with the vesicle and produce enzymes to destroy the captured bacteria.

Mitochondria convert energy into useful forms for the cell. Each mitochondrion is encased in a double-layered membrane. The inner membrane is folded into fingerlike projections called **cristae**. The middle region of the mitochondrion is called the **matrix**. To provide cellular energy, organic molecules from the breakdown of food are passed along a series of metabolic enzymes in the cristae to produce ATP, the energy currency of cells. Mature red blood cells lack mitochondria, whereas muscle and nerve cells have a huge population of mitochondria to supply energy for contraction or to generate nerve impulses.

Centrioles are paired organelles composed of microtubules. Centrioles are involved in cell division. Other microtubules occur in cells; all cells have a **cytoskeleton** for structural support. Many cells of the respiratory and reproductive systems have **cilia**, short hairlike projections extending from the plasma membrane. One type of human cell, spermatozoa, has a single, long **flagellum** for locomotion.

Laboratory Activity   IDENTIFICATION OF CELL ORGANELLES

PROCEDURES

1. Label the organelles in Figure 4.1.
2. Identify the organelles on a laboratory cell model.
3. Complete Table 4.1 by filling in the blank boxes.

C. Cell Division

Cells must reproduce for growth and tissue repair. During cell reproduction, a cell equally divides its chromosomes and splits into two identical cells. The divisional process includes **mitosis**, which is the division of the genetic material in the nucleus, and **cytokinesis**, the separation of the cytoplasm into equal parts to produce **daughter cells**. The daughter cells have the same number of chromosomes as the original "mother" cell. Figure 4.3 highlights the process of mitosis.

TABLE 4.1	**Summary of Cell Organelles and Their Functions**	
Organelle	*Organelle Structure*	*Organelle Function*
Mitochondria		
		Contains digestive enzymes
	Phospholipid bilayer with proteins	
Endoplasmic reticulum		
		Exocytosis of cellular products
	Composed of two subunit molecules	
		Regulates cell activities

Most of the time, a cell is not dividing and is in **interphase**. During interphase, the cell does its various functions. Before division, the cell must replicate its genetic material, DNA, so that the resulting cells will each contain a complete set of genes. After replication, each chromosome has two strands of DNA, the original strand and the identical copy. Each strand is called a **chromatid**, and matching chromatids are held together by a **centromere**.

Mitosis starts with **prophase** as chromosomes become visible in the nucleus. During early prophase, the chromosomes are long and disorganized. As prophase progresses, the chromosomes coil to shorten and move toward the equator, or middle, of the cell. In the cytoplasm, a pair of **centrioles** separate to opposite poles (sides) of the cell. Between the centrioles, microtubules called **spindle fibers** span across the cell.

Metaphase occurs as the chromosomes line up in the middle of the cell at the **metaphase** (equatorial) **plate**. Spindle fibers extend across the cell from one pole to the other and attach to centromeres of the chromosomes.

Anaphase is the division of the cell's nuclear material. Chromatids of each chromosome separate as the spindle fibers pull them apart and move them toward opposite poles of the cell. The individual chromatids are now considered chromosomes. The cell membrane begins to pinch inward to partition the cytoplasm and chromosomes for the two developing cells. This process, called cytokinesis, continues into the next stage of mitosis.

During the last phase, **telophase**, chromosomes in each pole unwind, and a nuclear membrane forms around them, producing a nucleus for each new cell, complete with a set of nuclear material and organelles. Telophase ends as cytokinesis completes the formation of the two **daughter cells**. The daughter cells are in interphase and, depending on the cell type, may divide again.

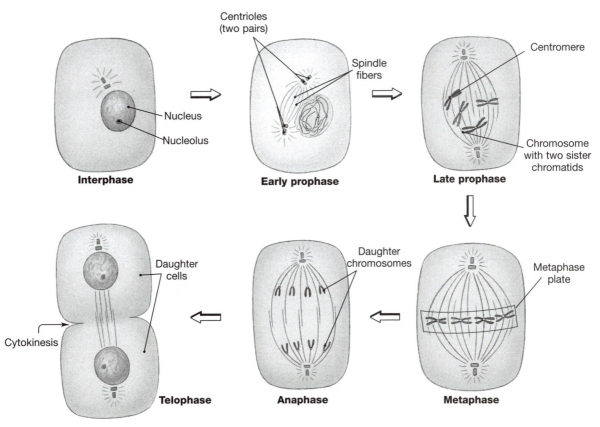

●**Figure 4.3**
Mitosis

Laboratory Activity   OBSERVATIONS OF MITOSIS _____

MATERIALS

Ascaris (round worm) slide or whitefish blastula slide
microscope

PROCEDURES

Ascaris is a parasitic nematode or round worm with only 6 chromosomes, making it a good choice for mitosis studies (human cells have 46 chromosomes). The slide is made from the oviduct of a pregnant female worm. In the oviduct are fertilized eggs, called zygotes, in their first mitotic division. Each zygote consists of a single cell surrounded by an egg membrane and an egg shell. The cells undergoing mitosis are inside the egg membranes. Your laboratory instructor may choose to use a whitefish blastula slide for showing the stages of mitosis. A blastula is an embryonic ball of cells that forms early in the development of all vertebrate animals. In a living blastula, the cells are dividing rapidly as the embryo grows. The thin blastula sections on the slides contain hundreds of cells in various stages of mitosis.

Notice the following on the *Ascaris* slide:

- Scan the slide at low magnification, and observe the many eggs and the outer wall of the oviduct. The slide usually has several oviduct sections.
- Increase magnification to medium power, and examine the contents of a single egg. Distinguish between the egg shell, egg membrane, and cell membrane. Inside a single cell, locate the nucleus, centrioles, and spindle fibers. The chromosomes appear as dark thick structures in the cell.

1. Using Figure 4.3 as reference, identify cells in:
   a. Interphase with a distinct nucleus
   b. Prophase with disorganized chromosomes
   c. Metaphase with equatorial chromosomes
   d. Anaphase with chromosomes separating toward opposite poles
   e. Telophase with nuclear membranes forming around new cell

2. Observe cytokinesis in late anaphase and telophase.
3. Draw and label each stage of mitosis in the space provided.

PART TWO—CELL FUNCTION

WORD POWER	MATERIALS
diffusion (diffu—apart)	microscope, slides, coverslips
equilibrium (equil—equal)	dropper bottle, food color dye
gradient (grad—a hill)	beakers—250 ml, 500 ml
osmosis (osmo—pushing)	dialysis tubing and tubing clips
solution (solu—to dissolve)	starch solution, Lugol's solution
solute (solu—dissolved)‹	blood (nonhuman)
dialysis (dialy—to separate)	wax pencil
isotonic (iso—same)	isotonic solution
hypertonic (hyper—excessive)	hypertonic solution
hypotonic (hypo—below)‹	hypotonic solution‹
hemolysis (...lys—loose)	

OBJECTIVES

On completion of this part of the exercise, you should be able to:

- Describe the two main processes by which substances move into and out of cells.
- Discuss diffusion, concentration gradients, and equilibrium.
- Explain the effect of isotonic, hypertonic, and hypotonic solutions on cells.

Cells are the functional living unit of your body. For cells to survive, materials must be transported across the plasma membrane. Cells import nutrients, oxygen, hormones, and other regulatory molecules. Wastes and cellular products are exported to the extracellular (interstitial) fluid. Plasma membranes are selectively permeable and regulate passage of certain materials. Proteins and other macromolecules are too big to pass through channels in the plasma membrane. Movement of these molecules requires the use of carrier molecules in a process called **active transport** that consumes cell energy. Smaller molecules, such as water and many ions, cross the membrane without assistance from the cell. This movement is called **passive transport** and requires no energy expenditure by the cell. Diffusion and osmosis are the primary passive processes in the body and will be studied in this section.

A. Diffusion

Diffusion is the net movement of substances from a region of greater concentration to a region of lesser concentration. Simply put, diffusion is the spreading out of substances owing to the collisions between vibrating molecules. Diffusion occurs throughout the body, in extracellular fluids, across cell membranes, and in the cytosol of cells. Examples of diffusion include oxygen moving from the lungs into pulmonary capillaries, odor molecules moving through the nasal lining to reach the olfactory cells, and movement of ions in and out of nerve cells to produce electrical impulses. Molecules diffuse through cells by two basic mechanisms: lipid-soluble molecules diffuse through the phospholipid bilayer of the membrane, and small molecules and ions diffuse through the channels of integral proteins.

Cells cannot directly regulate diffusion; it is a passive transport process. If a substance is unequally distributed, a **concentration gradient** exists, and one region will have a greater concentration of the substance than other regions. The substance will diffuse until an equal distribution occurs, a point called **equilibrium**. Figure 4.4 details the process of diffusion. Notice in step 1 that the molecules in the ink drop are concentrated

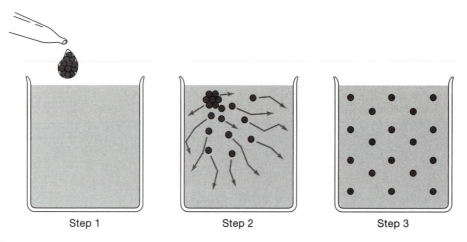

●**Figure 4.4**
Diffusion

Step 1: Placing an ink drop in a glass of water establishes a strong concentration gradient because there are many ink molecules in one location and none elsewhere. **Step 2:** As diffusion occurs, the ink molecules spread through the solution. **Step 3:** Eventually diffusion eliminates the concentration gradient, and the ink molecules are distributed evenly. Molecular motion continues, but there is no directional movement.

together before they are placed in water. Once in the water, step 2, the ink molecules disperse as they bump into other ink molecules and water molecules. Eventually, the molecules will be evenly distributed in equilibrium, as illustrated in step 3.

Laboratory Activity   DIFFUSION OF A LIQUID IN A LIQUID

MATERIALS

    250 ml or larger beaker
    food coloring dye

PROCEDURES

1. Fill the beaker 3/4 full with tap water. Leave the beaker undisturbed for several minutes to let the water settle.
2. Carefully add several drops of food coloring to the water.
3. Observe diffusion of the dye for several minutes.

RESULTS

1. Why did the dye diffuse in water?

    _____

    _____

2. How would cool or warm water affect the diffusion rate?

    _____

    _____

B. Osmosis

Osmosis is the net movement of water through a selectively permeable membrane, from a region of greater concentration to a region of lesser concentration. With an understanding of diffusion, one may define osmosis as the diffusion of water through a selectively permeable membrane. Osmosis occurs when solutions of differing solute concentration are separated by a selectively permeable membrane. A **solution** is the result of dissolving a **solute**, such as salt, in a **solvent**, such as water. A 2% salt solution is, by volume, 2% salt and 98% water. Solute levels determine solvent concentrations. This establishes the concentration gradient for osmosis. As solute concentration increases, the solvent concentration will decrease. For example, fill a beaker completely with water so that if you attempt to add one more drop of water it will overflow. The beaker now has 100% water. If a solute such as salt is added to the water, the solute will occupy space in the beaker and therefore cause some water to be displaced and overflow the beaker. If the salt occupies 2% of the total beaker volume, then a 2% salt solution has been prepared with 98% water and 2% salt.

Study Tip

Only water moves across the membrane during osmosis. If the solutes could move across the membrane, they would diffuse to equilibrium and eliminate the concentration gradient of water, and osmosis would not occur.

Examine Figure 4.5 presenting osmosis. The figure illustrates a beaker divided by a selectively permeable membrane. Each side of the beaker has identical molecules but in different concentrations. The small dots represent water, the larger circles are solutes that cannot cross the membrane. Side A of the beaker has more water than side B, the side with more solute and less water. In step 1, the water level between the sides of the membrane is equal. As water moves from side A to side B,

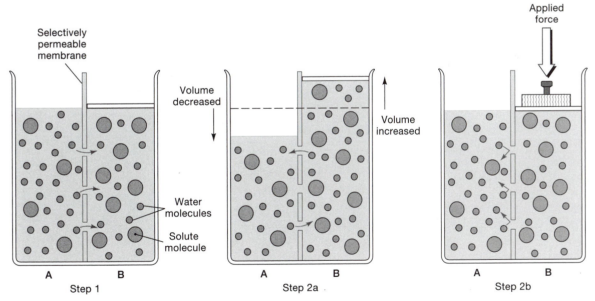

●**Figure 4.5**
Osmosis

Step 1: Two solutions containing different solute concentrations are separated by a selectively permeable membrane. Water molecules (small dots) begin to cross the membrane toward the solution with the higher concentration of solutes (larger circles) (solution B).
Step 2a: At equilibrium the solute concentrations on the two sides of the membrane are equal. The volume of solution B has increased at the expense of solution A. **Steb 2b:** Osmosis can be prevented by resisting the volume change. The osmotic pressure of solution B is equal to the amount of hydrostatic pressure required to stop the osmotic flow.

the volume of side B increases until equilibrium is reached in step 2. Water and solute concentrations are equal between the sides of the membrane.

Solutions have an **osmotic pressure** due to the presence of solutes. The greater the solute concentration, the greater the osmotic pressure of the solution. The solution with the greatest osmotic pressure will cause water to osmose toward it. In effect, "water follows solute and osmotic pressure is a pulling pressure." Notice in step 2b of Figure 4.5 that a force applied to side B will stop osmosis if that force is equal to the osmotic pressure causing the osmosis. Drinking water is often purified by reverse osmosis, applying pressure to water on one side of a membrane to force water through the membrane while leaving the solutes behind.

Laboratory Activity   OBSERVATION OF OSMOSIS _____

Experiment I—Osmosis

This exercise uses an artificial membrane called dialysis tubing. The dialysis tubing, like cell membranes, is selectively permeable. The tubing has small pores that allow the passage of small molecules yet restricts the passage of large molecules. The dialysis tubing will be folded into a bag and filled with a concentrated starch solution (see Figure 4.6). The bag will then be placed in a beaker containing water with iodine. To detect if water has osmosed, the dialysis bag will be weighed before and after the experiment. Any change in weight will be contributed to a change in water volume in the bag. To detect if diffusion of starch and iodine has occurred, a simple starch test will be performed. Starch in the presence of iodine produces a dark blue color.

●**Figure 4.6**
Osmosis Setup Using Dialysis Membrane

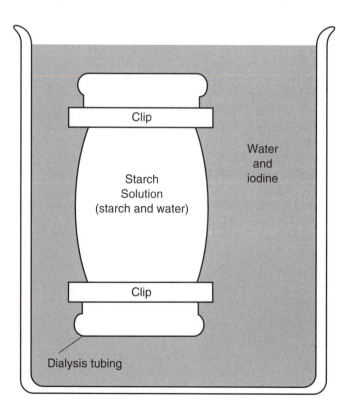

MATERIALS

1 strip of dialysis tubing 15 cm long
2 dialysis tubing clips (or string)
500 ml beaker
distilled water (DW)
5% starch solution (5 gm starch dissolved in 95 ml DW)
Lugol's (iodine) solution

PROCEDURES

1. Add approximately 100 ml of distilled water to the beaker. Soak the dialysis tubing in the water for 3–4 minutes.
2. Remove dialysis tubing from the beaker.
3. Fold one end of the tubing over and close it securely with a tubing clip. Rub the other side of the tubing between your fingers to open the membrane.
4. Fill the bag approximately 3/4 full with starch solution. Fold and clip the bag securely without trapping too much air in the bag.
5. Rinse off the outer surface of the bag to remove any spilled starch solution. Dry and weigh the full dialysis bag; record this data in Table 4.2, then return the bag to the beaker of water. Add more water if the bag is not completely submerged. Add just enough Lugol's solution (iodine) to discolor the water. Record other observations in Table 4.2.
6. After 60 minutes, remove, dry, and weigh the bag. Record your data and observations in the "Results" section.

RESULTS

Complete Tables 4.2 and 4.3, and answer the following questions.

1. Which solution had the greater osmotic pressure, the starch solution in the bag or the iodine solution in the beaker?

   _____

   _____

2. Use osmotic pressure to explain the osmosis of water in this experiment.

   _____

   _____

3. How did you decide if starch and iodine moved?

   _____

   _____

TABLE 4.2	**Dialysis Experiment Observations**	
Dialysis Bag	*Initial Observations*	*Final Observations*
Weight of bag		
Shape of bag		
Color of starch solution		
Color of beaker water		

TABLE 4.3	Dialysis Experimental Results	
Substance	*Movement* *(in, out, one)*	*Process* *(diffusion, osmosis)*
Water		
Starch		
Iodine		

Experiment II—Observation of Osmosis in Red Blood Cells

To observe osmosis in living cells, solutions of various concentrations will be added to blood. Solutions with the same solute concentrations as a cell are **isotonic solutions**. If solute concentrations are the same, solvent concentrations will be equal, too. Solutions with more solute (and therefore less solvent) than the cell are **hypertonic solutions**, and solutions having less solute than the cell are **hypotonic solutions**. Note that the cell is the reference point; solutions are compared to the cell. If a cell loses water owing to osmotic movement, the cell will shrink or **crenate**. In a hypotonic solution, a cell will gain water and perhaps burst or **lyse**. Blood cells undergo **hemolysis** in hypotonic solutions. Figure 4.7 illustrates the effect of various solutions on red blood cells.

Your laboratory instructor may choose to use plant cells rather than blood for observing osmosis. Leaves from the aquatic plant *Elodea* are placed in different solutions to promote osmosis. Plant cells have a thick outer cell wall that provides structural support of the plant. Pushed against the inner surface of the cell wall is the cell membrane. To study osmosis in plant cells, observe the distribution of the cell's organelles, and attempt to locate the cell membrane. In hypotonic solutions, for example, the plant cell will lose water, and the cell membrane will shrink away from the cell wall.

●**Figure 4.7**
Osmotic Flow across Cell Membranes

White arrows indicate the direction of osmotic water movement. **(a)** Because these red blood cells are immersed in an isotonic saline solution, no osmotic flow occurs and the cells have their normal appearance. **(b)** Immersion in a hypotonic saline solution results in the osmotic flow of water into the cells. The swelling may continue until the cell membrane ruptures. **(c)** Exposure to a hypertonic solution results in the movement of water out of the cells. The red blood cells shrivel and become crenated. (SEM × 833)

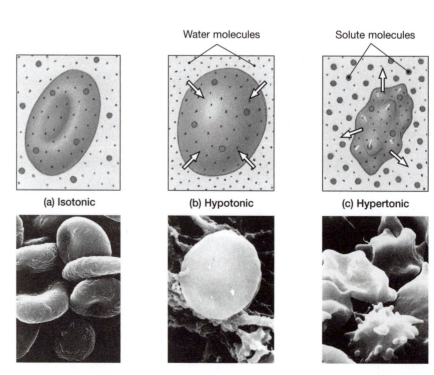

(a) Isotonic   (b) Hypotonic   (c) Hypertonic

MATERIALS

blood (nonhuman blood if possible) or live *Elodea*
microscope slides and coverslips, microscope, eyedroppers, wax pencil
0.90% saline solution (isotonic)
2.0% saline solution (hypertonic)
distilled water (hypotonic)
gloves and safety glasses

PROCEDURES

1. With the wax pencil, label (on the ends) three microscope slides "Iso," "Hypo," and "Hyper."
2. Put on safety glasses and disposable gloves before handling any blood. Place a small drop of blood on three microscope slides. *Do not touch the blood*. If you are using plant cells to observe osmosis, place a single *Elodea* leaf flat on the slide.
3. Place a coverslip over each slide.
4. Add a drop of isotonic solution to the outer edge of the coverslip of the Iso slide. Repeat with the other slides and solutions.
5. Observe changes in cell shape as osmosis occurs. Draw cells from each slide in the space provided. Compare your results with the cells in Figure 4.7.

CELL STRUCTURE AND FUNCTION
Exercise 4 Laboratory Report

A. Matching

Match each cellular structure listed on the left with the correct description on the right.

1. _____ cell membrane	A.	copy of a chromosome
2. _____ centrioles	B.	component of cell membranes
3. _____ ribosome	C.	short, hair-like cellular extensions
4. _____ smooth ER	D.	part of phospholipid molecule
5. _____ chromatid	E.	intracellular fluid
6. _____ lysosomes	F.	involved in mitosis
7. _____ integral protein	G.	folds of inner mitochondrial membrane
8. _____ cytoplasm	H.	composed of a phospholipid bilayer
9. _____ cristae	I.	stores calcium ions in most cells
10. _____ hydrophilic heads	J.	site for protein synthesis
11. _____ cytosol	K.	vesicle with digestive enzymes
12. _____ cilia	L.	intracellular fluid and the organelles

B. Fill in the Blanks

1. Replication of genetic material results in chromosomes composed of two strands, each called a _____.
2. A cell in metaphase has chromosomes located in the _____ of the cell.
3. Division of the cytoplasm to produce two daughter cells is called _____.
4. Double-stranded chromosomes separate during _____ of mitosis.
5. Microtubules called _____ attach to chromatids and pull them apart.
6. Chromosomes become visible during _____ of mitosis.
7. The last stage of mitosis is _____.
8. Matching chromatids are held together by a _____.

C. Short Answer Questions

1. What is the purpose of cell division?

2. Which organelles are involved in protein synthesis?

3. What is the function of the spindle fibers during mitosis?

4. What structures in the cell membrane regulate ion passage?

5. Why is a concentration gradient necessary for passive transport?

6. Describe the components of a 2% sugar solution.

7. A blood cell is placed in a 3% salt solution. Will osmosis, diffusion, or both occur across the blood cell's membrane? Why?

8. After a long soak in the tub, you notice that your skin has become wrinkled and your fingers and toes feel bloated. Describe why this occurred.

D. Define

1. osmotic pressure

2. dialysis

3. hypertonic solution

4. hemolysis

E. Examples

Give an example of each of the following transport processes:
1. Diffusion of a substance in the air

2. Phagocytosis

3. A hypertonic solution

F. Completion

1. Complete the Concept Map.

Concept Map

Using the following terms, fill in the circled, numbered, blank spaces to complete the concept map. Follow the numbers that comply with the organization of the map.

Ribosomes   Nucelolus   Membranous   Centrioles
Lysosomes   Lipid bylayer   Proteins   Organelles
Fluid component

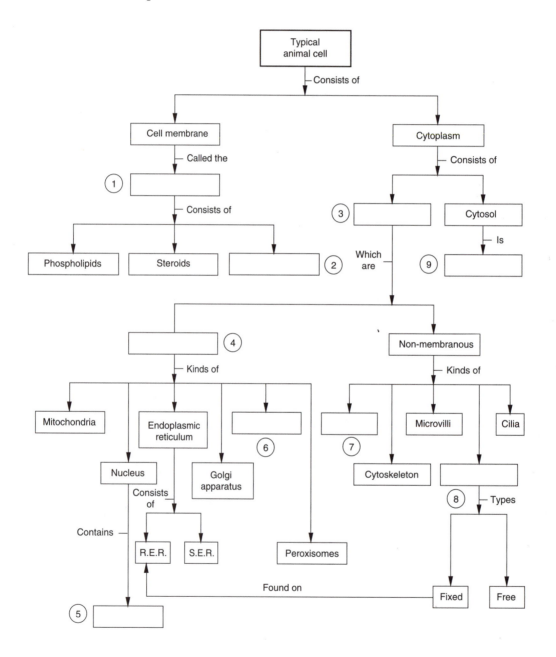

5 SURVEY OF HUMAN TISSUES

Histology is the study of tissues. A **tissue** is a group of similar cells working together to accomplish a specialized function. Figure 5.1 illustrates an overview of tissues of the body. Molecules and atoms combine to form cells that secrete materials into their surrounding extracellular fluid. The cells and their secretions compose the various tissues of the body. There are four major categories of tissues in the body: **epithelial, connective, muscle**, and **neural** tissues. Each category includes specialized tissues with specific locations and functions. Groups of specialized tissue form an organ such as the stomach, a muscle, or a bone. Organs with common functions compose the organ systems of the body.

It may be difficult for us to relate how cells contribute to the life of an individual as a whole, yet we readily see the effects of fat tissue and muscle tissue on our body. To appreciate how organs function, you must have a clear understanding of their histological organization. During your microscopic observations of tissues, it is important to scan the entire slide to examine the tissue at low magnification. Then, increase magnification and observe individual cells in the tissue. Remember, a slide often has more than one tissue. When examining adipose tissue, for example, you will also see blood vessels and perhaps a nerve. Take your time when studying a tissue; a quick glance through the microscope is not sufficient to learn the tissue for identification on a laboratory examination.

PART ONE   EPITHELIAL TISSUE

WORD POWER

epithelium (epi—above; ...thel—tender)
lamina (lamin—thin sheet)
squamous (squam—a scale)
stratified (strat—a layer)
pseudostratified (pseudo—false)

MATERIALS

compound microscope
prepared microscope slides
    simple squamous epi. (lungs or frog skin)
    simple cuboidal epi. (kidney)
    simple columnar epi. (digestive tract)
    stratified squamous epi. (skin)
    pseudostratified epi. (trachea)
    transitional epi. (urinary bladder, relaxed)

OBJECTIVES

On completion of this part of the exercise, you should be able to:

- List the characteristics used to classify epithelium.
- Describe how epithelium is attached to the body.
- Describe the microscopic appearance of each type of epithelium.
- List the location and function of each epithelial tissue.
- Identify each tissue under the microscope.

●**Figure 5.1**
An Orientation to the Tissues
of the Body

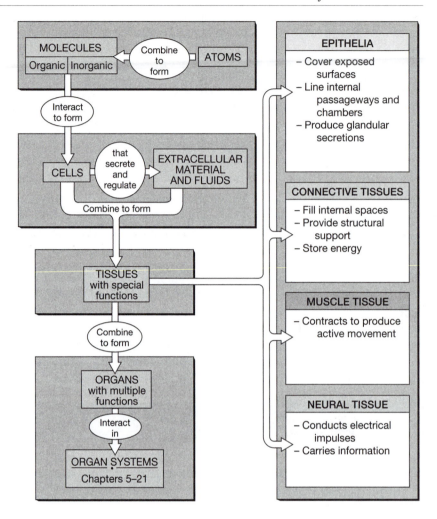

Epithelial tissue, or **epithelium**, is a lining and covering tissue. It is the only tissue you can directly see on your body. Epithelium covers the body surface and lines passageways open to the external environment such as the nose, mouth, throat, anus, and lungs. Glide your tongue over the inner surface of your cheeks and feel the epithelium exposed to your oral surface. Below this epithelium is a connective tissue layer.

Epithelial cells are packed closely together, leaving little room for other structures between the cells. The cells on the surface of epithelium are exposed to a body cavity or the external environment. The bottom surface of epithelium is always attached to a deeper connective tissue layer by a double-layered **basement membrane** (see Figure 5.2). The membrane acts as a barrier to prevent materials from the underlying connective tissue from seeping into the epithelium. Connective tissue cells produce protein fibers that anchor the epithelium in place.

The epithelium has a wide range of functions. The epithelium protects exposed surfaces of the body from excessive friction, and prevents dehydration and the invasion of microbes and chemicals into the body. All substances must pass through a layer of epithelium to enter the internal environment. Digested food in the intestines, for example, passes through a lining of simple columnar epithelium. Many sensory organs are found in the epithelium of the skin. Epithelium also forms the secretory cells of glands. Some epithelium has fine hairlike **cilia** that sweep mucus across the free surface of the tissue.

Epithelium is classified according to cell shape and cell layers. Together, these characteristics are used to name a tissue. **Squamous** cells are irregularly shaped, flat, and scalelike. **Cuboidal** cells are squarish and have a large central nucleus. **Columnar**

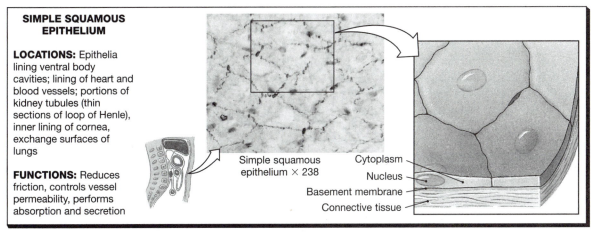

SIMPLE SQUAMOUS EPITHELIUM

LOCATIONS: Epithelia lining ventral body cavities; lining of heart and blood vessels; portions of kidney tubules (thin sections of loop of Henle), inner lining of cornea, exchange surfaces of lungs

FUNCTIONS: Reduces friction, controls vessel permeability, performs absorption and secretion

Simple squamous epithelium × 238

Cytoplasm
Nucleus
Basement membrane
Connective tissue

(a)

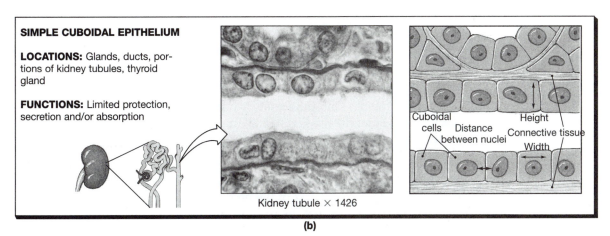

SIMPLE CUBOIDAL EPITHELIUM

LOCATIONS: Glands, ducts, portions of kidney tubules, thyroid gland

FUNCTIONS: Limited protection, secretion and/or absorption

Kidney tubule × 1426

Cuboidal cells
Distance between nuclei
Height
Connective tissue
Width

(b)

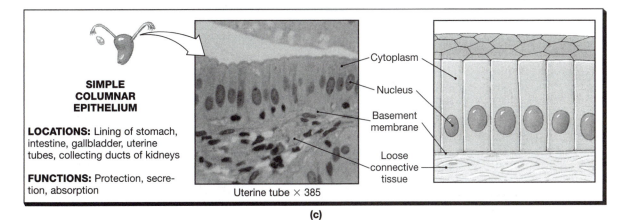

SIMPLE COLUMNAR EPITHELIUM

LOCATIONS: Lining of stomach, intestine, gallbladder, uterine tubes, collecting ducts of kidneys

FUNCTIONS: Protection, secretion, absorption

Uterine tube × 385

Cytoplasm
Nucleus
Basement membrane
Loose connective tissue

(c)

●**Figure 5.2**
Simple Epithelia

(a) A superficial view of the simple squamous epithelium that lines the periotoneal cavity. The three-dimensional drawing shows the epithelium in superficial and sectional view. **(b)** A section through the cuboidal epithelial cells of a kidney tubule. The diagrammatic view emphasizes structural details that permit the classification of an epithelium as cuboidal. **(c)** Micrograph showing the characteristics of simple columnar epithelium. In the diagrammatic sketch, note the relationships between the height and width of each cell: the relative size, shape, and location of nuclei; and the distance between adjacent nuclei. Compare with Figure 5.3.

cells are taller than they are wide, like a column of a building. Epithelium is also named by the number of cell layers. **Simple** epithelium consists of a single layer of cells. All cells in simple epithelium touch the basement membrane and are exposed to the upper surface of the tissue. **Stratified** epithelium has layers of cells, one stacked on another. Only the upper layer of cells in stratified epithelium is exposed to the free surface of the tissue. **Pseudostratified** epithelium, as its name implies, is "falsely stratified." Cells grow to different heights, and the taller cells grow over and cover the shorter cells. The layering of nuclei gives the tissue a stratified appearance. **Transitional** epithelium lines organs that expand and shrink. This epithelium comprises a variety of cell types and shapes to accommodate stretching.

A. Simple Squamous Epithelium

Simple squamous epithelium (Figure 5.2a) is found in serous membranes, the lining of blood vessels and the heart, and in the air sacs (alveoli) of the lungs. When viewing a superficial preparation of this epithelium under the microscope, tissue appears as a sheet of cells, similar to ceramic tiles on a floor. Functions of simple squamous epithelium include secretion and diffusion.

B. Simple Cuboidal Epithelium

Simple cuboidal epithelium comprises cube-shaped cells with large central nuclei. The tissue forms ducts and tubules, and the secretory portions of glands such as the salivary glands and the thyroid gland. Note in Figure 5.2b the cube-shaped cells arranged in a ring to make a tubule. Simple cuboidal epithelium functions in secretion and absorption.

C. Simple Columnar Epithelium

Simple columnar epithelium lines most of the uterine tubes, the digestive tract, and the gallbladder. Notice in Figure 5.2c that the nuclei are uniformly organized at the base of the cells. Simple columnar epithelium functions in protection, secretion, and absorption.

D. Pseudostratified Columnar Epithelium

Pseudostratified columnar epithelium lines the trachea and bronchi of the respiratory system, and parts of the male reproductive system. This tissue is always ciliated to propel materials along the tissue surface. Notice in Figure 5.3a that the nuclei are unevenly distributed, creating a stratified appearance. All cells are attached to the basement lamina, yet some do not reach the free tissue surface. Between the simple columnar cells are **goblet cells** (not shown in the figure). In the respiratory system, mucus produced by goblet cells traps dust and other particles in the inhaled air. Cilia on the columnar cells sweep the mucus up to the throat for disposal.

E. Transitional Epithelium

Transitional epithelium covers internal surfaces that stretch and shrink, such as the urinary bladder and the gallbladder. This tissue has a variety of cell shapes and sizes and can stretch in all directions (see Figure 5.3b). Most transitional tissue slides are prepared with relaxed transitional tissue, and the cells are stacked one on another.

F. Stratified Epithelium

Stratified epithelium is a multilayered tissue with only the bottom cellular layer in contact with the basement lamina and only the upper cells exposed to the free tissue surface. A variety of cell shapes occurs in stratified epithelium, and the type of cell on the free surface layer is used to describe and classify these tissues. Exposed surfaces,

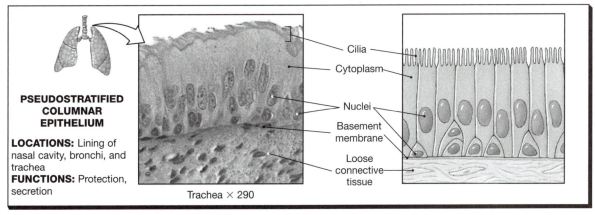

PSEUDOSTRATIFIED COLUMNAR EPITHELIUM

LOCATIONS: Lining of nasal cavity, bronchi, and trachea
FUNCTIONS: Protection, secretion

Cilia
Cytoplasm
Nuclei
Basement membrane
Loose connective tissue

Trachea × 290

(a)

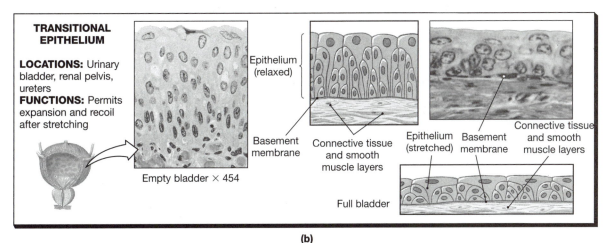

TRANSITIONAL EPITHELIUM

LOCATIONS: Urinary bladder, renal pelvis, ureters
FUNCTIONS: Permits expansion and recoil after stretching

Empty bladder × 454

Epithelium (relaxed)

Basement membrane
Connective tissue and smooth muscle layers

Epithelium (stretched)   Basement membrane   Connective tissue and smooth muscle layers

Full bladder

(b)

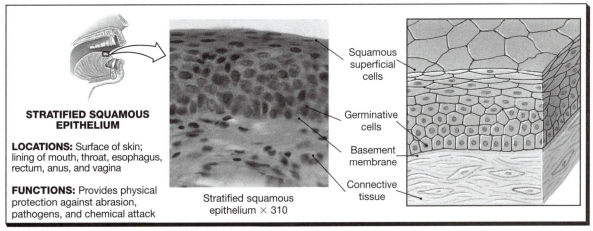

STRATIFIED SQUAMOUS EPITHELIUM

LOCATIONS: Surface of skin; lining of mouth, throat, esophagus, rectum, anus, and vagina

FUNCTIONS: Provides physical protection against abrasion, pathogens, and chemical attack

Stratified squamous epithelium × 310

Squamous superficial cells
Germinative cells
Basement membrane
Connective tissue

(c)

●**Figure 5.3**
Stratified Epithelia

(a) The pseudostratified, ciliated, columnar epithelium of the respiratory tract. Note the uneven layering of the nuclei. **(b)** At left, the lining of the empty urinary bladder, showing transitional epithelium in the relaxed state. At right, the lining of the full bladder, showing the effects of stretching on the arrangement of cells in the epithelium. **(c)** A sectional view of the stratified squamous epithelium that covers the tongue.

such as the upper layer of the skin and the lining of the mouth, protect against abrasion by building up multiple cell layers of stratified squamous epithelium. Figure 5.3c highlights stratified squamous of the mouth.

When viewing stratified squamous epithelium of the skin, the tissue layer is usually stained red or purple on the slide. Close observation within the stratified layer reveals that not all cells are squamous and that many cuboidal and columnar cells are distributed in middle layers. Cells near the surface of the tissue, however, are squamous. The bottom of the tissue is attached to connective tissue by the basement membrane.

Laboratory Activity   MICROSCOPIC OBSERVATIONS OF EPITHELIUM _____

MATERIALS

Compound microscope
prepared microscope slides
  simple squamous epithelium (lungs)
  simple cuboidal epithelium (kidney)
  simple columnar epithelium (intestine)
  stratified squamous epithelium (skin)
  pseudostratified columnar epithelium (trachea)
  transitional epithelium (urinary bladder)

PROCEDURES

1. Use the micrographs in the lab manual and your textbook as reference during your observations. On each slide, identify as many structures detailed in the diagrams as possible.

2. Examine the tissues under the microscope at various magnifications, and observe the following for each tissue:

Simple Squamous Epithelium

Scan the slide of simple squamous epithelium at low and medium power. What is the shape of the cells, and how are they arranged in the tissue?

_____

_____

Draw a section of simple squamous epithelium as viewed at medium power in the space provided.

Simple Cuboidal Epithelium

Scan the slide of simple cuboidal epithelium at low and medium power. What type of structure do the cuboidal cells form? Is the basement membrane visible?

_____

_____

Draw a tubule, and detail the simple cuboidal epithelium in the space provided.

Simple Columnar Epithelium

Examine the intestine slide at low power, and locate the lumen or open cavity of the digestive tract. Lining the lumen is simple columnar epithelium. What is the function of the columnar epithelium?

_____

_____

Draw a section of the intestine and detail simple columnar epithelium in the space provided.

Pseudostratified Columnar Epithelium

The trachea is lined with pseudostratified columnar epithelium. Focus on the edge of the lumen of the trachea to locate the tissue. How are the nuclei distributed in this tissue? Are goblet cells present?

_____

_____

Draw a section of the trachea, and detail the pseudostratified columnar epithelium in the space provided.

Transitional Epithelium

Observe transitional epithelium lining the inner wall of the urinary bladder. Describe the shape and arrangement of cells in this tissue.

_____

_____

Sketch a portion of transitional epithelium in the space provided.

Stratified Squamous Epithelium

Observe the skin slide, and locate the superficial stratified squamous epithelium. Do all cells touch the basement membrane? Which cells are exposed to the tissue surface?

_____

_____

Draw a section of skin, and detail the stratified squamous epithelium in the space provided.

PART TWO   CONNECTIVE TISSUE

WORD POWER	MATERIALS
fibroblast (...blast—a bud or sprout)	compound microscope
adipocyte (...cyte—a cell)	prepared microscope slides:
osteoclast (...clast—to break)	loose connective tissue
macrophage (...phage—to eat)	adipose
areolar (areol—small open space)	reticular tissue
canaliculus (canali—a canal)	dense connective tissue (regular)
chondroblast (chondro—cartilage)	dense connective tissue (irregular)
perichondrium (peri—around)	hyaline cartilage
osteon (osteo—a bone)	elastic cartilage
lamella (lamella—small plate)	fibrocartilage
	compact bone

OBJECTIVES

On completion of this part of the exercise, you should be able to:

- List the major groups and characteristics of connective tissue.
- Discuss the composition of the matrix of each connective tissue.

- List the location and function of each connective tissue.
- Identify each connective tissue and its cell and matrix structure under the microscope.

Connective tissue provides the body with structural support and attachment of other tissues. Unlike epithelium, cells of connective tissue are loosely arranged and scattered in the tissue (see Figure 5.4). The most abundant cells in connective tissue are **fibroblasts**. These cells secrete a **matrix** of protein **fibers** and a **ground substance**. The ground substance is usually a syrupy mixture of glycoproteins and other molecules. The fibers made by connective tissue cells include white **collagen** fibers that give the tissue strength and yellow **elastic** fibers that provide flexibility. White material in uncooked beef and chicken is collagen. Leather is mostly collagen fibers from the dermis of animal skin tanned and preserved. As we age, cells secrete fewer protein fibers into the matrix resulting in brittle bones and wrinkled skin. **Reticular** fibers are found in reticular connective tissue and provide a framework for support of internal soft organs such as the liver and spleen.

Connective tissue is classified into three broad groups distinguished primarily by their cellular composition and characteristics of the extracellular matrix. **Connective tissue proper** is a group of connective tissues that have a matrix similar to the consistency of watery Jello. Connective tissue proper includes loose (areolar), adipose, reticular, and dense connective tissues. These tissues have a matrix comprising a viscous (thick) ground substance with many extracellular fibers. As shown in Figure 5.4, connective tissue proper has a variety of cell types.

Fluid connective tissue has a liquid matrix and includes blood and lymph. The liquid matrix allows these tissues to flow freely in blood vessels or lymphatic vessels. Blood and lymph will be studied in a later exercise. Cartilage and bone tissues are **supporting connective tissues**. Cartilage has a gelatinous type of matrix and is very flexible and resistant to breaking. Bone has a solid matrix and provides a rigid framework to support other tissues of the body.

A. Connective Tissue Proper

Loose (areolar) connective tissue is distributed throughout the body. This tissue fills spaces between structures for support and protection, similar to packing material around an object in a box. Muscles, for example, are separated from surrounding anatomy by loose connective tissue. This tissue is very flexible and lets the muscle move freely under the skin. Figure 5.4a details the organization of loose connective tissue. A variety of cells is scattered in the tissue with collagen and elastic fibers clearly visible in the matrix.

Adipose or fat tissue is also distributed throughout the body and is abundant under the skin, in the buttocks, breasts, and abdomen. Fat cells are called **adipocytes** and are packed closer together than other proper connective tissues (see Figure 5.4b). The distinguishing feature of adipose tissue is the displacement of the cytoplasm owing to the storage of lipids. Adipocytes fill their vacuoles with so much lipid that the organelles and cytosol are pushed to the periphery of the cells.

Dense regular connective tissue, illustrated in Figure 5.4c, has thick parallel bundles of collagen fibers with small fibroblasts between the collagen bundles. This strong tissue forms tendons between muscles and bones and ligaments between bones. Flat layers of dense connective tissue called fascia protect and isolate muscles from surrounding structures and allow for muscle movement. The profusion of collagen fibers makes this tissue appear white under the microscope. The dermis of the skin is composed of **dense irregular connective tissue** characterized by the presence of interlacing collagen fibers. The irregular arrangement of the fibers weaves a strong capsule that supports organs such as the urinary bladder and the kidneys, and lines the cavities of joints.

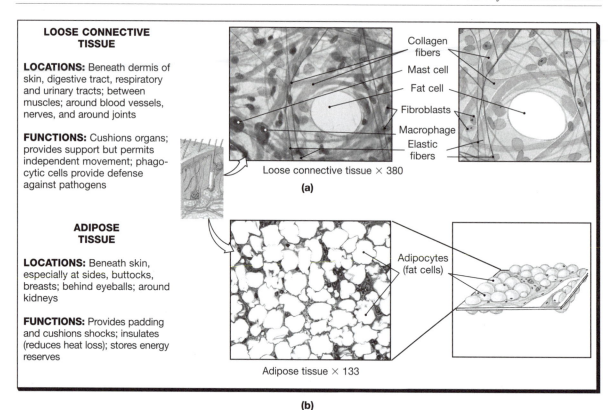

LOOSE CONNECTIVE TISSUE

LOCATIONS: Beneath dermis of skin, digestive tract, respiratory and urinary tracts; between muscles; around blood vessels, nerves, and around joints

FUNCTIONS: Cushions organs; provides support but permits independent movement; phagocytic cells provide defense against pathogens

Collagen fibers
Mast cell
Fat cell
Fibroblasts
Macrophage
Elastic fibers

Loose connective tissue × 380

(a)

ADIPOSE TISSUE

LOCATIONS: Beneath skin, especially at sides, buttocks, breasts; behind eyeballs; around kidneys

FUNCTIONS: Provides padding and cushions shocks; insulates (reduces heat loss); stores energy reserves

Adipocytes (fat cells)

Adipose tissue × 133

(b)

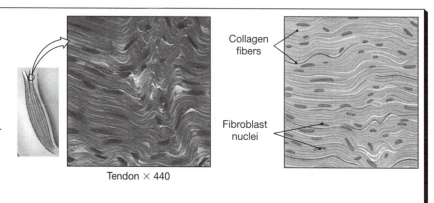

DENSE CONNECTIVE TISSUES

LOCATIONS: Between skeletal muscles and skeleton (tendons); between bones (ligaments); covering skeletal muscles; capsules of visceral organs

FUNCTIONS: Provide firm attachment; conduct pull of muscles; reduce friction etween muscles; stabilize relative positions of bones; help prevent overexpansion of organs such as the urinary bladder

Collagen fibers

Fibroblast nuclei

Tendon × 440

(c)

●**Figure 5.4**
Loose and Dense Connective Tissue

(a) All of the cells of connective tissue proper are found in loose connective tissue. **(b)** Adipose tissue is loose connective tissue dominated by adipocytes. In standard histological preparations the tissue looks empty because the lipids in the fat cells dissolve during the sectioning and staining procedures. **(c)** The dense regular connective tissue in a tendon. Notice the densely packed, parallel bundles of collagen fibers. The fibroblast nuclei can be seen flattened between the bundles.

Laboratory Activity   MICROSCOPIC OBSERVATIONS OF CONNECTIVE TISSUE PROPER_____

MATERIALS

compound microscope
prepared microscope slides
    loose (areolar) connective tissue
    adipose tissue
    dense regular connective tissue
    dense irregular connective tissue

PROCEDURES

1. Use the micrographs in the lab manual and your textbook as reference during your observations. On each slide, identify as many structures detailed in the diagrams as possible.

2. Examine the tissues under the microscope at various magnifications, and observe the following for each tissue:

Loose (Areolar) Connective Tissue

Scan the slide of loose connective tissue at low and medium magnifications. What types of fibers do you see?

_____

_____

Draw a section of loose connective tissue with the microscope at medium magnification in the space provided. Label the extracellular fibers.

Adipose Tissue

Scan the slide of adipose tissue at low and medium magnifications. How are the cells arranged? Describe the matrix of the tissue.

_____

_____

At high magnification, observe individual adipocytes. Draw several cells in the space provided.

Dense Regular Connective Tissue

Examine dense connective tissue at low and medium magnifications. Describe the composition of the matrix. How are the fibroblasts and collagen fibers arranged in this tissue?

_____

_____

Draw and label dense regular connective tissue in the space provided.

Dense Irregular Connective Tissue

Examine dense irregular connective tissue at low and medium magnifications. Describe the arrangement of fibroblasts and collagen fibers in this tissue. Contrast this arrangement with the organization of dense regular connective tissue.

_____

_____

Draw and label dense irregular connective tissue in the space provided.

B. Supportive Connective Tissue

Supportive connective tissue includes **cartilage** and **bone**. These tissues have a strong matrix of fibers to support the body's weight and handle mechanical stress. Cartilage is an avascular tissue lacking blood vessels. Materials diffuse through a gelatinous matrix to reach the cartilage cells. Bone has a solid matrix comprising calcium phosphate and calcium carbonate salts.

Supportive connective tissues are surrounded by a membrane that protects the tissue and supplies new tissue-producing cells. The **perichondrium** surrounds cartilage and produces **chondroblasts**, which secrete the fibers and ground substance of the cartilage matrix. Eventually, chondroblasts become trapped in matrix within small spaces called **lacunae** and lose the ability to produce additional matrix. These cells are now called **chondrocytes** and function in maintenance of the mature tissue. Examine Figure 5.5, and locate the features of cartilage.

HYALINE CARTILAGE

Hyaline cartilage is the most common cartilage tissue in the body. It is located in most joints of the skeletal system, the nasal septum, the larynx, and lower respiratory

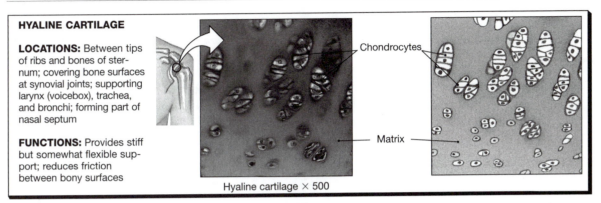

HYALINE CARTILAGE

LOCATIONS: Between tips of ribs and bones of sternum; covering bone surfaces at synovial joints; supporting larynx (voicebox), trachea, and bronchi; forming part of nasal septum

FUNCTIONS: Provides stiff but somewhat flexible support; reduces friction between bony surfaces

Chondrocytes

Matrix

Hyaline cartilage × 500

(a)

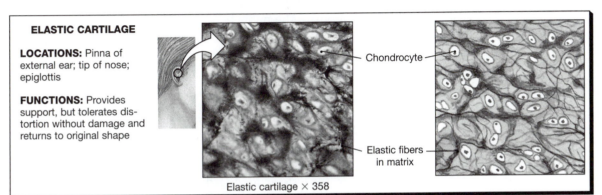

ELASTIC CARTILAGE

LOCATIONS: Pinna of external ear; tip of nose; epiglottis

FUNCTIONS: Provides support, but tolerates distortion without damage and returns to original shape

Chondrocyte

Elastic fibers in matrix

Elastic cartilage × 358

(b)

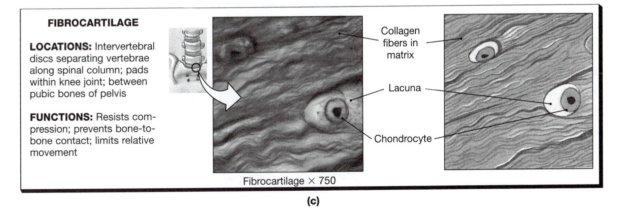

FIBROCARTILAGE

LOCATIONS: Intervertebral discs separating vertebrae along spinal column; pads within knee joint; between pubic bones of pelvis

FUNCTIONS: Resists compression; prevents bone-to-bone contact; limits relative movement

Collagen fibers in matrix

Lacuna

Chondrocyte

Fibrocartilage × 750

(c)

●Figure 5.5
Types of Cartilage

(a) Hyaline cartilage. Note the translucent matrix and the absence of prominent fibers.
(b) Elastic cartilage. The closely packed elastic fibers are visible between the chondrocytes.
(c) Fibrocartilage. The collagen fibers are extremely dense, and the chondrocytes are relatively far apart.

passageways. The gelatinous matrix provides flexible support and reduces friction between bones in a joint. Examine the micrograph of hyaline cartilage in Figure 5.5a. The tissue is distinguishable from other cartilages by the lack of stained fibers in the matrix. Hyaline cartilage does contain extracellular elastic and collagen fibers, yet they do not pick up commonly used staining solutions. Along the outer perimeter of the cartilage is a perichondrium with chondroblasts. Chondrocytes occur deeper in the tissue and are surrounded by lacunae.

ELASTIC CARTILAGE

Elastic cartilage, shown in Figure 5.5b, has many elastic fibers in the matrix and is therefore easy to distinguish from hyaline cartilage. Chondrocytes are trapped in lacunae in the middle of tissue, whereas the perichondrium and chondroblasts are found on the periphery. The elastic fibers allow considerable binding and twisting of the tissue. The pinna (flap) of the ear, the larynx, and the tip of the nose contain elastic cartilage. Bend the pinna of one of your ears. Notice how the elastic cartilage returns to the original shape after distortion.

FIBROCARTILAGE

Fibrocartilage contains collagen fibers that are visible in the matrix and chondrocytes stacked one on another (see Figure 5.5c). This cartilage is very strong and durable, and functions to cushions joints and limit bone movement. The intervertebral discs of spine, the anterior pelvis joint where the pubis bones join, and pads in the knee joint are all exposed to stress forces of the skeleton. Fibrocartilage functions as a shock absorber in these areas.

BONE

Bone tissue, Figure 5.6, is surrounded by the **periosteum** that contains **osteoblasts** for bone growth and repair. As with chondroblasts, osteoblasts secrete organic components of the matrix, become trapped within lacunae, and mature into **osteocytes**. Compact bone is characterized by the presence of columns of tissue called **osteons**. Each osteon surrounds a **central canal** that contains blood vessels. Layers of matrix, **lamellae**, surround the central canal. **Canaliculi** perforate the osteon for the diffusion of nutrients and wastes. Carbonic acid is secreted by **osteoclasts** to dissolve portions of the bone matrix and release calcium ions into the blood for various chemical processes. Bone functions in the support of the body, attachment of skeletal muscles, and protection of internal organs.

Laboratory Activity   MICROSCOPIC OBSERVATIONS OF SUPPORTIVE CONNECTIVE TISSUE___

MATERIALS

compound microscope
prepared microscope slides
   hyaline cartilage   fibrocartilage
   elastic cartilage   bone

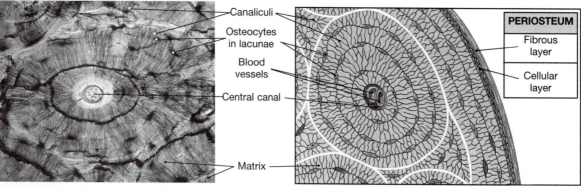

●**Figure 5.6**
Bone

The osteocytes in bone are usually organized in groups around a central space that contains blood vessels. For the photomicrograph, a sample of bone was ground thin enough to become transparent. Bone dust filled the lacunae and the central canal, making them appear dark.

PROCEDURES

1. Use the micrographs in the lab manual and your textbook as reference during your observations. On each slide, identify as many structures detailed in the diagrams as possible.
2. Examine the tissues under the microscope at various magnifications, and observe the following for each tissue:

Hyaline Cartilage

Scan the slide of hyaline cartilage at low and medium magnifications. Do you see the perichondrium and chondroblasts?

_____

_____

Examine the deeper middle region of the cartilage. How do these cartilage cells differ from the cells in the perichondrium?

_____

_____

Draw a section of hyaline cartilage in the space provided, and label the perichondrium, chondroblasts, chondrocytes, and lacunae.

Elastic Cartilage

Identify the perichondrium with many small chondroblasts and the chondrocytes trapped in lacunae deeper in the tissue. What is visible in the matrix of this cartilage?

_____

_____

Draw a section of elastic cartilage in the space provided, and label the perichondrium, chondroblasts, chondrocytes, lacunae, and elastic fibers in the matrix.

Fibrocartilage

Fibrocartilage comprises groups of chondrocytes stacked up in lacunae. Collagen fibers occur in the matrix, yet they may be difficult to see microscopically. How does the arrangement of cells in this cartilage differ from the cell distribution in hyaline and elastic cartilages?

_____

_____

Draw a section of fibrocartilage in the space provided, and label the chondrocytes.

Bone (Osseous) Tissue

Do not use the high power objective on this slide. Bone tissue slides are much thicker than most slide preparations. Use care to avoid forcing an objective lens onto the slide and damaging the lens or cracking the slide.

Scan the bone slide and observe the pattern of osteons. Increase magnification and observe the detailed structure of a single osteon. Compare bone tissue to cartilage.

Draw several osteons in the space provided, and label the central canal, canaliculi, lamellae, and osteocytes.

PART THREE   MUSCLE TISSUE

WORD POWER	MATERIALS
sarcolemma (sarco—flesh)	compound microscope
striated (stria—streaked)	prepared microscope slides
cardiac (cardia—heart)	skeletal muscle
cardiocytes	cardiac muscle
visceral (viscera—internal organ)	smooth muscle

OBJECTIVES

On completion of this part of the exercise, you should be able to:

- List the three types of muscle tissue and a function of each.
- Describe the histological appearance of each type of muscle tissue.
- Identify each muscle tissue in microscope preparations.

Muscle tissue specializes in contraction. Muscle cells shorten during contraction and produce force or tension that causes a variety of movements. Muscular valves control the movement of materials through the digestive system. Rapid skeletal muscle contractions generate body heat. The heart beats and pushes blood through the vascular system.

There are three types of muscle tissue, each named for their location in the body. **Skeletal** muscle is attached to bone and provides movement such as walking and moving your head. **Cardiac** muscle forms the walls of the heart. During the contraction phase of a heart beat, blood is pumped into blood vessels. The pressure generated by the heart's contraction forces blood to flow through the vascular system and supply cells with oxygen, nutrients, and other essential materials. **Visceral** or **smooth** muscle tissue is found inside hollow organs such as the stomach, intestines, blood vessels, and uterus. Visceral muscle controls the movement of materials through the digestive system, the diameter of blood vessels, and the contraction of the uterus during labor.

A. Skeletal Muscle

Figure 5.7a highlights skeletal muscle tissue. The tissue comprises large, long cellular structures called **muscle fibers**. Each fiber is a composite of many fused myoblasts creating a **multinucleated** fiber. The nuclei are clustered under the **sarcolemma**, the muscle fiber's cell membrane. Muscle fibers are **striated** with a distinct banded pattern due to a repeating organization of internal contractile proteins called **myofilaments**. Skeletal muscle tissue may be stimulated to contract consciously and is therefore under **voluntary** control.

B. Cardiac Muscle

Cardiac muscle tissue is striated like skeletal muscle tissue. Each cardiac muscle fiber (cardiocyte) has a single nucleus (**uninucleated**) and branches. Cardiocytes are interconnected by **intercalated discs**, which conduct the stimulus of contraction to neighboring cells. Locate an intercalated disc in Figure 5.7b. Unlike your skeletal muscles, the control of cardiac muscle is **involuntary**. For example, when you exercise, autonomic nerves increase the heart rate.

C. Visceral (Smooth) Muscle

Figure 5.7c shows smooth muscle tissue. Smooth muscle fibers are **nonstriated**, which means they lack the bands found in skeletal and cardiac muscle tissue. Each fiber is **uninucleated** and is spindle shaped, thick in the middle and tapered at the ends. The tissue usually occurs in double sheets of tissue organized at right angles. This arrangement enables the tissue to shorten structures and reduce their diameter. Visceral muscle is under **involuntary** control.

Laboratory Activity   MICROSCOPIC OBSERVATIONS OF MUSCLE TISSUE _____

MATERIALS

compound microscope
prepared microscope slides
    skeletal muscle
    cardiac muscle
    smooth muscle

SKELETAL MUSCLE TISSUE

LOCATIONS: Combined with connective tissues and nervous tissue in skeletal muscles, organs such as the skeletal muscles of the limbs

FUNCTIONS: Moves or stabilizes the position of the skeleton; guards entrances and exits to the digestive, respiratory, and urinary tracts; generates heat; protects internal organs

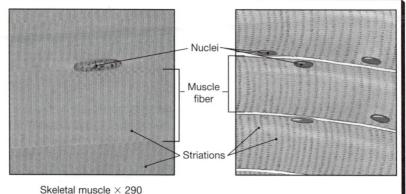

Nuclei

Muscle fiber

Striations

Skeletal muscle × 290

(a)

CARDIAC MUSCLE TISSUE

LOCATION: Heart

FUNCTIONS: Circulates blood; maintains blood (hydrostatic) pressure

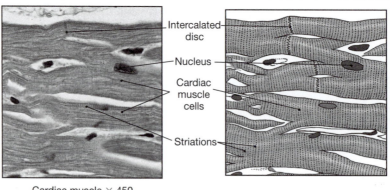

Intercalated disc

Nucleus

Cardiac muscle cells

Striations

Cardiac muscle × 450

(b)

SMOOTH MUSCLE TISSUE

LOCATIONS: Encircles blood vessels; found in the walls of digestive, respiratory, urinary, and reproductive organs

FUNCTIONS: Moves food, urine, and reproductive tract secretions; controls diameters of respiratory passageways; regulates diameter of blood vessels; and contributes to regulation of tissue blood flow

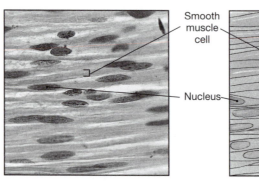

Smooth muscle cell

Nucleus

Smooth muscle × 300

(c)

●**Figure 5.7**
Muscle Tissue

(a) Skeletal muscle fibers. Note the large fiber size, prominent banding pattern, multiple nuclei, and unbranched arrangement. **(b)** Cardiac muscle cells. Cardiac muscle cells differ from skeletal muscle fibers in three major ways: size (cardiac muscle cells are smaller), organization (cardiac muscle cells branch), and number and location of nuclei (a typical cardiac muscle cell has one centrally placed nucleus). Both contain actin and myosin filaments in an organized array that produces striations. **(c)** Smooth muscle fibers. Smooth muscle fibers are small and spindle-shaped, with a central nucleus. They do not branch, and there are no striations.

PROCEDURES

1. Examine skeletal muscle tissue under the microscope at various magnifications. Use the figures in the lab manual and your textbook for reference. Compare the appearance of the tissue in cross section and longitudinal section. Draw and label the microscopic structure of skeletal muscle tissue in the space provided.

2. Examine the heart muscle under the microscope at various magnifications. Use the figures in the lab manual and your textbook for reference. Examine individual cardiocytes at a high magnification. Are there striations and branches? What structure interconnects adjacent cells? _____

   _____

   _____

   Draw and label the microscopic structure of cardiac muscle tissue in the space provided.

3. Examine smooth muscle under the microscope at various magnifications. Where is the nucleus located in the cell? Are striations visible?_____

   _____

   Draw and label the microscopic structure of visceral muscle tissue in the space provided.

PART FOUR   NERVE TISSUE

WORD POWER	MATERIALS
neuron (neuro—a nerve)	compound microscope
soma (soma—a body)	prepared microscope slide:
dendrite (dendro—tree)	multipolar neuron
axon (axo—an axle)	
synapse (synap—a union)	

OBJECTIVES

On completion of this part of the exercise, you should be able to:

- List the basic functions of neural tissue.
- Identify a neuron and its basic structure under the microscope.
- Describe how neurons communicate with other cells.

To maintain homeostasis, the body must constantly evaluate internal and external conditions and respond quickly and appropriately to environmental changes. The nervous system processes information from sensory organs and responds with motor instructions to the body's effectors, the muscles and glands. Cells responsible for retrieving, interpreting, and sending the electrical signals of the nervous system are called **neurons**.

Neurons are excitable; that is, they can respond to environmental changes by processing stimuli into electrical impulses called **action potentials**. These impulses may be conducted to adjacent neurons or to muscle and glandular tissues for homeostatic adjustments.

A typical neuron has distinct cellular regions. Examine Figure 5.8, and locate the **soma** surrounding the **nucleus**. This area also contains most of the organelles of the cell. Many fine extensions called **dendrites** conduct impulses toward the soma that then passes the signal into a single **axon** that conducts information away from the soma, toward another neuron or to a muscle or gland. The end of the axon is an enlarged **synaptic knob** that contains membranous structures called **synaptic vesicles**. These vesicles contain **neurotransmitter** molecules, a chemical messenger used to excite or inhibit other cells. The axon communicates with an adjacent neuron, muscle fiber, or gland across a specialized junction called the **synapse**. Notice in Figure 5.8 the three axons on the left synapsing with the neuron shown in full on the right. At the synapse, cells do not touch; a small **synaptic cleft** separates the cells. When an action potential reaches the end of the axon, synaptic vesicles release neurotransmitter molecules that diffuse across the synaptic cleft and excite or inhibit the postsynaptic cell.

Laboratory Activity   MICROSCOPIC OBSERVATIONS OF NERVE TISSUE

MATERIALS

compound microscope
prepared microscope slide
multipolar neuron

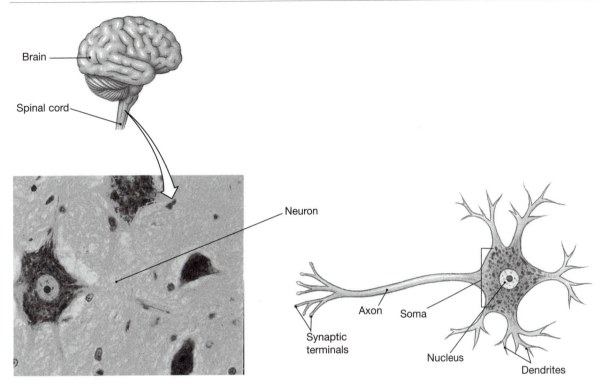

●Figure 5.8
Neural Tissue

Study Tip

On most neuron slides, it is difficult to distinguish the axon from the dendrites. Locate a neuron isolated from others, and examine the soma for a large extension. This branch is most likely the axon.

PROCEDURES

1. Start at low magnification, and select a neuron to examine more closely. Center the neuron in the field and increase magnification. Adjust the light setting of the microscope if necessary.
2. Identify the following structures of a neuron:
   soma, nucleus, dendrites (thin extensions), axon (thicker extension)
3. Draw and label several neurons in the space provided.

Name _____   Date _____

Section _____   Number _____

SURVEY OF HUMAN TISSUES
Exercise 5 Laboratory Report

A. **Short Answer Questions**

1. Which part of a neuron conducts an impulse toward the soma?

2. How are skeletal and cardiac muscle tissues similar and dissimilar?

3. Which types of muscle tissues are striated?

4. List two locations of smooth muscle tissue in the body.

5. What are the functions of intercalated discs in cardiac muscle?

6. Which muscle tissues are controlled involuntarily?

7. What is the matrix of a tissue? Do all tissues have a matrix?

8. How are epithelia attached to the body?

9. List the three major groups of connective tissues, and give an example of each.

10. How are the various epithelial tissues named?

B. **Matching**

Match each term on the left with the correct description on the right. Each choice may be used more than once.

1. _____ muscle fiber membrane   A. sarcolemma
2. _____ electrical conduction in heart B. intercalated disc
3. _____ striated, uninucleated   C. cardiac muscle
4. _____ muscle of tip of tongue   D. skeletal muscle
5. _____ muscle tissue of artery   E. smooth muscle
6. _____ voluntary muscle
7. _____ nonstriated
8. _____ involuntary, striated

C. Identify

Identify the connective tissue described in each statement.

1. _____ Cells with nucleus and cytoplasm displaced
2. _____ Solid matrix, lamellae surrounding central canals
3. _____ Found in intervertebral discs, chondrocytes stacked inside lacunae
4. _____ Parallel bundles of collagen fibers with fibroblasts between fibers
5. _____ Gelatinous matrix, chondrocytes in lacunae, elastic fibers in matrix
6. _____ Numerous cells in a network of reticular fibers.

D. Matching

Match each term on the left with the correct description on the right.

1. _____ simple cuboidal   A. lines urinary bladder
2. _____ stratified squamous   B. forms thin serous membranes
3. _____ transitional   C. contains goblet cells
4. _____ simple columnar   D. lines the mouth cavity
5. _____ simple squamous   E. forms tubules in kidneys

E. Fill in the Blanks

1. Epithelium that occurs in a single layer is _____ epithelium _____.
2. Epithelium is attached to the body by _____.
3. Epithelium that stretches and relaxes is _____ epithelium.
4. Cells that secrete mucus are called _____.
5. The tissue attached to the skeleton is called _____.
6. Strong fibers called _____ give connective tissue strength.
7. The tissue group with widely scattered cells is _____.
8. _____ muscle tissue comprises organs such as blood vessels and intestines.

F. Completion

Identify the tissue in each of the following illustrations:

1. Epithelial tissue in Figure 5.9.

1. _____

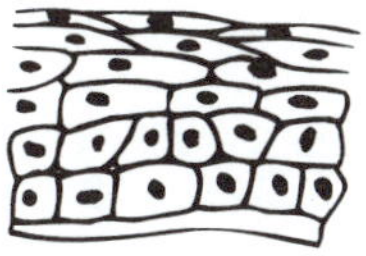

2. _____

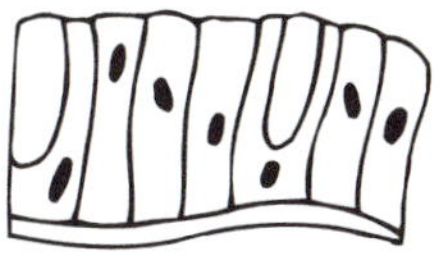

3. _____

4. _____

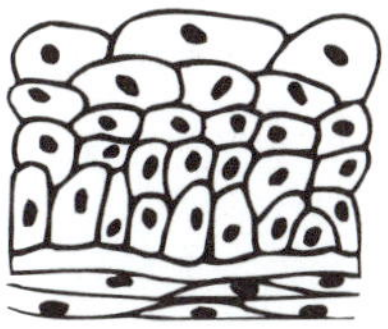

5. _____

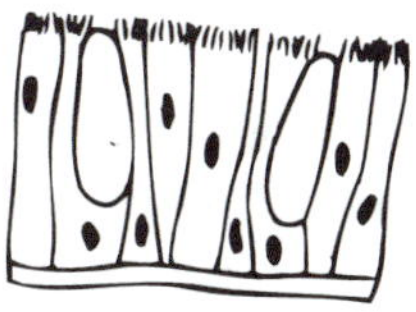

6. _____

●Figure 5.9
Types of Epithelial Tissue

   2. Connective tissue in Figure 5.10.
   3. Muscle tissue in Figure 5.11.

Connective Tissue Proper

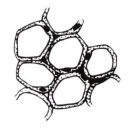

1. Loose Connective

2. Adipose

3. _____

Supporting Connective Tissue Proper

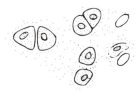

5. _____

6. dense Connective

7. _____

●**Figure 5.10**
Types of Connective Tissue

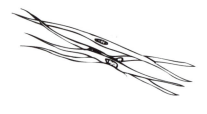

1. _____

2. _____

3. _____

●**Figure 5.11**
Types of Muscle Tissue

6 THE INTEGUMENTARY SYSTEM

WORD POWER

epidermis (epi—over, upon)
stratum (strat—a layer)
melanin (melan—black)
spinosum (spin—a thorn)
granulosum (gran—grain)
lucidum (luci—clear)
corneum (corn—a horn)

dermis (derm—the skin)
papillary (papill—a nipple)
sebaceous (seb—grease)
apocrine (apo—off, from)
eccrine (ec—out, away)
pili (pil—hair)

MATERIALS

skin models
skin charts
compound microscope
prepared microscope slides
   scalp cross section
   palm or sole cross section

OBJECTIVES

On completion of this exercise, you should be able to:

- Identify the two main layers of the skin.
- Identify the layers of the epidermis.
- Distinguish between the papillary and reticular layers of the dermis.
- Identify a hair follicle, the parts of a hair, and an arrector pili muscle.
- Distinguish between sebaceous and sudoriferous glands.
- Describe three sensory organs of the integument.

The integumentary system is the most visible organ system of the human body. Most individuals spend more time grooming their skin and skin derivatives such as hair and nails than exercising their muscular system. Our impressions of others are sometimes influenced by their skin. If their skin is wrinkled, we think of them as old. Tanned skin is associated with health even though the sun can damage the skin. The skin seals your body in a protective barrier that is flexible yet resistant to abrasion and evaporative water loss. We contact the external environment with the skin. To caress a baby's head, to feel the texture of granite, or to test the temperature of your bath water, all these involve sensory organs in the skin. Sweat glands in the skin cool the body to regulate body temperature. Vitamin D, an essential compound in calcium and phosphorus balance, is synthesized by the skin on exposure to sunlight. In this exercise, you will study the structure of the skin.

The skin is organized into two tissue layers, a superficial layer of epithelium, called the **epidermis**, and a deeper layer of connective tissue, called the **dermis**. Locate these two main layers in Figure 6.1. Isolating the dermis from underlying structures is a fatty layer of connective tissue, the **hypodermis or subcutaneous** layer.

A. Epidermis

The epidermis comprises stratified squamous epithelium organized into many distinct cell bands or strata, shown in Figure 6.2. Thick-skinned areas such as the palms and soles have five strata, thin-skinned areas have four. Cells are produced in the basal region of

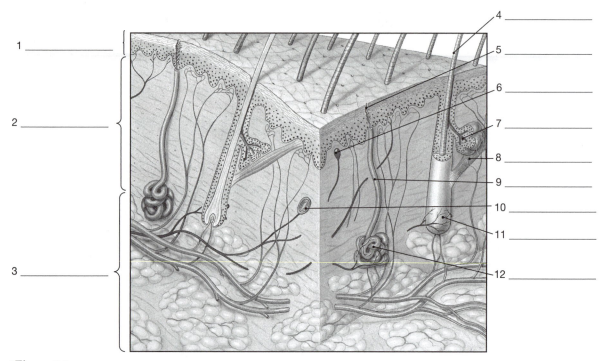

1 _____

2 _____

3 _____

4 _____

5 _____

6 _____

7 _____

8 _____

9 _____

10 _____

11 _____

12 _____

●**Figure 6.1**
Components of the Integumentary System

Relationships among the major components of the integumentary system (with the exception of nails, shown in Figure 6.4).

the epidermis and are pushed upward toward the surface of the skin. During this migration, cells synthesize and accumulate the protein keratin that reduces water loss across their cell membrane. The uppermost layer of the epidermis comprises dead, dry, scalelike cells.

The **stratum germinativum**, or **stratum basale**, is a single cell layer between the base of the epidermis and the upper dermis. Locate this layer in Figure 6.2. The cells of this stratum are in a constant state of mitosis to replace cells that have rubbed off the epidermal surface. As new cells are produced, they push previously formed cells toward the surface. It takes from 15 to 30 days for a cell to migrate from the stratum basale to the top of the stratum corneum.

Many different cell types populate the stratum basale. **Melanocytes** produce a pigment called **melanin** that protects deeper cells from the harmful effects of ultra-violet radiation from the sun. Prolonged exposure to UV light causes an increase in melanin synthesis, resulting in a darkening or tanning of the skin. Other epidermal cells produce a yellow-orange pigment called **carotene**. This pigment is common in light-skinned individuals.

Superficial to the stratum basale is the **stratum spinosum**, which consists of five to seven rows of cells interconnected by thickened cell membranes called **desmosomes**. During the slide preparation process, cells in this layer often shrink, whereas the desmosomes bridges between cells remain intact. This results in cells with a spiny out-line, therefore, the layer's name, spinosum.

Superior to the stratum spinosum is a layer of darker cells composing the **stratum granulosum**. As cells from the stratum basale are pushed upward, they synthesize the protein **keratohyalin**, which increases durability and reduces water loss from the skin. Keratohyalin granules stain dark and give this layer its color.

The uppermost layer of the epidermis is the **stratum corneum**, with many layers of flattened, dead cells. As cells from the stratum granulosum move upward, keratohyalin

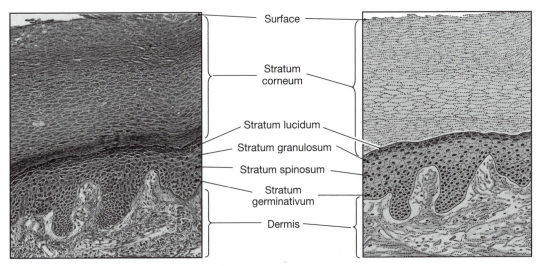

Surface

Stratum corneum

Stratum lucidum

Stratum granulosum

Stratum spinosum

Stratum germinativum

Dermis

●Figure 6.2
Layers of the Epidermis

A light micrograph through a portion of the epidermis, showing the major stratified layers of epidermal cells. (LM × 90)

granules form into the fibrous protein **keratin**, which kills the cells and waterproofs them with a hard covering. If the skin is located in an area of high abrasion, like the palms or soles, a thin transparent layer of cells, called the **stratum lucidum**, lies between the stratum granulosum and the stratum corneum. In the stratum lucidum, keratohyalin is converted into keratin. Cells of the stratum corneum are constantly being shed or worn off and replaced by new cells produced and pushed up from deeper layers.

B. Dermis

Below the stratum germinativum is the **dermis**, a band of connective tissues that anchors the epidermis in place (see Figure 6.1). The dermis is divided into two regions, the **papillary** and **reticular** layers. Although no distinct boundary occurs between these layers, the upper fifth of the dermis composes the papillary region. At the interface between the dermis and the epidermis is the papillary layer, consisting of loose connective tissue with many collagen and elastic fibers. Here the dermis is folded and projects into the epidermis. These dermal ridges fold the surface of the epidermis, such as the swirls of fingerprints. These patterns on the skin are genetically based and do not change during an individual's life. Within the dermal projections are small sensory receptors for light touch, called **Meissner's corpuscles**.

Deep to the papillary region is the reticular region, distinguished by widely scattered cells and dense irregular connective tissue with interlaced collagen fibers. By comparison, collagen fibers in the papillary region are less distinct. Large blood vessels, sweat glands, and adipose tissue are also visible in the reticular region. Sensory receptors, called **Pacinian corpuscles**, detect deep pressure. Attaching the dermis to underlying structures is the **subcutaneous layer** or **hypodermis**, profuse with adipose tissue and loose connective tissue.

C. Accessory Structures of the Skin

During embryonic development, the epidermis produces accessory structures called **epidermal derivatives**, which include oil and sweat glands, hair, and nails. These structures are exposed on the surface of the skin and project deep into the dermis.

SEBACEOUS GLANDS

Sebaceous glands secrete **sebum**, an oily substance, which coats hair shafts and the epidermal surface. Observe in Figures 6.1 and 6.3 the infolding of epithelium to form the hair follicle and sebaceous gland. Notice how the glandular cells empty onto the follicle. The gland is a **holocrine** exocrine gland. The cells fill with secretory materials and burst to deliver the product into a duct. In sebaceous glands, oil-filled sebaceous cells are pushed toward the lumen of the duct. Once on the free surface of the lumen, their cell membranes rupture, and sebum is discharged into the duct. Sebum coats the hair shaft to reduce brittleness and prevents excessive drying of the scalp.

Sebaceous **follicles** secrete sebum onto the surface of the skin. These glands lack hair and are distributed on the face, most of the trunk, and the male reproductive organs. Secretions from these glands lubricate the skin and provide antibacterial action.

Most teenagers have dealt with the "embarrassment" of skin blemishes called acne. When sebaceous ducts become blocked, sebum becomes trapped in the ducts. As the sebum accumulates, it causes the skin to raise and make a pimple. The white head of the pimple shows the duct is infected with bacteria that are feeding on the sebum.

SUDORIFEROUS (SWEAT) GLANDS

Sudoriferous glands are scattered throughout the dermis of most of the skin. Sweat glands are exocrine glands that secrete their fluid into sweat ducts that lead directly to the surface of the skin or into ducts that empty into hair follicles, shown in Figure 6.1.

Two types of sweat glands occur in the dermis, apocrine and eccrine glands. **Apocrine glands** are found in the groin, nipples, and axillae. Apocrine sweat glands secrete a thick sweat into ducts associated with hair follicles. Bacteria on the hair metabolize the sweat and produce the characteristic body odor of, for example, axillary sweat. When body temperature increases, **eccrine glands** secrete sweat containing electrolytes, proteins, urea, and other compounds onto the body surface. The sweat absorbs body heat and evaporates from the skin, cooling the body. This sweat also con-

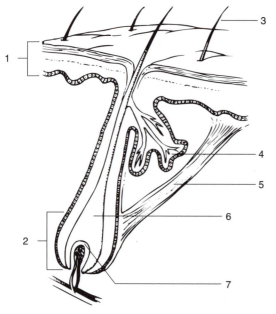

1. _____

2. _____

3. _____

4. _____

5. _____

6. _____

7. _____

●Figure 6.3
Hair Follicle, and Associated Structures

tributes to body odor because of the presence of waste products, such as urea, in the sweat. Eccrine glands are not associated with hair follicles and are distributed throughout most of the skin.

HAIR

Most of the skin is covered with hair. Only the lips, nipples, portions of the external genitalia, and the palmar and plantar surfaces of the fingers and toes are without hair. Hair generally serves a protective function. It cushions the scalp and prevents foreign objects from entering the eyes, ears, and nose. Hair also serves as a sensory receptor. Wrapped around the base of each hair is a **root hair plexus**, a sensory neuron sensitive to movement of the hair.

Examine Figure 6.3, and note that each hair is embedded in a **hair follicle**, a pocket of epidermis that extends deep into the dermis. At the base of the hair follicle is the **hair root**. At the root is a **papilla** containing nerves and blood vessels and the living, proliferative part of the hair, the **matrix**. Cells in the matrix undergo mitotic divisions that cause elongation and growth of the hair. Above the matrix, keratinization of the hair cells causes them to harden and then to die. The resulting **hair shaft** contains an outer **cortex** and inner **medulla**. Figure 6.3 also depicts an **arrector pili muscle** attached to each hair follicle. When furry animals such as dogs and cats are cold, these muscles contract to raise the hair and trap a layer of warm air by the skin. In humans, the muscle has no known thermoregulatory use; humans do not have enough hair to gain an insulation benefit. The muscles contract when we are cold and produce "gooseflesh." The arrector pili muscles also respond to emotional stimuli such as beautiful music or chalk scratching on a board. The arrector pili muscles in animals also have an emotional response. Animals raise their fur to look bigger when they feel threatened.

NAILS

To protect the tips of the fingers and toes, hard nails cover their dorsal surface. Nails comprise tightly packed keratinized cells. Figure 6.4 illustrates the parts of a typical nail. The visible elongated body of the nail, called the **nail body**, protects the underlying nail bed of the skin. Blood vessels underneath the nail body give the nail its pinkish coloration. The **free edge** of the nail body extends past the end of the digit. The **nail root** is at the base of the nail where new growth occurs. The **lunula** is a whitish portion of the proximal nail body where blood vessels do not show through. The cuticle around the nail is called the **eponychium**. It is composed of a band of epidermis that seals the **nail groove** to the epidermis. Under the free edge of the nail is the **hyponychium**, a thicker region of the epidermis.

●**Figure 6.4**
Structure of a Nail

The prominent features of a typical fingernail as viewed from the surface.

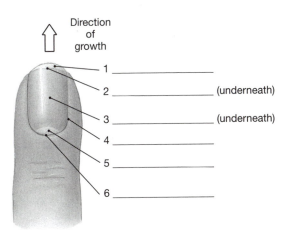

Direction of growth

1 _____

2 _____ (underneath)

3 _____ (underneath)

4 _____

5 _____

6 _____

Laboratory Activity   MICROSCOPIC OBSERVATION OF THE SKIN_____

MATERIALS

compound microscope
prepared slide of the scalp in cross section

PROCEDURES

1. Label Figures 6.1, 6.3, and 6.4
2. Scan the scalp slide at low magnification, and identify the epidermis, dermis, and hypodermis.
3. Increase magnification and examine the epidermis. Locate the epidermal layers beginning with the deepest layer, the stratum germinativum.
4. Study the dermis; identify the papillary and reticular regions. In the papillary region, look for Meissner's corpuscles positioned where the dermis folds upward along the epidermis.
5. Locate a hair follicle. Identify the hair shaft, cortex, and medulla. Identify a sebaceous gland associated with the hair follicle.
6. Scan the dermis of the slide for a sudoriferous gland. Trace the duct from the gland to the surface of the skin.
7. Study the models, charts, and other material available in the lab.

THE INTEGUMENTARY SYSTEM
Exercise 6 Laboratory Report

A. Matching

Match each skin structure listed on the left with the correct description on the right.

1. _____	sebaceous gland	A. layer in thickened areas of epidermis
2. _____	apocrine sweat gland	B. deep to the dermis, contains adipose
3. _____	keratin	C. sweat gland associated with hair follicle
4. _____	arrector pili	D. produces new epidermal cells
5. _____	stratum corneum	E. deep layer of the dermis
6. _____	papillary layer	F. protein that reduces water loss
7. _____	stratum germinativum	G. functions in thermoregulation
8. _____	reticular layer	H. surface layer of epidermis
9. _____	subcutaneous layer	I. muscle attached to hair follicle
10. _____	stratum lucidum	J. folded layer of dermis next to epidermis
11. _____	eccrine sweat gland	K. produces sebum

B. Short Answer Questions

1. Describe the layers of epidermis in an area where the skin is thick.

2. Why is the epidermis keratinized?

3. How does the skin tan when exposed to sunlight?

4. List the types of sweat glands associated with the skin.

5. How are cells replaced in the epidermis?

C. Drawing

Draw and label a transverse section of the scalp that includes a hair follicle.

D. **Completion**

  Complete the Concept Map of the skin. Use your textbook as reference if necessary.

Concept Map

  Using the terms below, fill in the circled, numbered blank spaces to complete the concept map. Follow the numbers that comply with the organization of the map.

Nerves   Epidermis   Collagen   Skin
Hypodermis   Connective   Fat   Granulosum
Papillary Layer

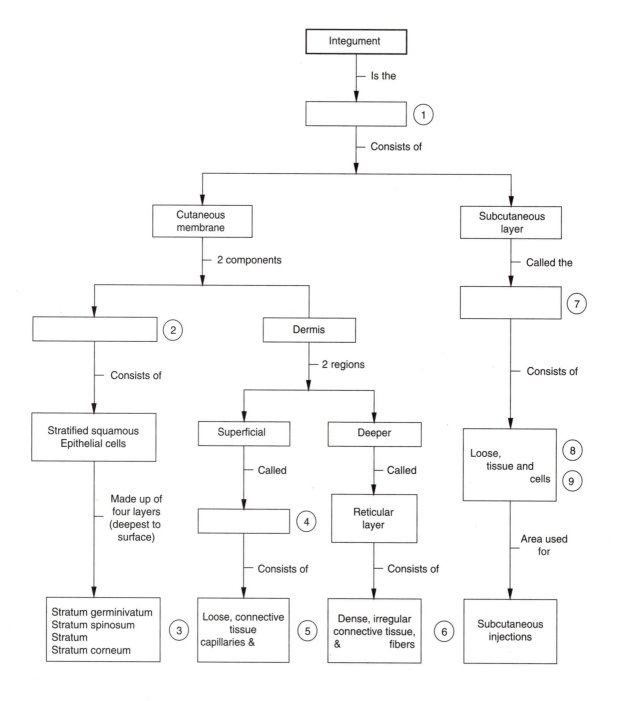

7 THE SKELETAL SYSTEM

The skeletal system serves many functions. Bones support soft tissues of the body and protect vital internal organs. Calcium, lipid, and other materials are stored in the matrix of bone tissue, and blood cells are manufactured in the red marrow of bones. Bones serve as levers for the muscular system to pull on and produce movement or maintain posture. In this exercise, you will study the gross structure of bone and the individual bones of the skeletal system.

PART ONE ORGANIZATION OF THE SKELETAL SYSTEM

OBJECTIVES

On completion of this part of the exercise, you should be able to:

- Describe the gross anatomy of a long bone.
- Describe the histological organization of compact and spongy bone.
- List the five shapes of bones, and give an example of each.

WORD POWER	MATERIALS
endosteum (endo—inner, osteo—bone)	articulated skeleton
metaphysis (meta—middle)	preserved long bone or fresh long bone
	gloves
	safety glasses
	blunt probe

A. Anatomy of a Long Bone

Long bones, such as the humerus and femur, have a shaft called the diaphysis with an epiphysis on each end (see Figure 7.1). The proximal epiphysis is on the superior end where the bone attaches to the skeleton, and the distal epiphysis is on the inferior end. The diaphysis contains compact bone, and the epiphyses contain spongy bone. A layer of hyaline cartilage, the **articular cartilage**, covers the epiphysis where it articulates with another bone. The interior of the diaphysis is hollow, forming the **medullary** or **marrow cavity** where lipid is stored as **yellow marrow**. An inner membrane called the **endosteum** lines the bone tissue facing the medullary cavity. The outer surface of the bone is a tough fibrous membrane called the **periosteum**. This membrane appears shiny and glossy in the living condition.

Between the diaphysis and an epiphysis is the **metaphysis**. In a juvenile's bone, the metaphysis is called the **epiphyseal plate** and consists of a plate of hyaline cartilage that allows the bone to grow in length. By early adulthood, the diaphysis fuses to the epiphysis leaving a bony remnant of the growth plate, the **epiphyseal line**.

●**Figure 7.1**
Structure of a Long Bone

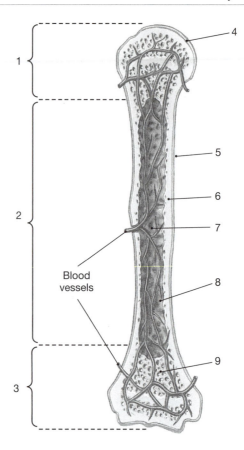

Blood
vessels

1. _____
2. _____
3. _____
4. _____
5. _____
6. _____
7. _____
8. _____
9. _____

Laboratory Activity   EXAMINATION OF A FRESH LONG BONE _____

MATERIALS

preserved long bone or a fresh long bone from a butcher shop
gloves and safety glasses
blunt probe

PROCEDURES

1. Examine the long bone, and locate the periosteum. Does it appear shiny? Are any
tendons or ligaments attached to it? Is cartilage present at the tips of the bone?

_____

_____

2. If the bone has been sectioned, observe the internal bone tissue of the diaphysis.
Is the bone tissue similar in all regions of the sectioned bone?

_____

_____

3. Look closely and locate the metaphysis. Most likely the bone has an epiphyseal
line rather than a epiphyseal plate. Why?

_____

_____

4. Label Figure 7.1.

B. Histological Organization of Bone

Two types of bone tissue are found in the skeleton: **compact** and **spongy** bone. Compact bone seals the outer surface of bones and is found in the shaft of long bones such as the femur. Spongy bone is found in the epiophyses and lining the medullary cavity of long bones.

Compact bone is characterized by supportive columns called **osteons** (see Figure 7.2). Each osteon consists of many rings of calcified matrix called **lamellae**. Between the lamellae are **osteocyte** cells in small spaces called **lacunae**. A **central canal** in the middle of the osteon surrounds blood vessels and nerves. Small diffusion channels called **canaliculi** radiate outward from the central canal through the lamellae.

Spongy bone is organized into a meshwork of bony struts called **trabeculae**. Filling the spaces between trabeculae is **red marrow**, the tissue that produces most blood cells.

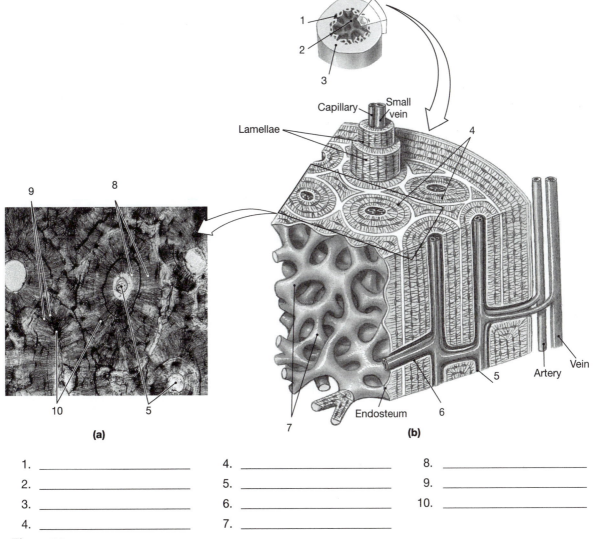

(a)   (b)

1. _____   4. _____   8. _____
2. _____   5. _____   9. _____
3. _____   6. _____   10. _____
4. _____   7. _____

●Figure 7.2
Structure of a Typical Bone

(a) A thin section through compact bone; in this procedure the intact matrix and central canals appear white, and the lacunae and canaliculi are shown in black. (LM × 272) (b) Diagrammatic view of the structure of a typical bone, the humerus.

Laboratory Activity   OBSERVATION OF BONE TISSUE _____

MATERIALS

bone tissue model

PROCEDURES

1. Label Figure 7.2.
2. Examine the bone model, and locate each structure shown in Figure 7.2.
3. Review the histology of bone tissue in Exercise 5 of the manual.

C. Overview of the Skeleton

The adult skeleton contains 206 bones. Each bone is an organ and comprises osseous tissue, cartilage, and other connective tissues. The skeleton is subdivided into the axial and appendicular divisions. The **axial division** includes the **skull**, the **vertebrae** of the spine, the **sternum**, 12 pairs of **ribs** and the **hyoid**, for a total of 80 bones. Locate these bones in Figure 7.3.

The **appendicular division** consists of the pectoral girdles, the upper limbs, the pelvic girdle, and the lower limbs. Each girdle articulates with the axial division and provides attachment and mobility of the limbs. The appendicular division of the adult skeleton has 126 bones.

Each bone of has certain anatomical features on the surface called **bone markings**. A particular bone marking may be unique to one bone or found throughout the skeleton. Table 7.1 summarizes a variety of bone markings.

| TABLE 7.1 | **Skeletal Terminology** |

General Description		Anatomical Term	Definition
Elevations and projections (general)	Process	Any projection or bump	
	Ramus	An extension of a bone making an angle to the rest of the structure	
Processes formed where tendons or ligaments attach	Trochanter	A large, rough projection	
	Tuberosity	A smaller, rough projection	
	Tubercle	A small, rounded projection	
	Crest	A prominent ridge	
	Line	A low ridge	
Processes formed for articulation with adjacent bones	Head	The expanded articular end of an epiphysis, separated from the shaft by a narrower neck	
	Condyle	A smooth, rounded articular process	
	Trochlea	A smooth, grooved articular process shaped like a pulley	
	Facet	A small, flat articular surface	
	Spine	A pointed process	
Depressions	Fossa	A shallow depression	
	Sulcus	A narrow groove	
Openings	Foramen	A rounded passageway for blood vessels and/or nerves	
	Fissure	An elongated cleft	
	Meatus	A canal leading through the substance of a bone	
	Sinus or antrum	A chamber within a bone, normally filled with air	

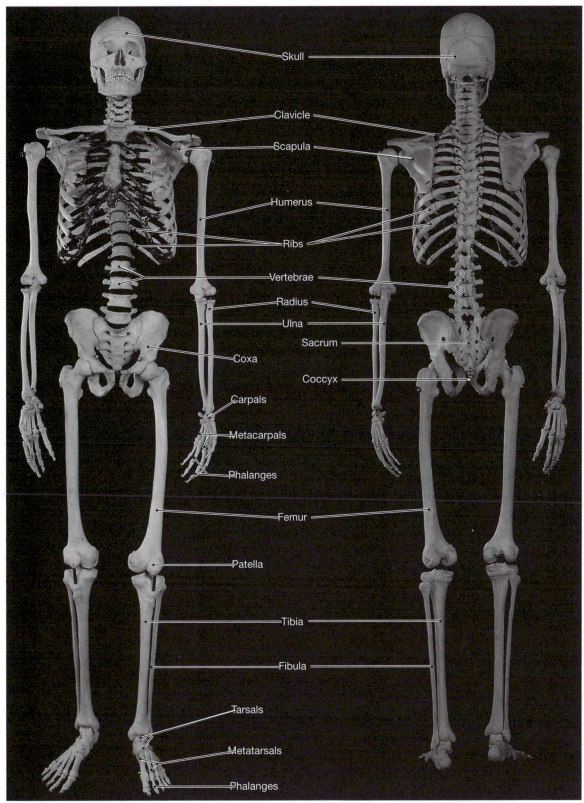

(a) **Anterior view**   (b) **Posterior view**

●**Figure 7.3**
The Skeleton
(a) Anterior and (b) posterior views of the skeleton.

PART TWO   THE AXIAL SKELETON

<table>
<tr><td colspan="2" align="center">**WORD POWER**</td><td align="right">**MATERIALS**</td></tr>
<tr>
<td>foramen (foram—opening)
condyle (condyl—a knuckle)
styloid (stylo—a point)
petrous (petro—a rock)
sella turcica (sella—saddle)
cribriform plate (cribr—a sieve)
crista galli (crist—crest;
  galli—rooster)</td>
<td>sinus (sinu—a hollow)
demifacet (demi—half; facette—small face)
manubrium (manu—a hand)
xiphoid (xiph—a sword)
dens (dens—a tooth)
tubercle (tuber—a knot)</td>
<td>articulated skeleton
disarticulated skeleton
sectioned skull</td>
</tr>
</table>

OBJECTIVES

On completion of this part of the exercise, you should be able to:

- List the components of the axial skeleton.
- Identify all cranial and facial bones on each view of the skull.
- Describe the unique features of each region of the vertebral column.
- Identify the features of a typical vertebra.
- Identify the components of the sternum.

The axial skeleton includes the skull, hyoid bone, spine, ribs, and sternum. The skull protects the brain and forms the framework for the face. Only the mandible, the lower jaw, moves on the skull for processing food and for speech. All other skull bones are immovable. Inferior to the mandible is the hyoid bone that serves as a bridge to support muscles of the pharynx and larynx. Organs of the thoracic cavity are protected by the sternum and ribs. The spine or vertebral column supports and balances the weight of the body and protects the spinal cord.

Use an articulated skeleton and a skull while you work through the sections of the axial skeleton.

Care of the Skeleton

During your laboratory studies of the skeleton, handle all bones with care and respect. The cleaning and preparation of study skeletons results in brittle bones. Many bones of the skull are thin and fragile and are easily damaged. Do not use your pencil or pen as a pointer since these will mark the skeleton. Do not force a probe or even a pipe cleaner into a hole or any other bone feature. Furthermore, do not mix bones from different skeletons. Following these simple guidelines will help the laboratory faculty maintain their skeleton inventory and provide future students with quality skeleton specimens.

A.   The Skull

The skull has 22 bones organized into 14 facial bones and 8 cranial bones that encase the brain. Six bones along the midline of the skull are unpaired, whereas the lateral bones of the skull are paired. Seven other bones are associated with the skull, six bones of the middle ear (three auditory ossicles per ear) and the hyoid bone. Therefore, the actual total number of skull bones is 29. The auditory ossicles will be presented in detail in the exercise on the anatomy of the ear.

THE CRANIUM

The cranium comprises eight bones: two temporal, two parietal, one occipital, one frontal, one sphenoid, and one ethmoid.

Locate each bone and structure on a study skull in the laboratory. The skull has most likely been sectioned transversely for observation of internal cranial structures.

Frontal Bone   The frontal bone forms the roof, walls, and floor of the anterior cranium (see Figures 7.4, 7.5, and 7.6). The frontal bone forms the superior portion of the eye orbit. It extends from the forehead posteriorly to the **coronal suture**, where it articulates with a pair of parietal bones. The **frontal squama** is the flattened expanse of the frontal bone commonly called the forehead. Directly superior to the orbit of the eye is a small foramen, the **supraorbital foramen**.

Parietal Bones   The parietal bones form the posterior crest of the skull and are joined by the **sagittal suture** at the superior midline of the skull (see Figure 7.4). The bones are smooth and have few surface features.

Occipital Bone   The occipital bone makes up the posterior floor and wall of the skull. Position your study skull as in Figure 7.4b, an inferior view, and locate each structure as it is discussed. The most conspicuous structure of the occipital bone is the **foramen magnum**, the large hole where the spinal cord enters the skull to join the brain. The lateral margins of the foramen magnum are surrounded by smooth, rounded projections, the **occipital condyles**, that articulate with the first vertebra of the spine. The **occipito-parietal (lambdoidal)** suture occurs where the occipital bone articulates with the parietal bones on the posterior skull.

Temporal Bones   The temporal bones form the lower lateral walls and part of the floor of the cranium. The temporal bone articulates with the parietal bone at the **squamosal suture**. The flat expanse of temporal bone below this suture is called the **squamous portion**. Figure 7.5 illustrates the lateral aspect of a temporal bone. Locate the **external auditory canal**, a tube for conducting sound waves toward the ear drum. Directly posterior to the external auditory canal is the **mastoid process**, a blunt, rough process used as an attachment site for muscles that move the head. Medial to the mastoid process is the needlelike **styloid process**, where muscles of the tongue and pharynx attach. Anterior to the external auditory canal is the **zygomatic arch**, a thin bridge of bone joining the temporal and zygomatic bones.

Position the study skull as in Figure 7.4b to view the inferior aspect of the temporal bones. Remove the mandible from your specimen if your lab instructor allows. Anterior to the external auditory canal is the **mandibular (condylar) fossa**, the shallow depression where the mandible articulates with the temporal. Between the mastoid process and the occipital condyle is the **jugular foramen**. Directly anterior to the jugular foramen is the **carotid canal**. The carotid canal is a passageway for the carotid artery, a major blood vessel supplying the brain. The jugular foramen is the exit site for the jugular vein, the major vein that drains blood from the brain.

Examine the floor of the temporal bones as shown in Figure 7.6a. The large crest arising from the floor is the **petrous portion** of the temporal bone. Inside this bony mass are the organs for hearing and equilibrium, as well as three small bones of the middle ear, the auditory ossicles. Identify the **internal acoustic canal** on the posterior medial surface of the petrous portion.

Sphenoid Bone   The sphenoid bone is the keystone of the cranium; all cranial bones articulate with the sphenoid. It is visible from all views of the skull: anteriorly, posteriorly, laterally, inferiorly, and internally. Use Figures 7.4, 7.5, and 7.6 to review the position of the sphenoid bone.

Locate the sphenoid bone in Figure 7.6a, and notice how it articulates with the frontal, temporal, and occipital bones. The anterior margin of the sphenoid is the bat-shaped

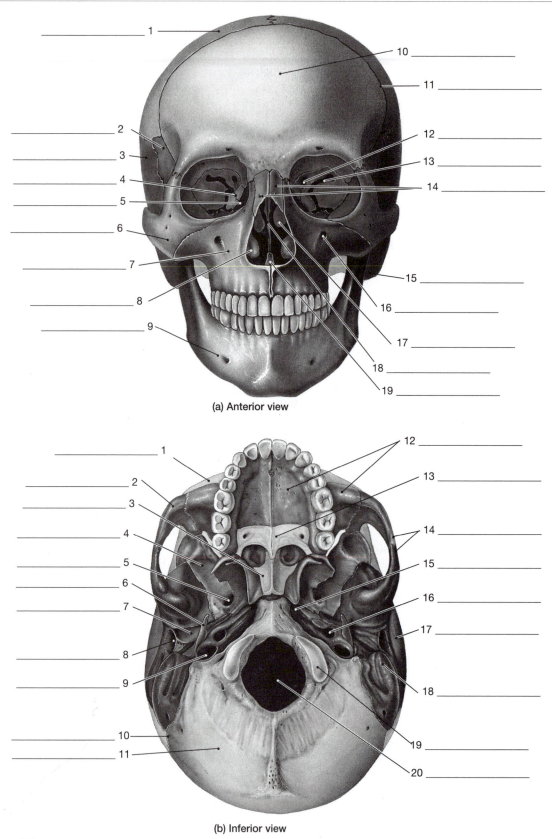

(a) Anterior view

(b) Inferior view

●Figure 7.4
The Adult Skull

(a) Anterior view and (b) inferior view.

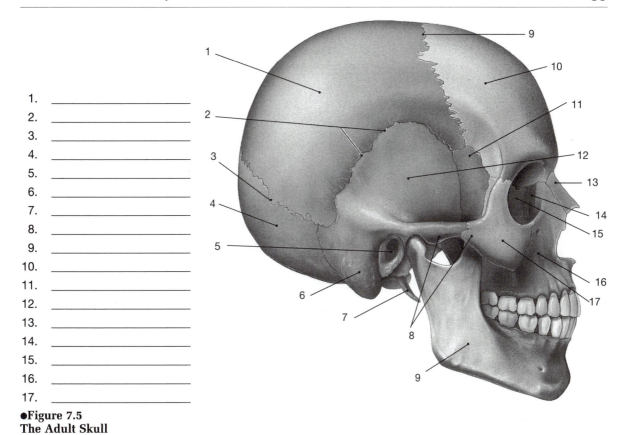

1. _____
2. _____
3. _____
4. _____
5. _____
6. _____
7. _____
8. _____
9. _____
10. _____
11. _____
12. _____
13. _____
14. _____
15. _____
16. _____
17. _____

●Figure 7.5
The Adult Skull
The adult skull in lateral view.

lesser wing. Just inferior to the medial portion of the lesser wing are the **optic foramina**, the passageways of the optic nerves from the eyes. The **greater wing** is posterior to the lesser wing and contributes to the floor of the cranium. In the middle of the sphenoid is a raised platform, the **sella turcica**, or Turk's saddle, which is where the pituitary gland rests.

Three foramina are aligned lateral to each side of the sella turcica and serve as passageways for blood vessels and nerves. The **foramen ovale** is the oval hole. Posterior to the foramen ovale is a small **foramen spinosum**. The round foramen anterior to the foramen ovale is the **foramen rotundum**. The **foramen lacerum** is medial to the carotid canal.

Ethmoid Bone   The **ethmoid** is a single rectangular bone deep in the eye orbit behind the bridge of the nose. Locate the ethmoid in Figure 7.4a, an anterior view of the skull. The ethmoid is directly posterior to the small lacrimal bone, the bone with a small foramen in the medial "corner" of the eye orbit. Observe in Figure 7.6a the vertical crest of bone called the **crista galli** and the surrounding sievelike **cribriform plate**. Fine extensions of the olfactory nerves pass though many foramina that perforate the cribriform plate. The inferior ethmoid has a thin sheet of bone, the **perpendicular plate**, which divides the nasal cavity. Thin shelves of bone extend from the lateral walls of the ethmoid as the **superior** and **middle conchae**.

BONES OF THE FACE

The face is fashioned by 14 bones. There are two nasal bones, two maxillae, two palatine, two lacrimal, two inferior nasal conchae, two zygomatic, one vomer, and one mandible.

Nasal Bones   The nasal bones form the bridge of the nose. The bones articulate superiorly with the frontal bone, laterally with the maxilla and posteriorly (internally) with the ethmoid bone. Review the nasal bones in Figures 7.4a and 7.5.

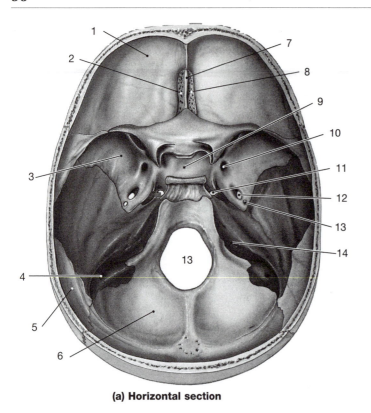

(a) Horizontal section

1. _____
2. _____
3. _____
4. _____
5. _____
6. _____
7. _____
8. _____
9. _____
10. _____
11. _____
12. _____
13. _____
14. _____

(c) Sagittal section

1. _____
2. _____
3. _____
4. _____
5. _____
6. _____
7. _____
8. _____
9. _____
10. _____
11. _____
12. _____
13. _____
14. _____
15. _____
16. _____

(b) Sagittal section

●**Figure 7.6**
Sectional Anatomy of the Skull

(a) Horizontal section through the skull, looking down on the floor of the cranial cavity.
(b) Saggital section through the skull

The Maxillae   The paired maxillae are the foundation of the face. The **alveolar processes** are the U-shaped processes where the upper teeth are embedded into the maxilla. From the inferior aspect (see Figure 7.4b), the **palatine process** of the maxilla is visible. This bony shelf forms the anterior hard palate of the mouth.

Lacrimal Bones   The lacrimal bones form the anterior portions of the medial orbital wall, illustrated in Figure 7.5. Tears flow medially across the eye and drain into the inferior **lacrimal fossa** that transports tears to the nasal cavity.

Zygomatic Bones   The check bones are called zygomatic bones. This bone also contributes to the floor and lateral wall of the orbit (see Figure 7.5). The posterior margin of the zygomatic bone narrows inferiorly to form part of the **zygomatic arch** with the temporal bone.

Palatine Bones   The palatine bones are small bones in the roof of the mouth posterior to the palatine processes of the maxillae. These bones, along with the maxillae, form the roof of the mouth and separate the oral cavity from the nasal cavity. Locate the palatine bones in Figure 7.4b.

The Vomer   The vomer separates the nasal chamber into right and left cavities. This thin sheet of bone is best viewed from the inferior aspect of the skull looking into the nasal cavities, as in Figure 7.4b.

The Inferior Nasal Conchae   The inferior nasal conchae, Figure 7.4a, are bony shelves that extend medially from the lower lateral portion of the nasal wall. They cause inspired air to swirl in the nasal cavity so that the mucous membrane lining can warm, cleanse, and moisten the air. Similar shelves of bone occur on the lateral wall of the ethmoid.

The Mandible   The mandible is the U-shaped bone of the lower jaw, as shown in Figures 7.4 and 7.5. Posteriorly, the bone curves upward at the **angle** to a raised projection, the **ramus**, which terminates at a U-shaped **mandibular notch**. Two processes extend from the notch, the anterior **coronoid process** and the posterior **condylar process**. The smooth condylar process articulates in the mandibular fossa on the temporal bone to form the temporomandibular joint (TMJ). Open and close your mouth to feel this articulation. The **alveolar process** is the crest of bone where the lower teeth articulate with the mandible. On the anterior mandible are two small holes, the **mental foramina**. Medial to each ramus is the **mandibular foramen**, a passageway for nerves to the lower teeth and lip.

Hyoid Bone   The hyoid bone is a small bone that is located inferior to the mandible. This bone is unique because it articulates with no other bone of the skeletal system. You cannot feel your hyoid bone because it is surrounded by ligaments and muscles of the throat and neck. Locate the hyoid bone on an articulated skeleton in the laboratory.

PARANASAL SINUSES OF THE SKULL

The skull contains four cavities called **paranasal sinuses** that interconnect with the nasal cavity and reduce the weight of the skull. Use your textbook to study the four sinuses of the skull.

Laboratory Activity   THE SKULL _____

MATERIALS
skull
skull with sagittal section

PROCEDURES
1. Label Figures 7.4, 7.5, and 7.6.
2. Locate each cranial bone from all views of the skull.
3. Identify the surface features on each cranial bone.

4. Locate each facial bone from all views of the skull.

5. Identify the surface features on each facial bone.

6. Locate the four paranasal sinuses on a sagittally sectioned skull.

B. Vertebral Column

The vertebral column, or spine, is a flexible column of 33 vertebrae. The vertebral column articulates with the skull superiorly, the pelvic girdle inferiorly, and the ribs laterally. Between articulating vertebrae is a cushion of fibrocartilage called the **intervertebral disc**. Vertebrae are grouped into four regions based on their location in the vertebral column and their anatomical features, illustrated in Figure 7.7. The first seven vertebrae are the **cervical** vertebrae of the neck. Twelve **thoracic** vertebrae articulate with the ribs. The lower back has five **lumbar** vertebrae and the **sacrum** that consists of five fused vertebrae. The "tail bone" includes three to five fused **coccygeal** vertebrae; most vertebral columns have four coccygeals.

Notice in Figure 7.7 that the vertebral column of an adult is not straight but is curved. These spinal curves are necessary to balance the body weight while standing erect. Read in your lecture textbook about the curvatures of the vertebral column.

VERTEBRAL ANATOMY

Figure 7.8 illustrates anatomical features of a vertebra from each region of the spine. Each vertebra is shown in a superior view. The posterior of the vertebrae is characterized by a medial projection called the spinous process. Each vertebra has the following surface features:

- **Body (centrum)**—The thick, disc-shaped anterior portion.
- **Spinous process**—A long, single extension of the posterior vertebral wall.

●**Figure 7.7**
The Vertebral Column

The major divisions of the vertebral column, showing the four spinal curves.

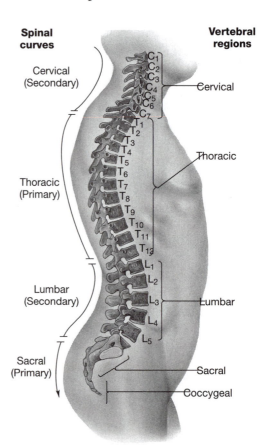

●Figure 7.8
Typical Vertebrae of the Cervical, Thoracic, and Lumbar Regions
Each vertebra is shown in superior view.

1. _____
2. _____
3. _____
4. _____
5. _____
6. _____
7. _____
8. _____

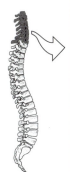
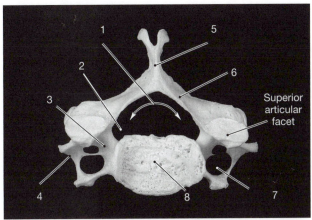

(a) Typical cervical vertebra

Superior articular facet

1. _____
2. _____
3. _____
4. _____
5. _____
6. _____

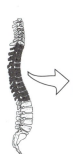
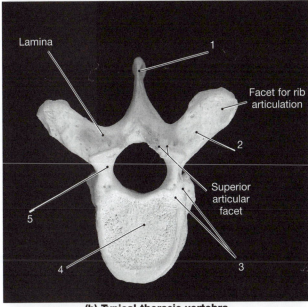

Lamina

Facet for rib articulation

Superior articular facet

(b) Typical thoracic vertebra

1. _____
2. _____
3. _____
4. _____
5. _____
6. _____

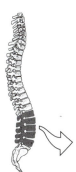

Superior articular facet

(c) Typical lumber vertebra

- **Transverse process**—A pair of extensions lateral to the spinous process.
- **Vertebral foramen**—The large hole posterior to the body. The spinal cord is protected in the spinal cavity, the tube formed by the vertebral foramina of vertebrae stacked in a column.
- **Pedicle**—The strut of bone extending posteriorly from the body to a transverse process. Each pedicle is notched to form part of an **intervertebral foramen**, a passageway for spinal nerves from the spinal cord.
- **Lamina**—A flat plate of bone between the transverse and spinous processes.

CERVICAL VERTEBRAE

Seven cervical vertebrae are located in the neck. Figure 7.8 illustrates the distinguishing feature of cervical vertebrae, the two **transverse foramina** located on the transverse processes.

The first two cervical vertebrae are modified for special articulations. The first cervical vertebra, C_1, is called the **atlas** and is the only vertebra that articulates with the skull. The atlas is unique in that it is the only vertebra that lacks a body. The **axis** is the second cervical vertebra. It is specialized to articulate with the atlas. A peglike **dens** or **odontoid process** arises superiorly from the body of the axis. The atlas pivots around the dens to rotate the head.

THORACIC VERTEBRAE

The 12 thoracic vertebrae are larger than the cervical vertebrae and increase in size as they approach the lumbar region. The thoracic vertebrae articulate with the 12 pairs of ribs. Most ribs attach at two sites on **articular facets** at the tip of the transverse process and on a **demifacet** located on the posterior of the body. Two demifacets usually are present on the same body, a **superior** and an **inferior demifacet**.

LUMBAR VERTEBRAE

The five lumbar vertebrae are large and heavy so as to support the weight of the head, neck, and trunk. Compared to thoracic vertebrae, lumbar vertebrae have a wider body, a blunt and horizontal spinous process, and shorter transverse processes.

SACRAL AND COCCYGEAL VERTEBRAE

The **sacrum** is a single bony element comprising five fused sacral vertebrae (see Figure 7.9). The sacrum articulates with the ilium of the pelvic girdle to form the posterior wall of the pelvis. Fusion of the sacral bones before birth consolidates the vertebral canal into a **sacral canal**. The **coccyx** articulates with the fifth fused sacral vertebra.

Laboratory Activity   THE VERTEBRAL COLUMN _____

MATERIALS

articulated skeleton
disarticulated vertebral column

PROCEDURES

1. Identify the four regions of the vertebral column on an articulated skeleton.
2. Describe the anatomy of a typical vertebra. Locate each feature on a vertebra. Label Figure 7.8.
3. Distinguish the anatomical differences between cervical, thoracic, and lumbar vertebrae.
4. Identify the unique features of the atlas and the axis. How do these two vertebrae articulate with the skull and each other?

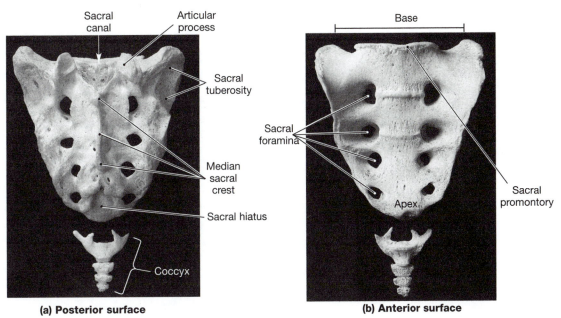

(a) Posterior surface   **(b) Anterior surface**

●**Figure 7.9**
The Sacrum and Coccyx
(a) Posterior view. **(b)** Anterior view.

C. The Ribs and Sternum

Twelve pairs of ribs occur in both males and females. The ribs articulate with the thoracic vertebrae posteriorly and the sternum anteriorly to enclose the thoracic organs in a protective rib cage. To breathe, muscles move the ribs to increase or decrease the size of the thoracic cavity and cause air to move into or out of the lungs.

THE RIBS

Ribs, also called **costa**, are classified by how they articulate with the sternum. Examine Figure 7.10, an anterior view of the rib cage. The first seven pairs are called **true ribs** because their cartilage, the costal cartilage, attaches directly to the sternum. The next five pairs of ribs are called **false ribs** because their costal cartilage does not directly connect with the sternum. Pairs 11 and 12 are called **floating ribs** and do not articulate with the sternum.

A rib has a **head** with two **facets** for articulating with the demifacets of thoracic vertebrae, and a **tubercle** that articulates with the costal facet of the transverse process. Between the head and tubercle is a slender **neck**.

THE STERNUM

The **sternum** comprises three bony elements, a superior **manubrium**, a middle **body**, and an inferior **xiphoid process**, all shown in Figure 7.10. The manubrium is triangular and articulates with the first pair of ribs and the clavicle (collar bone). The body is elongated and receives the costal cartilage of rib pairs 2–7.

Laboratory Activity   ANATOMY OF THE RIBS AND STERNUM _____

MATERIALS

articulated skeleton
disarticulated ribs and sternum

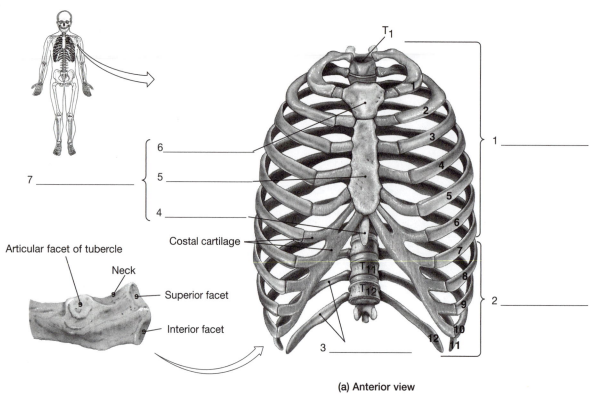

(a) Anterior view

●**Figure 7.10**
The Ribs and Sternum

PROCEDURES

1. Label Figure 7.10.
2. Discuss the anatomy of a typical rib. How many pairs of ribs do males and females have?
3. Describe the differences between true, false, and floating ribs.
4. Describe the anatomical features involved in the articulation of a rib on a thoracic vertebra.
5. Identify the different articular facets along the thoracic region, and relate this to how each rib articulates with the vertebrae.
6. Identify the bony elements of the sternum.

PART THREE   THE APPENDICULAR SKELETON

WORD POWER		MATERIALS
scapula (scapul—shoulder blade)	carpus (carp—the wrist)	articulated skeleton
glenoid (glen—a pit)	pisiform (pisi—a pea, form—shape)	disarticulated skeleton
coracoid (cora—a crow)	hamate (hamat—a hook)	
acromion (acromi—shoulder blade point)	coxa (cox—the hip)	
tubercle (tuber—a knot)	acetabulum (acetabul—a vinegar cup)	
capitulum (capit—the head)	trochanter (trochanter—a runner)	
trochlea (trochle—a pulley)	malleolus (malle—a hammer)	
coronoid (coron—a crown)	talus (talus—the ankle)	
olecranon (olecran—the elbow)	cuneiform (cuni—a wedge)	

OBJECTIVES

On completion of this part of the exercise, you should be able to:

- Describe the bones of the appendicular skeleton.
- Identify the bones and surface features of the pectoral girdle.
- Identify the bones and surface features of the upper limb.
- Identify the bones and features of the pelvic girdle and lower limb.

The appendicular division of the skeletal system hangs off the bones of the axial division. The shoulder or pectoral girdle is loosely attached to the sternum to allow a wide range of movement of the arm and shoulder. The arm or upper limb is used to handle and carry objects. The pelvic girdle is securely attached to the sacrum of the spine. The lower limb of the leg supports the body and is moved for locomotion. Review the components of the appendicular skeleton in Figure 7.3.

A. The Pectoral Girdle and Upper Limb

The pectoral girdle of each shoulder has a shoulder blade or **scapula** and a collar bone or **clavicle**. Each upper limb includes the bones of the arm, wrist, and hand. A total of 30 bones are in each limb, and all but 3 of the bones are in the wrist and hand. The upper arm bone is the **humerus**. Inferior to the humerus are the lateral **radius** and the medial **ulna**. The ulna and radius articulate with the humerus at the elbow. Eight **carpal** bones form the wrist. The hand contains 5 long, slender bones in the palm called **metacarpals** and 14 **phalanges** of the fingers.

THE SCAPULA

Locate a scapula in your study skeleton. Orient the bone as in Figure 7.11c to view the posterior surface. Most conspicuous is a prominent ridge of bone, the **spine**, extending across the scapula. At the lateral tip of the spine is the **acromion process** that hangs over the **glenoid fossa** where the humerus articulates. Can you feel the spine and acromion process on your scapula?

Superior to the glenoid cavity is the beakshaped **coracoid process**. At the base of the coracoid process, locate the **scapular notch**, an indentation in the superior border. Rotate the scapula to the lateral view, and notice how the coracoid process extends over the anterior surface of the scapula.

THE CLAVICLE

The S-shaped **clavicle** is the only connection between the pectoral girdle and the axial skeleton. The round **sternal end** articulates with the sternum, and the flat **acromial end** articulates with the scapula (see Figure 7.12).

THE HUMERUS

Refer to Figure 7.13 and examine the proximal end of a humerus. The **head** of the humerus articulates with the glenoid cavity of the scapula. Lateral to the head is the **greater tubercle**, and medial to the head is the **lesser tubercle**, both sites of muscle attachment. Trace a finger distally from the greater tubercle to feel the rough **deltoid tuberosity** where the large deltoid muscle of the shoulder attaches.

The distal end of the humerus accommodates two joints, the hingelike elbow joint and a pivot joint of the antebrachium. Locate the **medial epicondyle** in Figure 7.13. Notice how this epicondyle extends outward more than the **lateral epicondyle** on the opposite side. Below the epicondyles is a single **condyle** with two distinct features, the round **capitulum** on the lateral side and the cylindrical **trochlea** located medially. Can you feel the epicondyles on your humerus? Above the trochlea are two depressions, the anterior **coronoid fossa** and the posterior triangular **olecranon fossa**.

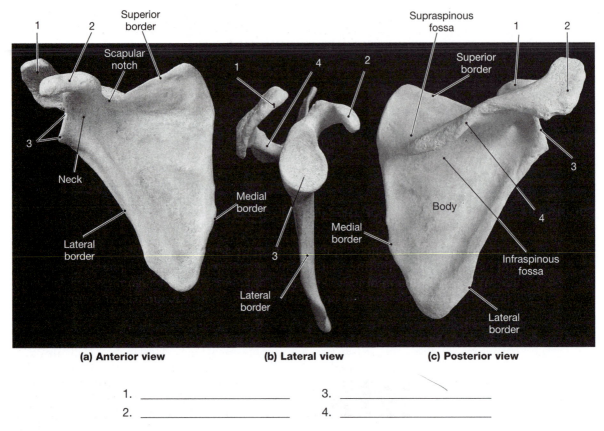

(a) Anterior view   (b) Lateral view   (c) Posterior view

1. _____   3. _____

2. _____   4. _____

●Figure 7.11
The Scapula
(a) Anterior, (b) lateral, and (c) posterior views of the right scapula.

THE ULNA

The forearm has two parallel bones, the ulna and the radius. The ulna is easy to identify by the conspicuous U-shaped **trochlear notch**, shown in Figure 7.14. The notch is like a C clamp with two processes that attach to the humerus, the superior **olecranon process** and the inferior **coronoid process**. On the lateral surface of the coronoid process is the small, curved **radial notch**. Below the notch is the rough **ulnar tuberosity**. The distal ulna has a pointed **styloid process**.

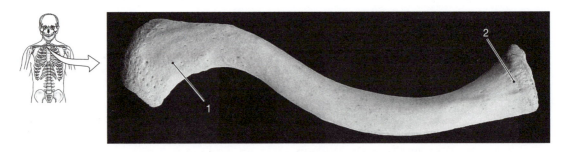

1. _____   2. _____

●Figure 7.12
The Clavicle
Anterior view of the right clavicle.

1. _____
2. _____
3. _____
4. _____
5. _____
6. _____
7. _____
8. _____
9. _____
10. _____
11. _____

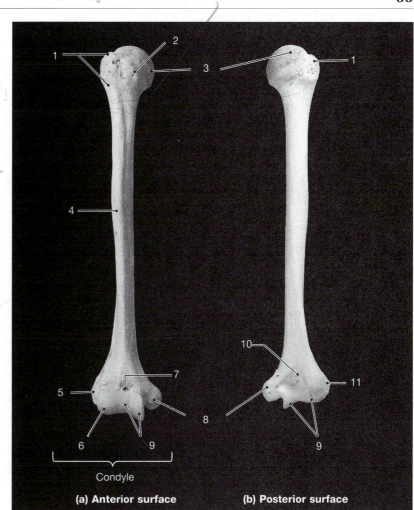

(a) Anterior surface   (b) Posterior surface

●**Figure 7.13**
The Humerus

(a) Anterior view of the right humerus. (b) Posterior view of the right humerus.

THE RADIUS

The radius, Figure 7.14, has a disk-shaped **head** that pivots in the radial notch of the ulna. Below the head is the **neck**. Distal to the neck is the **radial tuberosity**. The **styloid process** of the radius is larger and not as pointed as the styloid process of the ulna.

THE WRIST AND HAND

Each wrist and hand contains 27 bones. The wrist has eight **carpals** arranged in two rows of four, the proximal and the distal carpals (see Figure 7.15). An easy method of identifying the carpals is to use the anterior wrist and start with the carpal next to the styloid process of the radius. From this reference point, moving medially, the sequence of proximal carpals are the **scaphoid**, **lunate**, **triangular**, and the small **pisiform**. Return to the lateral side, and identify the four distal carpals, the **trapezium**, **trapezoid**, **capitate**, and **hamate**.

The five long bones of the palm are **metacarpals**. Each metacarpal is designated by a roman numeral starting with the thumb metacarpus as I. The 14 bones of the fingers are called **phalanges**. Each finger has three individual bones, and the thumb, or **pollex**, has two phalanges.

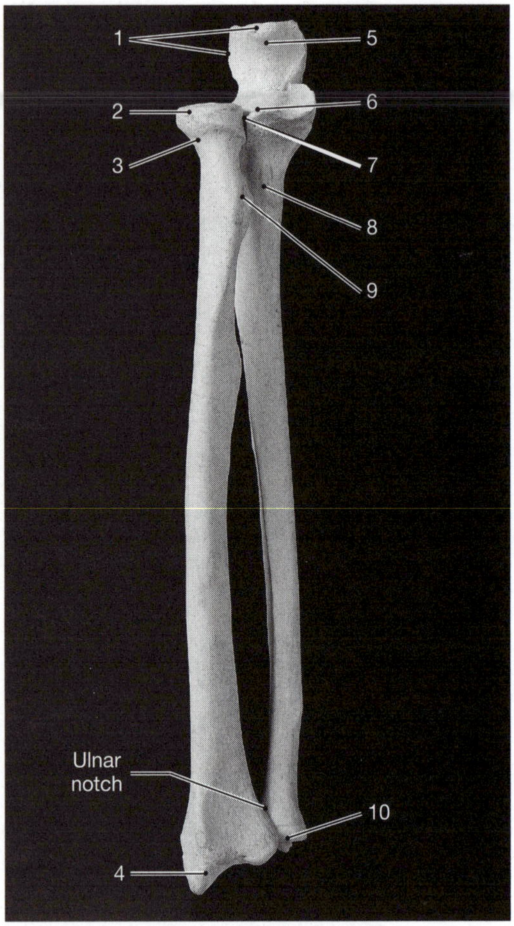

1. _____
2. _____
3. _____
4. _____
5. _____
6. _____
7. _____
8. _____
9. _____
10. _____

Ulnar notch

(a) Anterior view

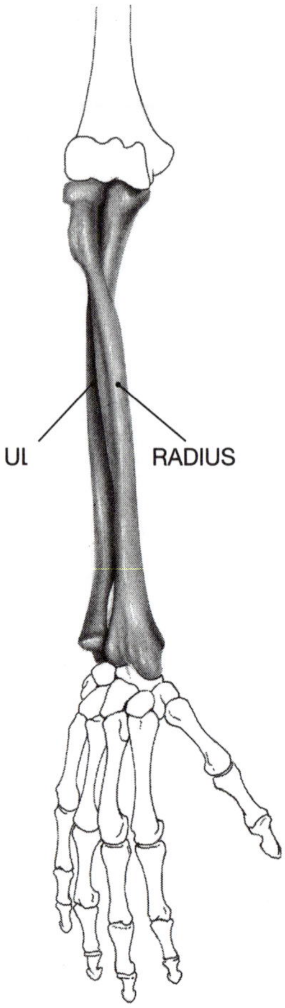

UL   RADIUS

(b) Pronation:
Anterior view

●Figure 7.14
The Radius and Ulna

(a) The radius and ulna are shown in anterior view. (b) Note the changes that occur during pronation.

Laboratory Activity   THE PECTORAL GIRDLE AND UPPER LIMB _____

MATERIALS

articulated skeleton
disarticulated skeleton

PROCEDURES

1. Locate the bones of the pectoral girdle and upper limb on your body.
2. Identify the surface features of the scapula and clavicle. Label Figures 7.11 and 7.12.
3. Identify the bone markings of the humerus, the ulna, and the radius. Label Figures 7.13 and 7.14.
4. Identify the bones of the hand and wrist. Label Figure 7.15.
5. Articulate the bones of the pectoral girdle and the upper limb.

●Figure 7.15
Bones of the Wrist and Hand
Posterior view of the right hand.

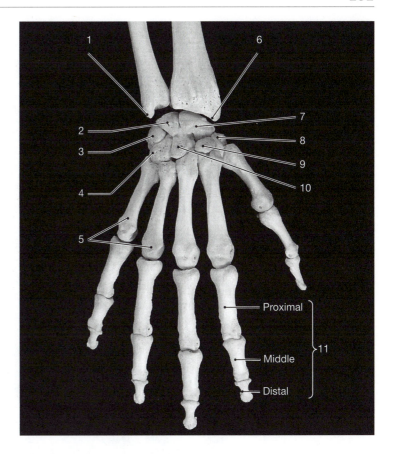

1. _____
2. _____
3. _____
4. _____
5. _____
6. _____
7. _____
8. _____
9. _____
10. _____
11. _____

B. The Pelvic Girdle and Lower Limb

The pelvic girdle forms the hips and is the anchoring point of the legs and spine. Each hip or **coxal bone** is formed by the fusion of three bones, the **ilium**, **ischium**, and **pubis**. The lower limb comprises the **femur** in the thigh, a kneecap or **patella**, and the **tibia** and **fibula** of the lower leg. Each ankle contains 7 **tarsal** bones; the foot has 5 **metatarsals** and 14 **phalanges**.

THE COXA

Locate a coxal bone in your study skeleton, and refer to Figure 7.16, a lateral view of the coxa. The **ilium**, **pubis**, and **ischium** fuse by early adulthood into a single **coxa**. The superior ridge of the ilium is the **iliac crest**. The crest narrows anteriorly to form the **anterior superior iliac spine**. You can probably palpate (feel) these structures on your ilia. The large socket in the middle of the coxa is the **acetabulum**, where the femur articulates with the hip. The large indentation on the posterior coxa is the **greater sciatic notch**. Trace the greater sciatic notch with your finger. The notch ends at a bony point, the **ischial spine**. The **ischial tuberosity** is the rough feature at the inferiormost portion of the ischium.

The pubis bone is on the anterior coxa. It has both a superior and an inferior ramus (branch) that join the ischium at the acetabulum and near the ischial tuberosity, creating the **obturator foramen**. Anteriorly, the pubis bones join at the **pubic symphysis**, a strong joint containing fibrocartilage.

Male and female pelvises are anatomically different. The female pelvis has a wider pelvic outlet, the space between the ischial spines. The female pelvis also has a wider

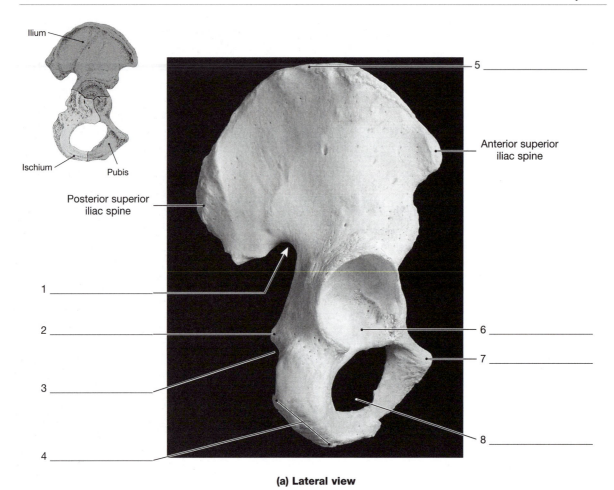

(a) Lateral view

●Figure 7.16
Lateral View of Coxa

U-shaped pubic angle, at the pubis symphysis, than the V-shaped male pelvis. The wider female pelvis provides a larger passageway for childbirth.

THE FEMUR

Locate a femur in your study skeleton. The femur is the largest bone in the skeleton. It supports the weight of the body and bears the stress from your legs. Use Figure 7.17 of the femur as a guide to locate the smooth, round **head** of the femur. The head fits into the acetabulum of the coxa and permits a wide range of movement. Lateral to the head is a large stump, the **greater trochanter**. The **lesser trochanter** is on the medial surface and is inferior and posterior to the greater trochanter. These large processes are attachment sites for powerful hip and thigh muscles. Along the middle of the posterior diaphysis is the **linea aspera**, where thigh muscles attach. At the distal end of the femur are the **lateral** and **medial condyles** that articulate with the tibia. Superior to the condyles are the **lateral** and **medial epicondyles**. The smooth **patellar surface** spans the lateral and medial condyles and serves as a gliding platform for the patella.

THE PATELLA

The patella is encased within the tendons of the anterior thigh muscles. The superior border of the bone is the flat **base**; the **apex** is at the inferior tip. Tendons attach to the rough anterior surface, and the smooth posterior **facets** glide over the condyles of the femur.

1. _____
2. _____
3. _____
4. _____
5. _____
6. _____
7. _____
8. _____
9. _____

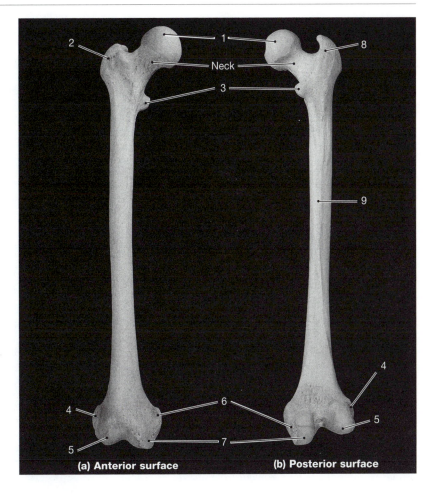

●**Figure 7.17**
The Femur
(a) Anterior view of the right femur. **(b)** Posterior view of the right femur.

THE TIBIA

The lower leg bones are illustrated in Figure 7.18. The tibia is the large medial bone of the lower leg. The proximal portion of the tibia flares to develop the **lateral** and **medial condyles** that articulate with the corresponding femoral condyles. Anteriorly below the condyles is the large **tibial tuberosity**, where thigh muscles attach. The distal tibia is constructed to articulate with the foot. A large wedge, the **medial malleolus**, helps stabilize the ankle joint.

THE FIBULA

The fibula is the slender bone lateral to the tibia, shown in Figure 7.18. The proximal and distal regions of the fibula appear very similar at first. Closer examination reveals the proximal **head** is diamond shaped and less massive than the distal wedge-shaped **lateral malleolus**.

THE ANKLE AND FOOT

The ankle is made up of seven **tarsal** bones. The **talus** sits on the heel bone, the **calcaneus**, and articulates with the tibia and the lateral malleolus of the fibula.

●Figure 7.18
The Tibia and Fibula

Anterior view of the right tibia and fibula.

1. _____
2. _____
3. _____
4. _____
5. _____
6. _____

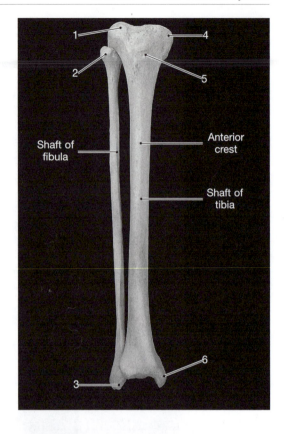

●Figure 7.19
Bones of the Ankle and Foot

Bones of the right foot as viewed from above. Note the orientation of the tarsals that convey the weight of the body to the heel and the plantar surfaces of the foot.

1. _____
2. _____
3. _____
4. _____
5. _____
6. _____
7. _____

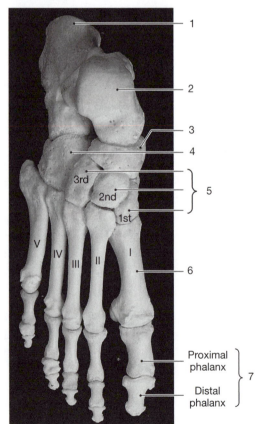

(a) Superior view, right foot

Anterior to the talus is the **navicular** that articulates with three **cuneiform** bones. Lateral to the third cuneiform is the **cuboid** that articulates posteriorly with the calcaneus. Most of the sole of the foot is formed by five **metatarsal** bones, numbered one through five from medial to lateral. As with the fingers of the hand, each toe has three **phalanges**, except the **hallux**, or big toe, which has two bones. Identify each bone of the foot in Figure 7.19.

Laboratory Activity   THE PELVIC GIRDLE AND LOWER LIMB _____

MATERIALS

articulated skeleton
diarticulated skeleton

PROCEDURES

1. Locate each bone of the pelvic girdle and lower limb on your body.
2. Identify the surface features of the coxa. Label Figure 7.16.
3. Identify the surface features of the femur, patella, tibia, and fibula. Label Figures 7.17 and 7.18.
4. Identify the bones of the ankle and foot. Label Figure 7.19.
5. Articulate both coxae and bones of the lower limb.

PART FOUR   ARTICULATIONS AND MOVEMENTS OF THE SKELETON

OBJECTIVES

On completion of this part of the exercise, you should be able to:

- Describe the basic anatomical features of a synovial joint.
- Identify and describe the joints of the skeleton.
- Demonstrate the movements of the skeleton.

A. Synovial Joints

Read about articulations and synovial joints in your textbook. Complete Table 7.2 on the types of synovial joints.

TABLE 7.2	**Types of Synovial Joints**
Joint Type	*Example*
Ball and socket	
Hinge	
Saddle	
Pivot	
Glide	
Condyloid	

B. Movements of the Skeleton

Read about the movements of the skeleton in your textbook. Label each type of movement in Figure 7.21.

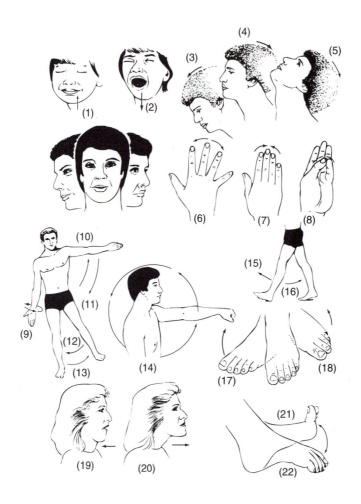

1. _____   9. _____   17. _____

2. _____   10. _____   18. _____

3. _____   11. _____   19. _____

4. _____   12. _____   20. _____

5. _____   13. _____   21. _____

6. _____   14. _____   22. _____

7. _____   15. _____

8. _____   16. _____

●**Figure 7.21**
Movements of the Skeleton

THE SKELETAL SYSTEM
Exercise 7 Laboratory Report

A. Discussion Questions

1. List the components of the axial and appendicular skeleton.

2. Give an example of five different types of surface marks of bones.

3. How many bones are found in the cranium and the face?

4. List the sutures of the skull and the bones that articulate at each.

5. Describe the unique features for each region of the vertebral column.

6. Describe how the bones of the pectoral girdle articulate with the axial skeleton.

7. Compare the bones of the hands and feet.

8. Describe the coxal bone.

9. How does the lower leg articulate with the coxal bone?

10. Compare the pelvises of males and females.

B. Matching

Match each bone with the correct division and part of the skeleton. Each question may have more than one answer, and each choice may be used more than once.

1. _____ scapula	A.	axial division
2. _____ coxal bone	B.	appendicular division
3. _____ patella	C.	pectoral girdle
4. _____ hyoid	D.	upper limb
5. _____ radius	E.	pelvic girdle
6. _____ metacarpal	F.	lower limb
7. _____ vertebra		
8. _____ clavicle		
9. _____ rib		
10. _____ femur		
11. _____ sternum		
12. _____ carpal		

C. **Matching**

Match the surface feature on the left with the correct bone on the right. Each choice on the right may be used more than once.

1. _____ acromion process	A. clavicle
2. _____ trochlea	B. patella
3. _____ glenoid fossa	C. fibula
4. _____ ulnar notch	D. humerus
5. _____ deltoid tuberosity	E. femur
6. _____ greater trochanter	F. scapula
7. _____ lateral malleolus	G. tibia
8. _____ capitulum	
9. _____ medial malleolus	
10. _____ base	

D. **Fill in the Blanks**

Complete each statement by filling the blank with the correct directional term, for example, the femur is <u>inferior</u> to the pelvis.

1. The humerus is _____ to the radius.
2. The fibula is _____ to the tibia.
3. The talus is _____ to the calcaneus.
4. The clavicle is _____ to the scapula.
5. The patella is _____ to the femur.
6. The ilium is _____ to the ischium.
7. The ulna is _____ to the radius.
8. The metacarpals are _____ to the carpals.

E. **Completion**

Describe the joints and movements involved in the following activities:

1. Walking _____
2. Throwing a ball _____
3. Turning a door knob _____
4. Crossing your legs while sitting _____
5. Shaking your head "no" _____
6. Chewing food _____

8 THE MUSCULAR SYSTEM

Every time a part of the body is moved, either consciously or unconsciously, muscles are used. Other functions of muscles include maintenance of **posture** and **body temperature**, **support** of soft tissues, and the passage of blood within blood vessels and food through the digestive tract. The body carefully coordinates these activities with many other organ systems, particularly the nervous system, to maintain homeostasis.

Three types of muscle tissue occur in the body: skeletal, cardiac, and smooth. Exercise 5 of this manual describes the major histological characteristics of each type of muscle tissue. This exercise is a study of skeletal muscles.

Skeletal muscles attach to bones and other tissues by tendons. Each muscle causes a movement, or **action**, that depends on many factors, especially the shape of the attached bones. For the muscle to produce a smooth, coordinated action, one end of the muscle must serve as a stable pulling point while the other end moves the attached bone. The less movable end of the muscle is called the **origin**. The opposite end that moves the bone is called the **insertion**. During the action, the insertion moves toward the origin to generate a pulling force. Usually, when a muscle pulls in one direction, an antagonistic muscle pulls in the opposite direction to produce resistance and promote a smooth movement.

A variety of methods is used to name muscles. Muscles are named after directional relations, body regions, muscle shape, muscle action, and muscle origins and insertions.

PART ONE   ORGANIZATION OF SKELETAL MUSCLES

WORD POWER

epimysium (epi—over; mys—muscle)
perimysium (peri—around)
fascicle (fasci—a bundle)
endomysium (endo—inner)
sarcolemma (sarco—flesh; lemma—a sheath)

MATERIALS

muscle model
muscle fiber model

OBJECTIVES

On completion of this part of the exercise, you should be able to:

- Identify the connective tissue components of muscles.
- Identify the major features of skeletal muscle fibers.

A. Connective Tissue Coverings

Skeletal muscles are surrounded and partitioned by three layers of connective tissue, each illustrated in Figure 8.1. The **epimysium** covers the surface of the muscle and separates the muscle from neighboring structures. The epimysium extends beyond the end of the muscle as a strong cord called the **tendon**. Deep to the epimysium is the **perimysium** that divides the muscle into bundles of muscle fibers called **fascicles**. The individual muscle fibers in the fascicles are separated by fibers of the **endomysium**.

B. **Organization of a Skeletal Muscle Fiber**

Skeletal muscles are made up of cellular structures called **muscle fibers** (see Figure 8.1). Each muscle fiber is a composite of many cells that fused during embryonic development. The cell membrane of a muscle fiber is called the **sarcolemma**. Connecting the sarcolemma to the interior of the muscle fiber are many **transverse tubules** (t-tubules), which function in passing contraction stimuli to deeper regions of the muscle fiber.

Inside the muscle fiber are proteins arranged into thousands of rods called **myofibrils** that extend the length of the fiber. Each myofibril is surrounded by membranes of the **sarcoplasmic reticulum**, a site of calcium storage. Transverse tubules stimulate the release of calcium ions from the sarcoplasmic reticulum during muscle contraction.

Myofibrils are groups of **myofilaments**, thick and thin protein molecules that interact to shorten the muscle during contraction. The thin myofilaments are mostly made of the protein **actin**; the thick myofilaments are made of **myosin**. The myofilaments are arranged into repeating patterns called **sarcomeres** along a myofibril. Distinct regions of the sarcomere are distinguishable. The thin myofilaments connect at the **Z lines** on each end of the sarcomere. Areas near the Z line that contain only thin myofilaments are the **I bands**. Between the I bands of a sarcomere is the **A band**, the area within the length of a thick myofilament. The edges of the A band overlap where the thick and thin myofilaments bind during muscle contraction.

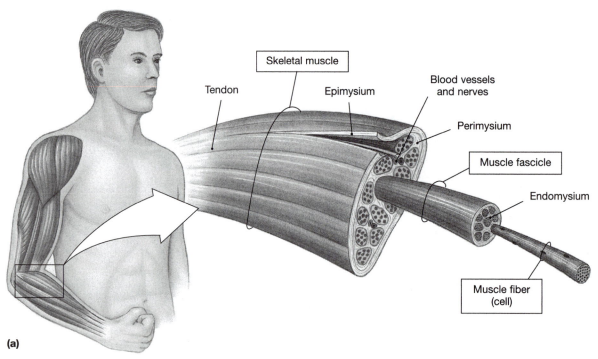

(a)

●**Figure 8.1**
Organization of Skeletal Muscles

(a) Gross organization of skeletal muscle.

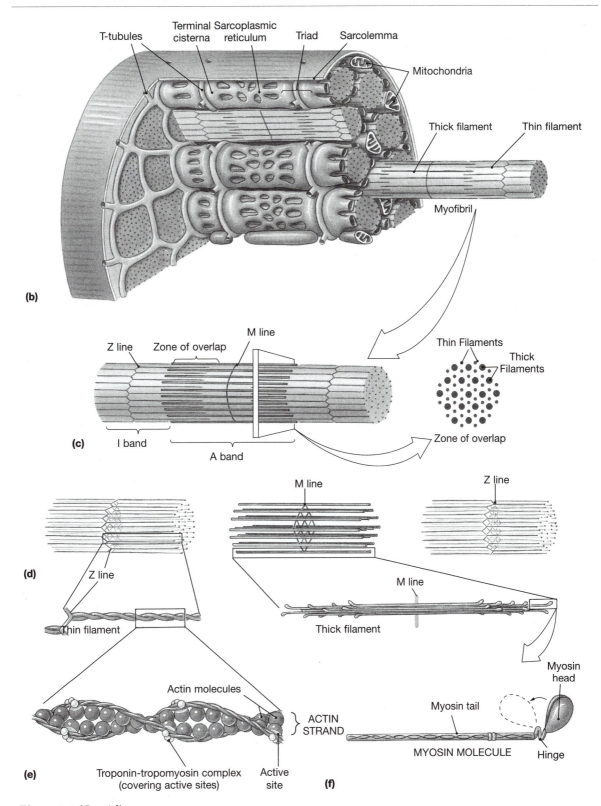

●**Figure 8.1** *(Cont'd)*
Organization of Skeletal Muscles

(b) Structure of a skeletal muscle fiber. **(c)** Diagrammatic organization of a sarcomere, part of a single myofibril. **(d)** The sarcomere in part (c), stretched to the point where thick and thin filaments no longer overlap (This cannot happen in an intact muscle fiber.) **(e)** Structure of a thin filament. **(f)** Structure of a thick filament.

Laboratory Activity   ORGANIZATION OF SKELETAL MUSCLES _____

MATERIALS

muscle model
muscle fiber model

PROCEDURES

1. Label the muscle structures in Figure 8.1.

2. Examine the muscle models in the laboratory. Identify the connective tissue coverings of muscles. Your laboratory instructor may have prepared a muscle demonstration from a round steak or similar cut of meat. Examine the meat for the various connective tissues. Are fascicles visible on the specimen?

3. Examine the skeletal muscle fiber models in the laboratory, and identify each feature. Describe the location of the sarcoplasmic reticulum, myofibrils, sarcomeres, and myofilaments.

4. Review the histology of skeletal muscle fibers in Exercise 5 of the manual.

PART TWO   MUSCLES OF THE HEAD AND NECK

WORD POWER

orbicularis oculi (orb—a circle; oculi—the eye)
buccinator (buccin—a trumpet)
platysma (platy—broad, flat)
masseter (masseter—a chewer)
temporalis (tempor—the temples)
sternocleidomastoid (sterno—breastbone; cleido—clavicle)

MATERIALS

head model
torso model

OBJECTIVES

On completion of this part of the exercise, you should be able to:

- Identify the muscles of facial expression and mastication.
- Identify the major muscles of the neck and head.

A. Muscles of Facial Expression

ORBICULARIS OCULI

The **orbicularis oculi** surrounds the eye (see Figure 8.2). It originates from the medial wall of the orbit and forms a band of muscle that wraps around the eye. On contraction, it acts as a sphincter to close the eye, as during an exaggerated wink.

ORBICULARIS ORIS

The **orbicularis oris** is a sphincter muscle that surrounds the mouth. It originates on the bones surrounding the mouth and inserts on the lips. This muscle shapes the lips for a variety of functions, such as speech, food manipulation, and facial expressions like puckering the lips. Identify this muscle on your face.

1. _____

2. _____

3. _____

4. _____

5. _____

6. _____

7. _____

8. _____

9. _____

10. _____

11. _____

12. _____

13. _____

14. _____

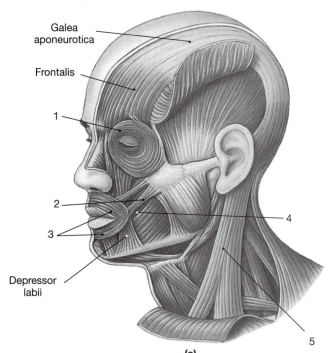

Galea aponeurotica

Frontalis

1

2

3

4

5

Depressor labii

(a)

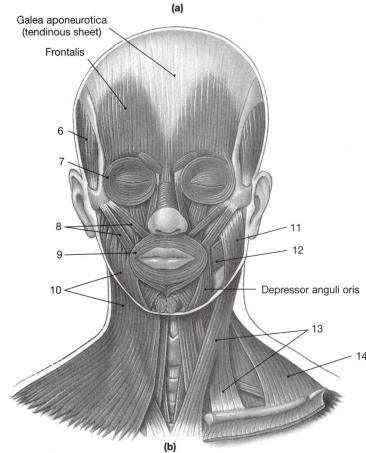

Galea aponeurotica (tendinous sheet)

Frontalis

6

7

8

9

10

11

12

Depressor anguli oris

13

14

(b)

●Figure 8.2
Muscles of the Head and Neck

(a) Anerior and lateral view. (b) Anterior view.

BUCCINATOR

The **buccinator** is the horizontal muscle spanning the jaws. While chewing, the buccinator moves food across the teeth from the cheeks. This muscle is also involved in tensing the cheeks as when blowing out or sucking on a straw. The name *buccinator* is from the Latin term trumpet. Trumpeters use the buccinator to pressurize their exhalations while playing.

ZYGOMATICUS

The **zygomaticus** arises from the temporal process of the zygomatic bone and inserts on the angle of the mouth. This muscle pulls the skin of the mouth upward and laterally to produce a smile. Place your fingers on your cheek bones and smile. Did you feel the zygomaticus contract?

PLATYSMA

The **platysma** is a thin, broad muscle covering the sides of the neck. It covers the pectoralis and deltoid muscles and extends upward to insert on the inferior edge of the mandible. The platysma draws down the mandible and the soft structures of the lower face, which results in an expression of horror and disgust.

B. Muscles of Mastication

MASSETER

The muscles for chewing, or mastication, move the mandible. The **masseter**, shown in Figure 8.2, is a short, thick muscle overlying the buccinator. It originates on the maxilla and the zygomatic arch and inserts on the mandible. When the masseter contracts, it elevates the lower jaw against the teeth and produces a tremendous force for chewing. The masseter also protrudes the jaw when it acts alone. Put your finger tips at the angle of your jaw and clench your teeth. You should feel the masseter bulge as it forces the teeth together.

TEMPORALIS

The **temporalis** covers most of temple region where it originates. It inserts on the coronoid process of the mandible. The temporalis works with the masseter to elevate the jaw during chewing. The temporalis also can retract the jaw. To feel your temporalis contract, put your fingers on your temple and move the mandible as if chewing food.

C. Muscles That Move the Head

STERNOCLEIDOMASTOID

The **sternocleidomastoid** muscle (see Figure 8.2) occurs on both sides of the neck and is named after its points of attachment on the sternum, clavicle, and mastoid process of the temporal bone. Because this muscle spans the head and sternum, when both sides contract, the neck flexes and the head is tucked down toward the sternum. If only one side contracts, the head is pulled toward the shoulder of the opposite side. Rotate your head until your chin almost touches your right shoulder, and locate your left sternocleidomastoid just above the manubrium of the sternum.

Laboratory Activity   MUSCLES OF THE HEAD AND NECK _____

MATERIALS

head model
torso model

PROCEDURES

1. Examine the torso model. Locate each muscle described in the manual, and study its shape.
2. Since most of the muscles presented in this section are superficial muscles, attempt to locate them on your body. Contract the muscle to study the movement it produces.
3. Label each muscle in Figure 8.2.

PART THREE   MUSCLES OF THE CHEST AND ABDOMEN

WORD POWER   **MATERIALS**

pectoralis major (pector—the chest; major—larger)   torso model
serratus (serrat—a saw)   arm model
diaphragm (phragm—a partition)
intercostal (costa—a rib)
rectus (rect—straight)
latissimus (lati—wide)

OBJECTIVES

On completion of this part of the exercise, you should be able to:

- Identify the major muscles of the anterior and posterior chest.
- Identify the muscles used in respiration.
- Identify the muscles of the abdominal wall.

A.   Anterior Muscles of the Chest

PECTORALIS MAJOR

The largest muscle of the chest is the **pectoralis major** muscle (see Figure 8.3). It covers most of the upper rib cage on each side of the chest. Locate these muscles on your chest. In females, the lower part of the muscle mass is covered by the breast. The pectoralis major acts to adduct the arm and flex and rotate the humerus at the shoulder.

PECTORALIS MINOR

Tucked under the pectoralis major is the **pectoralis minor** muscle. This small, triangular-shaped muscle has a different function from its larger cousin. The minor originates along the edges of the third, fourth, and fifth ribs and inserts into the coracoid process of the scapula. It acts to pull the top of the scapula forward and depresses the shoulders.

SERRATUS ANTERIOR

The **serratus anterior**, Figure 8.3, appears as wedges of muscle on the side of the ribs. This gives the muscle a sawtooth appearance similar to a bread knife. The origin of the serratus anterior is on the upper edge of the first eight ribs. The muscle inserts on the vertebral border of the scapula. The serratus anterior functions to pull the scapula forward (protract) and rotate it laterally.

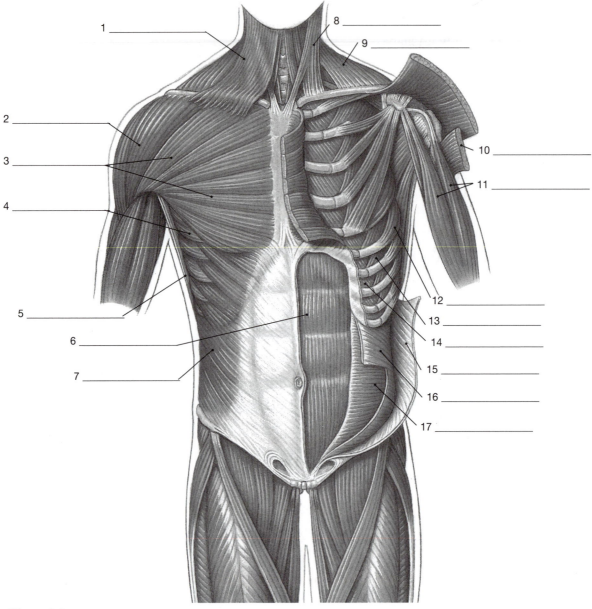

●Figure 8.3
Muscles of the Anterior Trunk

B. Muscles of Respiration

DIAPHRAGM

The **diaphragm** is a sheet of muscle that forms the thoracic floor and separates the thoracic and abdominopelvic cavities. It originates at many points along its edges, and the muscle fibers meet at a central tendon. When the muscle contracts, it pulls down on the central tendon and lowers the thoracic floor. This increases the volume of the thoracic cavity and results in a pressure decrease in the lungs. Contracting the diaphragm to expand the thoracic cavity is the muscular process for inhalation of air into the lungs.

EXTERNAL AND INTERNAL INTERCOSTALS

The intercostal muscles are found between the ribs and assist the diaphragm during breathing (see Figure 8.3). These muscles are difficult to palpate because they are deep to other muscles. The **external intercostals** span the region between each rib. Because their insertion is inferior to the origin, the ribs are elevated on contraction. This increases the volume of the thoracic cavity, decreasing the pressure there for inhalation.

The **internal intercostals** lie deep to the external intercostals. They act to depress the rib cage for forced expiration during exercise. Respiratory volumes and rates increase during exercise. The internal intercostals become active at this time and act to quickly depress the rib cage and force air out of the lungs.

Study Tip

Notice the difference in the orientation of the muscle fibers between the external and internal intercostals. The external fibers flare laterally as they are traced from bottom to top, whereas the internal fibers are directed medially. This tip is also useful for the external and internal oblique muscles of the abdomen.

C. Abdominal Muscles

RECTUS ABDOMINIS

The **rectus abdominis**, Figure 8.3, is the straight muscle along the abdomen between the pubic symphysis and the xiphoid process of the sternum. The muscle is divided by a midsagittal fibrous line called the **linea alba**. A well-developed rectus abdominis muscle has a "wash-board" appearance because transverse bands of tendons divide it into several segments. When the rectus abdominis contracts, it flexes the pubic symphysis and the xiphoid process toward each other, the movement that occurs in situps.

EXTERNAL AND INTERNAL OBLIQUE MUSCLES

Lateral to the rectus abdominis is the **external oblique**, a thin membranous muscle that covers the sides of the abdomen (see Figure 8.3). The external oblique originates on the inferior borders of the lower eight ribs and inserts on the anterior portion of the iliac crest and the linea alba. The **internal oblique** lies deep to the external oblique. The muscle fibers sweep upward from the lower abdomen and insert on the lower ribs, the xiphoid process, and the linea alba. Both the external and internal oblique muscles act with the rectus abdominis to flex the vertebral column to compress the abdomen. They also increase the pressure in the abdomen during defecation, urination, and childbirth.

TRANSVERSE ABDOMINIS

The **transverse abdominis**, Figure 8.3, is deep to the internal oblique. It originates on the lower ribs, the iliac crest, and the lumbar vertebrae and inserts on the linea alba and the pubic symphysis. The transverse abdominis muscle contracts with the other abdominal muscles to compress the abdomen.

D. Posterior Muscles of the Chest

TRAPEZIUS

The large diamond-shaped muscle of the upper back is the **trapezius** (see Figure 8.4b). It spans the gap between the scapulae and extends from the lower thoracic vertebrae to the back of the head. Locate this large muscle on your body. The trapezius has a complex action. It extends the neck and head, and moves the scapula.

LATISSIMUS DORSI

The **latissimus dorsi** is the large muscle of the lower back, located inferior and deep to the trapezius. Find this muscle in Figure 8.4b and on your back. The latissimus dorsi

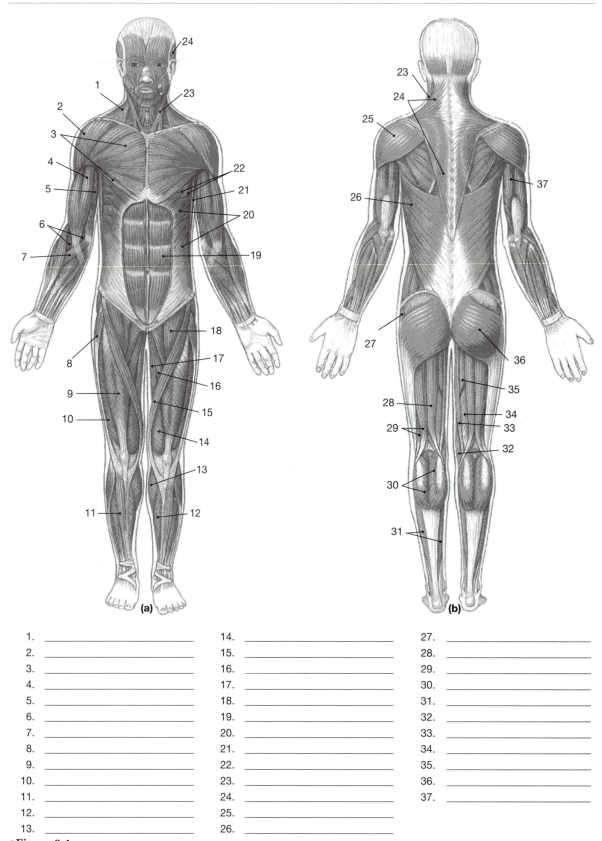

1. _____
2. _____
3. _____
4. _____
5. _____
6. _____
7. _____
8. _____
9. _____
10. _____
11. _____
12. _____
13. _____
14. _____
15. _____
16. _____
17. _____
18. _____
19. _____
20. _____
21. _____
22. _____
23. _____
24. _____
25. _____
26. _____
27. _____
28. _____
29. _____
30. _____
31. _____
32. _____
33. _____
34. _____
35. _____
36. _____
37. _____

●**Figure 8.4**
An Overview of the Major Skeletal Muscles

(**a**) Anterior view. (**b**) Posterior view.

has a broad origin on vertebrae T_6 and below and on the iliac crest. The muscle sweeps upward and inserts on the humerus near the insertion of the pectoralis major. The latissimus dorsi acts to adduct the arm and move it downward, as when hitting a nail with a hammer.

Laboratory Activity   MUSCLES OF THE CHEST AND ABDOMEN

MATERIALS

torso model

PROCEDURES

1. Examine the anterior chest of the torso model. Locate each muscle described, and study its shape.
2. Attempt to locate these muscles on your body. Contract the muscle to study the movement it produces.
3. Label each muscle in Figure 8.3 and 8.4.

PART FOUR   MUSCLES OF THE SHOULDER, ARM, AND HAND

The function of the muscles of this region of the body is support and movement of the arm, hand, and fingers. The muscles of the shoulder attach the humerus to the scapula and help stabilize the scapula on the posterior chest.

WORD POWER	MATERIALS
deltoid (delt—triangular)	torso model
biceps (bi—two; ceps—head)	arm model
supinator (supin—lying on the back)	
pronator (pron—bent forward)	

OBJECTIVES

On completion of this part of the exercise, you should be able to:

- Identify the major muscles of the shoulder region.
- Identify the major muscles of the arm.
- Identify the location of the flexors and extensors of the wrist and hand.

A. Muscles of the Shoulder and Upper Arm

DELTOID

The **deltoid**, Figure 8.4a, is the triangular muscle that covers the lateral shoulder. Locate this muscle on the upper part of your humerus, just below the shoulder. The muscle tapers and inserts on the humerus at the rough deltoid tuberosity. The major action of the deltoid is abduction of the humerus.

BICEPS BRACHII

The **biceps brachii**, Figure 8.4a, muscle makes up the fleshy mass of the anterior humerus when the arm is flexed. Locate the biceps brachii on your arm. The term *biceps* refers to the presence of two origins, both occurring on the scapula. The heads join a single tendon and insert on the radial tuberosity. The biceps brachii is a principal flexor of the forearm.

BRACHIALIS

The **brachialis** also flexes the forearm. It is located under the distal end of the biceps brachii (see Figure 8.4a). You can feel part of it if you flex your arm and palpate the area just lateral to the tendon of the biceps. It originates on the distal humerus and inserts onto the ulnar tuberosity. Since the ulna is limited to a hingelike movement, the action of the brachialis is flexion of the forearm.

TRICEPS BRACHII

The **triceps brachii** is the muscle of the posterior arm. Locate the horseshoe-shaped triceps in Figure 8.4a and on the back of your upper arm. Its name suggests three origins, two on the humerus and one on the scapula. The heads merge into a common tendon at the middle of the muscle that inserts on the olecranon process of the ulna. The triceps acts to extend the forearm and is therefore an antagonist to the biceps brachii and the brachialis.

B. Muscles of the Forearm

SUPINATOR

The **supinator**, shown in Figure 8.5, is found on the lateral side of the forearm, deep to several muscles. Its origin is on the lateral epicondyle of the humerus. The muscle crosses the antecubital region and inserts on the lateral side of the radius. When the supinator contracts, it rotates the radius into a parallel position with the ulna resulting in supination of the forearm.

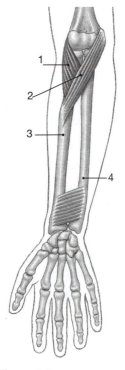

1. _____

2. _____

3. _____

4. _____

●Figure 8.5
Muscles of Pronation and Supination

PRONATOR TERES

The principal pronator muscle of the arm is the **pronator teres** (see Figure 8.5). Since pronation is the opposite action of supination, the origin and insertion of the pronator teres must be opposite of the supinator's. The pronator teres originates on the medial epicondyle of the ulna and inserts on the lateral side of the radius, about midway down the shaft of that bone. When the forearm is in a supine position, the pronator teres pulls the lateral edge of the radius toward the ulna and medially rotates the radius. This results in pronation of the forearm.

SUPERFICIAL FLEXORS AND EXTENSORS OF THE WRIST AND HAND

We will discuss the muscles of the wrist and hand as two groups, the flexors and the extensors. To flex the wrist and hand, the **flexor** muscles are located on the anterior surface of the forearm. They all share a similar origin on the common flexor tendon attached to the medial epicondyle of the humerus. The **extensors** originate on the common extensor tendon arising on the posterior humerus. Your laboratory instructor may assign specific flexors and extensors. Use your lecture textbook and Figures 8.4a and 8.4b to help you identify those muscles.

Locate the medial epicondyle on your right humerus, and place the fingers of your left hand on the muscle distal to the epicondyle. Now **flex** your right fingers and hand. You should feel the muscles contract near the epicondyle. Remember that the common **flexor** tendon originates on this epicondyle. Do the same for the lateral epicondyle, but instead place your fingers on the dorsal surface of the forearm and **extend** your right fingers and hand. The common **extensor** tendon arises here, and you should feel the extensor muscles contract and bulge just distal to the epicondyle in the forearm.

Laboratory Activity   MUSCLES OF THE SHOULDER, ARM, AND HAND _____

MATERIALS

arm model
torso model

PROCEDURES

1. Examine the arm model. Locate each muscle described, and study its shape.
2. Locate as many of these muscles on your body as possible. Contract the muscle to study the movement it produces.
3. Label each muscle in figures 8.4 and 8.5.

---

*P*ART FIVE   MUSCLES OF THE PELVIS, LEG, AND FOOT

The muscles of the pelvis help support the mass of the body and help stabilize the pelvic girdle. The leg has muscles that flex, extend, abduct, or adduct. Abductors, which move a structure away from the midline of the body or limb, are generally found on the lateral side of the structure being moved. Adductors are found on the medial side.

WORD POWER		MATERIALS
psoas (psoa—the loin)	sartoris (sartori—a tailor)	torso model
maximus (maxim—largest)	vastus (vast—huge)	leg model
gracilis (gracil—slender)		
tensor fasciae latae		
(tens—stretched; fasci—band; lat—wide)		

OBJECTIVES

On completion of this part of the exercise you should be able to:

- Identify the major muscles of the pelvis and gluteal regions.
- Identify the major muscles of the anterior and posterior thigh.
- Identify the major muscles of the lower leg.

A. Muscles of the Pelvis and Gluteal Region

ILIOPSOAS (PSOAS MAJOR AND ILIACUS)

The **iliopsoas** consists of two muscles, the psoas major and the iliacus (see Figure 8.6). Although these muscles are located in the lower abdomen, they act on the femur. The **psoas major** originates on the body and transverse processes of the vertebrae T_{12} through L_5. The muscle sweeps downward, passing between the femur and the ischium, and inserts on the lesser trochanter of the femur. The **iliacus** originates on the medial side of the ilium and joins the tendon of the psoas. Collectively, the iliopsoas flexes the thigh, bringing the anterior surface of the thigh toward the abdomen.

GLUTEUS MAXIMUS

The **gluteus maximus**, Figure 8.6, is the prominent muscle of the gluteal region. It is a large, fleshy muscle and is easily located on your body as the major muscle of the buttocks. It originates along the iliac crest and passes laterally to insert into a thick tendon called the **iliotibial tract**. The gluteus maximus extends the leg and is involved in the lateral rotation of the femur.

TENSOR FASCIAE LATAE

Tensor fasciae latae originates on the outer surface of the anterior ilium. The muscle projects downward and inserts into the iliotibial tract. As the name implies, the tensor fascia latae tenses the fascia of the thigh and helps stabilize the pelvis on the femur. The muscle also acts to abduct and medially rotate the thigh. Locate this muscle in Figure 8.6.

B. Muscles of the Upper Leg

SARTORIUS

The **sartorius** is a thin, ribbonlike muscle that originates on the anterior iliac spine. It crosses the knee joint to insert on the medial surface of the tibia, near the tibial tuberosity. This muscle is a flexor of the knee and hip joint and helps cross the legs by rotating the thigh (see Figure 8.6).

QUADRICEPS FEMORIS GROUP

The extensors of the lower leg are collectively called the **quadriceps femoris group** and include the **rectus femoris**, **vastus medialis**, **vastus lateralis**, and **vastus intermedius**. They make up the bulk of the anterior mass of the thigh and are easy to locate on your leg.

The **rectus femoris**, Figure 8.6, is located along the midline of the anterior surface of the thigh. It originates from the iliac spine and the acetabulum. The muscle blends into the tendon of the quadriceps group and inserts onto the patella. All the muscles of the quadriceps group converge onto this tendon, which inserts onto the tibial tuberosity. Because the rectus femoris crosses two joints, the hip and knee, it produces flexion of the hip joint and extension of the lower leg.

The vastus muscles are shown in Figure 8.6. Covering almost the entire medial surface of the femur is the **vastus medialis**. The **vastus lateralis** occurs on the lateral side

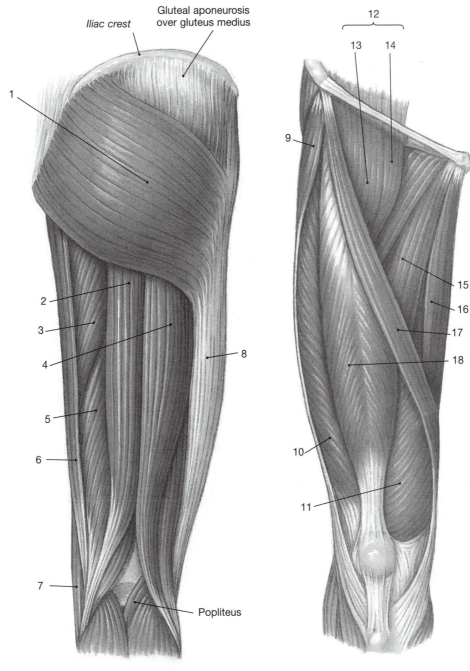

Iliac crest

Gluteal aponeurosis
over gluteus medius

Popliteus

(a) Posterior view

(b) Anterior view

1. _____
2. _____
3. _____
4. _____
5. _____
6. _____

7. _____
8. _____
9. _____
10. _____
11. _____
12. _____

13. _____
14. _____
15. _____
16. _____
17. _____
18. _____

●**Figure 8.6**
Muscles that Move the Leg

(a) Posterior view of thigh. **(b)** Anterior view of thigh.

of the rectus femoris. Directly under the rectus and between the lateralis and medialis is the **vastus intermedialis**. The vastus muscles originate on the anterior femur and posteriorly on the linea aspera of the femur. They converge onto the tendon of the quadriceps muscle. The vastus muscles act together to extend the lower leg.

GRACILIS

The muscles on the medial thigh act to adduct the leg. The **gracilis** is located midway along the medial thigh and is the most superficial of the thigh adductors (see Figure 8.6). It originates near the pubic symphysis and extends downward along the medial surface of the thigh and inserts near the tibial tuberosity. Since it passes over the knee joint, it acts to flex the lower leg and adduct the thigh.

ADDUCTOR MUSCLES

The major adductors of the thigh are the **adductor longus** and the **adductor magnus**, shown in Figure 8.6. These muscles originate along the pubis and are located on either side of the gracilis muscle. The adductor longus is anteriolateral to the gracilis. The largest adductor is the adductor magnus, which is located posterior to the gracilis. This muscle can be easily observed on the leg model if the superficial muscles are removed. The adductor muscles insert on the posterior femur and are powerful adductors of the thigh. They also cause medial rotation of the thigh.

THE HAMSTRING GROUP

The major muscles of the posterior upper leg are collectively called the **hamstrings**. They consist of the biceps femoris, the semitendinosus, and the semimembranosus. They all have a common origin on the ischial tuberosity and therefore act to extend the thigh. Figure 8.6 illustrates this muscle group.

The **biceps femoris** is the lateral muscle of the posterior thigh. It has two heads and two origins, one on the ischial tuberosity and a second on the femur. The two heads merge to form the belly of the muscle and insert on the lateral condyle of the tibia and the head of the fibula. Because this muscle spans the hip and the knee joints, it can flex the leg and extend the thigh.

Medial to the biceps femoris is the **semitendinosus**. It is a long muscle that passes behind the knee to insert on the posteriomedial surface of the tibia. The **semimembranosus** is medial to the semitendinosus and inserts on the medial tibia. These muscles cross both the hip and knee joints and act to extend the thigh and flex the leg. The hamstrings are therefore antagonists to the quadriceps. If the thigh is flexed and drawn up toward the pelvis, the hamstrings act to extend the thigh.

Flex your knee to feel the tendons of the semimembranosus and semitendinosus located just above the back of the knee on the medial side. Similarly, on the lateral side of the knee, just above the fibular head, you can palpate the tendon of the biceps femoris.

C. Muscles of the Lower Leg

TIBIALIS ANTERIOR

The **tibialis anterior**, Figure 8.7a, is located on the anterior side of the lower leg and comprises the fleshy mass just lateral to the anterior border (the "shin") of the tibia. The muscle has two origins, one of which is found on the anteriolateral side of the tibia and the other on a membrane between the tibia and fibula. The tendon passes over the dorsal surface of the foot and inserts onto the medial cuneiform and the first metatarsal. This muscle dorsiflexes your foot. Place your fingers on your tibialis, and dorsiflex your foot to feel the muscle contract.

1. _____
2. _____
3. _____
4. _____

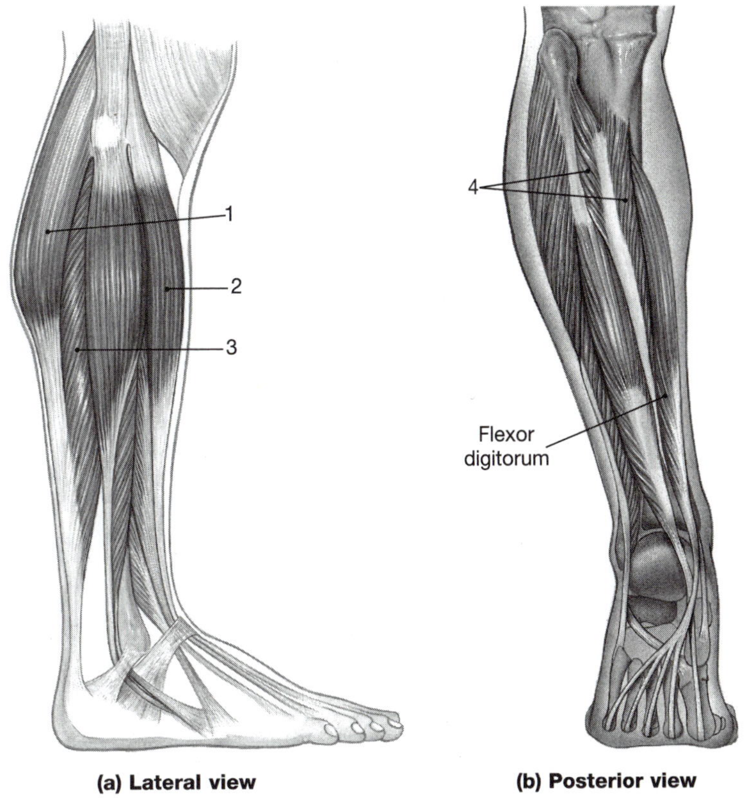

(a) Lateral view   **(b) Posterior view**

●Figure 8.7
Muscles that Move the Foot

GASTROCNEMIUS

The posterior lower leg is covered by the prominent calf muscles, the gastrocnemius, and the soleus (see Figure 8.7). These muscles share the calcaneal (Achilles') tendon that inserts onto the calcaneus of the foot. The **gastrocnemius**, Figure 8.7a, the most superficial calf muscle, originates on the lateral and medial epicondyles of the femur. The two heads cross over the back of the knee and form a fleshy mass consisting of two bellies before blending about halfway down the lower leg. The gastrocnemius acts to plantar flex, adduct, and invert the foot. Since the gastrocnemius also crosses the knee, it also flexes the lower leg.

SOLEUS

Deep to the gastrocnemius is the **soleus**, which has an origin on the upper third of the tibia and on the fibula. The belly of the muscle passes beneath the gastrocnemius and eventually converges into the calcaneal tendon. The soleus contracts with the gastroc-nemius to plantar flex, adduct, and invert the foot.

TIBIALIS POSTERIOR

The origin of the **tibialis posterior** is found midway along the shaft of the tibia and the fibula. The tendon passes medially to the calcaneus and inserts on the plantar surface of the navicular bone, the cuneiform bones, and the second, third, and fourth metatarsals. The tibialis posterior adducts and inverts the foot.

Laboratory Activity   MUSCLES OF THE PELVIS, LEG, AND FOOT _____

MATERIALS

leg model
torso model

PROCEDURES

1. Examine the torso model. Locate each muscle described, and study its shape.
2. Attempt to locate these on your leg. Contract each muscle to study the movement it produces.
3. Label each muscle in Figures 8.6 and 8.7.

THE MUSCULAR SYSTEM

Exercise 8 Laboratory Report

A. Matching

Match each muscle of the head and neck listed on the left with its correct action on the right. Some choices on the right may be used more than once.

1. _____ orbicularis oculi	A. purses lips
2. _____ buccinator	B. elevates corners of mouth
3. _____ masseter	C. closes eye
4. _____ orbicularis oris	E. turns corners of mouth down
5. _____ temporalis	F. elevates and retracts mandible
6. _____ platysma	G. flexes neck and head
7. _____ zygomaticus	H. tenses cheeks
8. _____ sternocleidomastoid	I. elevates and protracts mandible

B. Completion

1. Label Figure 8.8.
2. Label Figure 8.9

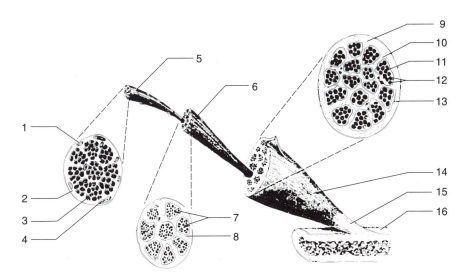

1. _____	7. _____	13. _____
2. _____	8. _____	14. _____
3. _____	9. _____	15. _____
4. _____	10. _____	16. _____
5. _____	11. _____	
6. _____	12. _____	

●Figure 8.8
Organization of Skeletal Muscles

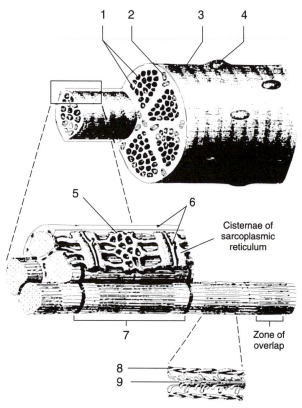

Cisternae of
sarcoplasmic
reticulum

Zone of
overlap

●Figure 8.9
Histological Organization of Skeletal Muscles

1. _____

2. _____

3. _____

4. _____

5. _____

6. _____

7. _____

8. _____

9. _____

C. Matching

Match each muscle of the chest and abdomen listed on the left with its correct action on
the right.

1. _____ pectoralis minor

2. _____ trapezius

3. _____ rectus abdominis

4. _____ transverse abdominis

5. _____ external oblique

6. _____ external intercostals

7. _____ serratus anterior

8. _____ internal intercostals

9. _____ pectoralis major

10. _____ latissimus dorsi

11. _____ internal oblique

12. _____ diaphragm

A. adducts arm posteriorly

B. elevates ribs

C. protracts scapula

D. lowers thoracic floor

E. second muscle layer of lateral abdomen

F. adducts arm anteriorly

G. extends neck and head

H. depresses ribs

I. deep muscle to internal oblique

J. depresses scapula

K. segmented by transverse tendon

L. superficial muscle of lateral abdomen

D. Matching

Match each muscle of the shoulder, arm, and hand listed on the left with its correct action on the right.

1. _____ deltoid   A. flexes arm, has two heads
2. _____ triceps brachii   B. rotates radius laterally
3. _____ supinator   C. extends arm
4. _____ brachialis   D. rotates radius medially
5. _____ pronator teres   E. abducts arm
6. _____ biceps brachii   F. flexes arm

E. Matching

Match each muscle of the pelvis and leg listed on the left with its correct action on the right.

1. _____ sartorius   A. large gluteal muscle
2. _____ tensor fasciae latae   B. extends lower leg
3. _____ gastrocnemius   C. flexes leg, adducts thigh
4. _____ psoas major   D. abducts thigh, iliotibial tract insertion
5. _____ rectus femoris   E. plantar flexes foot
6. _____ gracilis   F. adducts thigh, flexes leg
7. _____ vastus lateralis   G. dorsiflexes foot
8. _____ biceps femoris   H. flexes the thigh
9. _____ gluteus maximus   I. crosses the leg
10. _____ tibialis anterior   J. flexes thigh, extends leg

F. Short Answer Questions

1. Describe the muscles involved in mastication.

2. List the muscles that act on the mouth for facial expressions.

3. Describe the actions of the sternocleidomastoid muscle.

G. Short Answer Questions

1. Describe the muscle activity involved in inhalation of air.

2. List the muscles of the abdomen from deep to superficial.

3. Discuss the muscles that adduct and abduct the arm.

H. Short Answer Questions

1. Discuss the muscles involved in flexion and extension of the arm.

2. Compare the origins and insertions between the supinator and pronator teres.

J. Short Answer Questions

1. Compare the actions of the quadriceps femoris group and the hamstring group.

2. List the muscles involved in adduction and abduction of the thigh.

9 THE NERVOUS SYSTEM

The nervous system orchestrates body functions to maintain homeostasis. To accomplish this control, the nervous system must perform three vital tasks. First, it must detect changes in and around the body. Sensory receptors monitor specific environmental conditions and encode information about environmental changes as electrical impulses. Second, the nervous system must process the incoming sensory information and generate an appropriate motor response to adjust the activity of muscles and glands. Third, all sensory and motor activities must be orchestrated or integrated to achieve the balance of homeostasis.

PART ONE ORGANIZATION OF THE NERVOUS SYSTEM

WORD POWER	MATERIALS
afferent (afferen—carrying toward)	compound microscope
efferent (efferen—carrying away)	prepared slide
sensory (sens—a feeling)	multipolar neuron
somatic (somat—a body)	nervous system chart
autonomic (auto—self)	nervous system model

OBJECTIVES

On completion of this part of the exercise, you should be able to:

- Outline the organization and functions of the nervous system.
- Compare sensory receptors and effectors.
- Compare the autonomic and somatic divisions of the efferent PNS.
- Describe the cellular anatomy of a neuron.
- Generally discuss how a neuron communicates with other cells.

The nervous system is divided into two main components, the central nervous system and the peripheral nervous system (see Figure 9.1). The **central nervous system** (CNS) consists of the brain and the spinal cord. The **peripheral nervous system** (PNS) includes cranial and spinal nerves that communicate with the CNS. The PNS is subdivided into afferent and efferent divisions. The **afferent division** receives sensory information from sensory receptors, cells and organs that detect changes in the body and the surrounding environment, and sends the sensory information to the CNS for interpretation. The CNS decides the appropriate response to the sensory information and sends motor commands to the **efferent division**, which controls the activities of **effectors**, the muscles and glands of the body. The **somatic nervous system** conducts motor responses to skeletal muscles. The **autonomic nervous system**, which consists of the sympathetic and parasympathetic branches, sends commands to smooth and cardiac muscles and glands.

●Figure 9.1
Functional Overview of the Nervous System

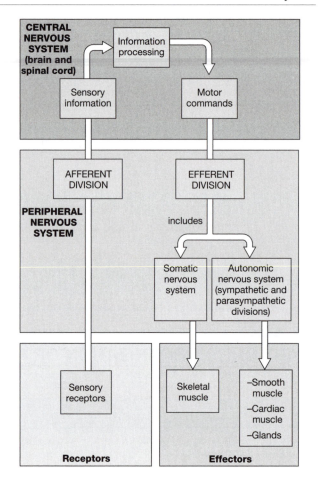

Two types of cells populate the nervous system, neurons and glial cells. Neurons are the communication cells of the nervous system and generate and transmit electrical impulses to respond to the ever-changing needs of the body. Glial cells have a supportive role to protect and maintain neural tissue. Refer to your textbook to study the various types of glial cells and their functions.

Use your textbook to study the organization of the nervous system. Know the basic pathways that sensory and motor information are conducted along in the nervous system.

A. Neurons

Neurons are the longest cells in your body. Some neurons are three feet long and extend from the spinal cord to your toes. A neuron has three distinguishable features: dendrites, a cell body, and an axon. Examine the neuron illustrated in Figure 9.2. Many dendrites carry information into the large, rounded cell body or **soma**, which contains the nucleus and organelles of the cell. One portion of the soma narrows into an **axon hillock** that extends into a single **axon**. The axon conducts impulses away from the soma toward other neurons or an effector. The axon may divide into several **collateral branches** that then subdivide into smaller branches called **telodendria**. Each telodendrium ends at a **synaptic knob** that contains chemical messengers called **neurotransmitters** stored in **synaptic vesicles**. When an electrical impulse arrives at the synaptic knob, the neuron releases the neurotransmitters that diffuse across a small space, called the **synaptic cleft**, and excites or inhibits the membrane of the neighboring neuron or effector. Notice that Figure 9.2 illustrates a single neuron synapsing with three different cell types, a neuron, a skeletal muscle, and gland cells. In reality, however, a neuron will synapse with a group of cells of only one cell type.

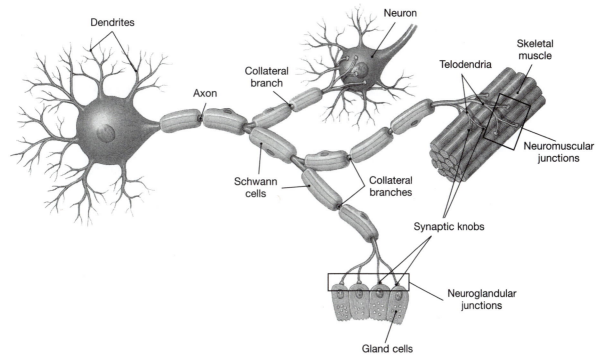

●Figure 9.2
Cellular Structure of a Neuron

The axon of a neuron is usually covered with a fatty **myelin** sheath that, in the peripheral nervous system, is produced by **Schwann cells**, a glial cell found in the peripheral nervous system. As a Schwann cell wraps around a small section of axon, it squeezes the cytoplasm out of its extensions and encases the axon in multiple layers of membrane. Notice that between the Schwann cells there are gaps called **nodes of Ranvier**. The membrane of the axon, the **axolemma**, is exposed at the nodes and permits a nerve impulse to rapidly arc from node to node.

Laboratory Activity   HISTOLOGY OF THE NERVOUS SYSTEM _____

MATERIALS
giant mutipolar neuron slide
compound microscope

PROCEDURES
1. Examine a prepared microscope slide of a giant multipolar neuron. Scan the slide at low magnification and locate several neurons.
2. Select a single neuron and increase magnification. Identify the soma and nucleus.
3. Can you distinguish between the dendrites and the axon?

_____

_____

4. Draw a neuron in the space provided below.

PART TWO   THE STRUCTURE OF THE BRAIN

WORD POWER

dura mater (dura—tough,
  mater—mother)
arachnoid (arachno—spider)
pia mater (pia—delicate)
folia (foli—a leaf)
vermis (verm—a worm)
aqueduct
  (aquaduct—water canal)
corpus callosum (corp—body,
  callo—thick skinned)

colliculus (collicul—a small hill)
peduncles (peduncul—little foot)
arbor vitae (arbor—tree, vitae—life)
pons (...pons—a bridge)
corpora (corp—a body)
quadrigemina (quadrigemina—four twins)
thalamus (thala—a chamber)

MATERIALS

brain models and charts
meningeal model

OBJECTIVES

On completion of this part of the exercise, you should be able to:

- Name the three meninges that cover the brain.
- Describe the extensions of the dura mater.
- Identify each major region of the brain and a basic function for each.
- Identify the surface features of each region of the brain.
- Identify the twelve pairs of cranial nerves.
- Identify the major internal features of the cerebrum, diencephalon, midbrain, brain stem and cerebellum.
- Trace a drop of cerebrospinal fluid through the ventricular system of the brain.

A. Protection and Support of the Brain

The brain and spinal cord of the central nervous system are protected by specialized membranes called **meninges**, a series of three membranes that anchor the CNS securely in place. Between certain meningeal layers is **cerebrospinal fluid** (CSF), which cushions the brain and prevents it from contacting the cranial bones during a head injury, much like a car's airbag prevents a passenger from hitting the dashboard. The cranial meninges include the following layers: the **dura mater**, **arachnoid**, and the **pia mater** (see Figure 9.3).

The **dura mater**, the outer meningeal covering, comprises two tissue layers. Between the layers are large blood sinuses, collectively called **dural sinuses**, which drain blood from cranial veins into the jugular veins. The **superior** and **inferior sagittal sinuses** are large veins in the dura between the two hemispheres of the cerebrum. Above the dura mater is the small **epidural space** filled with fat tissue and blood vessels.

The dura mater has extensions to further stabilize the brain. A midsagittal fold in the dura forms the **falx cerebri** and separates the right and left hemispheres of the cerebrum. Posteriorly, the dura mater folds again as the **tentorium cerebelli** and separates the cerebrum from the cerebellum. The **falx cerebelli** is a dural fold that separates the right and left hemispheres of the cerebellum. Locate these structures in Figure 9.3.

Deep to the dura mater is the **arachnoid**, named after the weblike connection this membrane has with the underlying pia mater. Between the dura and the arachnoid is the **subdural space**. The arachnoid forms a smooth covering over the brain and is visible in Figure 9.4.

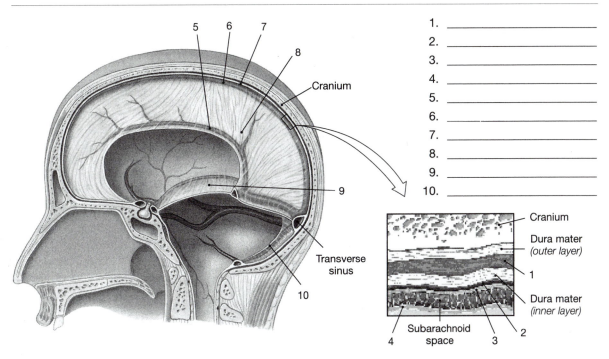

1. _____
2. _____
3. _____
4. _____
5. _____
6. _____
7. _____
8. _____
9. _____
10. _____

Cranium

Transverse
sinus

Cranium

Dura mater
(outer layer)

1

Dura mater
(inner layer)

Subarachnoid
space   2

●Figure 9.3
Organization of the Cranial Meninges
(a) Lateral view of the brain, showing the relationship of the meninges with the cranium.
(b) Detail of cranial meninges in lateral view.

Directly on the surface of the brain is the **pia mater**, which contains many blood vessels for the brain. Between the arachnoid and pia mater is the **subarachnoid space**, where **cerebrospinal fluid** (CSF) circulates. The fluid cushions the brain through buoyancy and protects it from contacting the hard bony wall of the cranium. Cerebrospinal fluid contains nutrients supplied by the blood to nourish the brain. The chambers in the brain, called ventricles, also contain CSF.

Laboratory Activity   PROTECTIVE COVERINGS OF THE BRAIN_____

MATERIALS
brain models
charts

PROCEDURES
1. Label and review the anatomy presented in Figure 9.3.
2. Locate the dura, arachnoid, and pia mater on available laboratory charts and models.

B. External Structures of the Brain

The brain is divided into five major regions: (1) the **cerebrum** or telencephalon, (2) the thalamus and hypothalamus, collectively called the **diencephalon**, (3) the **midbrain** or mesencephalon, (4) the **pons** and **cerebellum** that compose the mesencephalon, and (5) the **medulla oblongata** or myelencephalon. Locate the cerebrum, pons, cerebellum, and medulla in Figure 9.4. The midbrain, diencephalon, and other regions are shown in Figure 9.6, a midsagittal section of the brain.

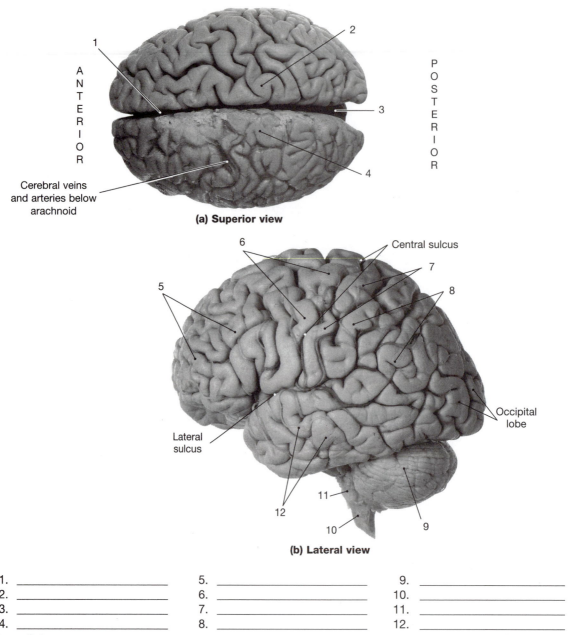

(a) Superior view

Central sulcus

Occipital lobe

Lateral sulcus

(b) Lateral view

1. _____	5. _____	9. _____
2. _____	6. _____	10. _____
3. _____	7. _____	11. _____
4. _____	8. _____	12. _____

●**Figure 9.4**
The Human Brain

(a) The paired cerebral hemispheres dominate the superior part of the brain. **(b)** Major
regions of the adult brain, lateral view.

THE CEREBRUM

The cerebrum is the most complex part of your brain. Your conscious thought, intel-
lectual reasoning, memory processing and storage occur in the cerebrum. Figure 9.4
shows the human brain from the superior and lateral views. The **cerebrum**, or telencephalon, is the largest portion of the brain. From the superior view, you can clearly
see that the cerebrum is divided into two lobes, the right and left **cerebral hemispheres**
by a deep fissure, the **longitudinal fissure**. The surface of each hemisphere is highly
convoluted and folded due to its rapid embryonic growth. Each small fold is called a
gyrus (pl. gyri), and each small groove is called a **sulcus** (pl. sulci). Deeper grooves,
such as the one separating the two hemispheres, are called **fissures**.

The cerebrum comprises five lobes, most named for the overlying cranial bone (see Figure 9.4). The anterior cerebrum is the **frontal** lobe. The prominent **central sulcus** located approximately midposteriorly separates the frontal lobe from the **parietal** lobe. The **occipital** lobe corresponds to the position of the occipital bone of the posterior skull. The **lateral sulcus** defines the boundary between the large frontal lobe and the **temporal** lobe of the lower lateral cerebrum. Cutting into the lateral sulcus and peeling away the temporal lobe reveals a fifth lobe, the **insula**.

THE CEREBELLUM

The cerebellum, Figures 9.4 and 9.5, is inferior to the occipital lobe of the cerebrum. Small folds on the cerebellar cortex are called **folia**. Right and left **cerebellar hemispheres** are

●**Figure 9.5**
The Cranial Nerves

(a) Inferior view of the brain. **(b)** Diagrammatic view, showing the attachment of the 12 pairs of cranial nerves.

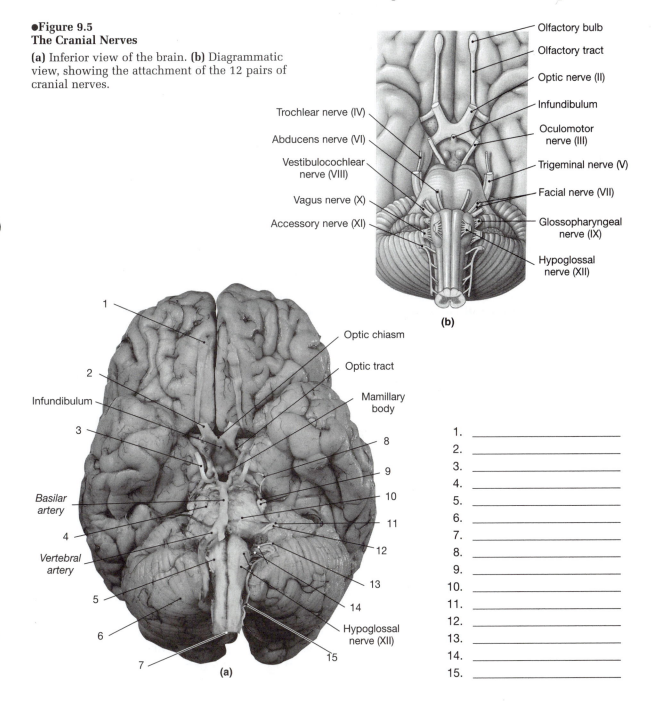

separated by a narrow **vermis**. Each cerebellar hemisphere comprises two lobes. A small **primary** (transverse) **fissure** separates the **anterior lobe** that is directly inferior to the cerebrum from the **posterior lobe**.

The cerebellum is primarily involved in the coordination of skeletal muscle contractions. Adjustments to postural muscles occur when impulses from the vestibulocochlear cranial nerve (VIII) of the inner ear pass into the cerebellum, where information concerning equilibrium is processed. Learned muscle patterns, such as those involved in serving a tennis ball or playing the piano, are stored in the cerebrum and processed in the cerebellum.

THE MEDULLA

The medulla connects the spinal cord with the brain. All ascending and descending messages between the brain and spinal cord pass through the medulla. The anterior surface of the medulla has two prominent folds called **pyramids** (see Figure 9.5). Since the right and left sides of the brain control the opposite sides of the body, the descending motor tracts decussate or cross over to the other side of the central nervous system in the pyramids. The medulla contains vital reflex centers for cardiovascular and respiratory functions.

THE PONS

The pons is superior to the medulla and functions as a relay station to direct sensory information to the thalamus and cerebellum. It also contains certain sensory, somatic motor, and autonomic cranial nerve nuclei. Locate the pons in Figures 9.4 and 9.5.

THE MIDBRAIN

The midbrain is located superior to the pons. Although covered by the cerebrum, the midbrain may be observed by gently pushing down on the superior surface of the cerebellum and looking below the cerebrum. Figure 9.6 shows the midbrain in midsagittal section. The **corpora quadrigemina** is a series of four bulges. The superior pair is the **superior colliculus**, which functions as a visual reflex center that moves the eyeballs and the head to keep an object centered on the retina of the eye. The lower pair is the **inferior colliculus**, which functions as an auditory reflex center and moves the head to locate and follow sounds.

THE DIENCEPHALON

The diencephalon is the upper portion of the brain stem that connects with the cerebrum and comprises the **thalamus** and the **hypothalamus** (Figure 9.6). All sensory information, except olfaction, pass through the thalamus, which then relays the impulse to the proper sensory cortex. Nonessential sensory data is filtered out by the thalamus and does not reach the sensory cortex. The **pineal gland**, an endocrine gland that secretes the hormone melatonin, is located on the roof of the diencephalon anterior to the corpora quadrigemina of the midbrain.

Only the hypothalamus of the diencephalon is exposed on the external surface of the brain. Examine the inferior view of the brain as in Figure 9.5, and locate the pons and the midbrain. Anterior to the mamillary bodies is the **infundibulum**, the stalk that attaches the **pituitary gland** or **hypophysis** to the hypothalamus.

CRANIAL NERVES

The brain and spinal cord of the central nervous system communicate to the rest of the body through cords of neurons called **nerves**. Two major groups of nerves occur: spinal nerves and cranial nerves. As the name of each group of nerves implies, the **spinal**

1. _____
2. _____
3. _____
4. _____
5. _____
6. _____
7. _____

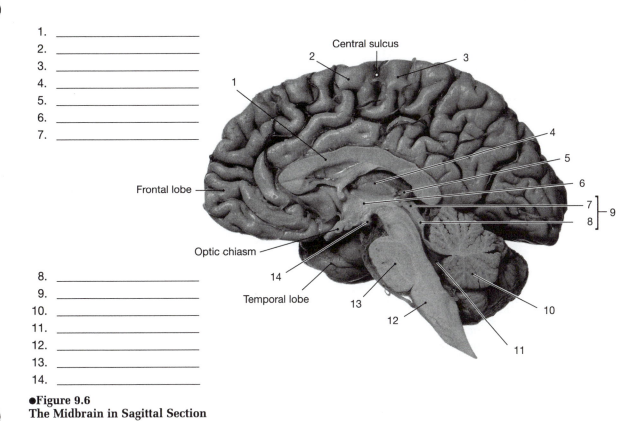

8. _____
9. _____
10. _____
11. _____
12. _____
13. _____
14. _____

●Figure 9.6
The Midbrain in Sagittal Section

nerves communicate to the CNS at the spinal cord, whereas the cranial nerves interface with the brain. The **cranial nerves** emerge from the brain at specific anatomical locations and pass through various foramina of the skull to get to the peripheral structures they innervate. All spinal and cranial nerves occur in pairs. There are 31 pairs of spinal nerves and 12 pairs of cranial nerves.

Each cranial nerve pair is identified by name; Roman numeral; and type of information the nerve conducts, sensory, motor, or both. Some cranial nerves are entirely **sensory nerves**, but most are **mixed nerves** with both motor and sensory neurons. Cranial nerves that primarily conduct motor commands are considered **motor nerves** though they have a few sensory fibers to inform the brain about muscle tension and position. Identify each cranial nerve shown in Figure 9.5. Use your textbook to study the function of each cranial nerve.

Laboratory Activity   EXTERNAL STRUCTURE OF THE BRAIN _____

MATERIALS

brain models and charts

PROCEDURES

1. Locate each major region of the brain on lab models and charts. Label each region in Figure 9.4.
2. Identify the surface features of each region on lab models.
3. Label the cranial nerves presented in Figure 9.5.
4. Locate each cranial nerve on lab models and charts.

C. Internal Structure of the Brain

CEREBRUM

Figure 9.6 shows the midsagittal anatomy of the human brain. The cerebral hemispheres are connected by a tract of white matter called the **corpus callosum**. It is easily identified as the curved white structure at the base of the cerebrum.

DIENCEPHALON

The diencephalon comprises the thalamus and the hypothalamus, both visible in Figure 9.6. The **thalamus** is shaped like a barbell with two masses on each end connected by an oval **intermediate mass** in the middle. The slight depression around the intermediate mass is the **third ventricle** that passes through the diencephalon. Below the thalamus and intermediate mass is the **hypothalamus**. Inferior to the posterior margin of the corpus callosum of the cerebrum is the conical **pineal gland**.

MIDBRAIN

The midbrain is clearly visible in the midsagittal sections of Figure 9.6. Posterior to the pineal gland is the **corpora quadrigemina**, which includes the **superior colliculus** and the **inferior colliculus**. A long canal called the **cerebral aqueduct** passes anterior to these masses. **Cerebral peduncles** are located within the walls that form the cerebral aqueduct. Their fibers extend to connect the pons and the cerebrum.

PONS AND MEDULLA

The enlarged area of the brain stem is the **pons**, located inferior to the cerebral peduncles. Inferior to the pons is the **medulla**, which connects the spinal cord to the rest of the brain. Notice in Figure 9.6 how the brain stem tapers from the pons to the spinal cord. The **fourth ventricle** lies posterior to the brain stem between the medulla and the cerebellum.

CEREBELLUM

The cerebellum is positioned posterior to the pons and medulla. In sagittal section, the white matter of the **cerebellum** is apparent. Because the tissue is highly branched, it is called the **arbor vitae**, the tree of life (see Figure 9.6). The fourth ventricle is anterior to the cerebellum.

VENTRICULAR SYSTEM OF THE BRAIN

Deep in the brain are four chambers called **ventricles**, detailed in Figure 9.7. Each ventricle contains modified capillaries called **choroid plexuses** that produce **cerebrospinal fluid** (CSF). This fluid flows through the ventricles and enters the subarachnoid space surrounding the brain. Cerebrospinal fluid protects and nourishes the tissues of the brain. A pair of **lateral ventricles** extends deep into the cerebrum as horseshoe-shaped chambers.

Cerebrospinal fluid circulates from the lateral ventricles through the **interventricular foramen** and enters the **third ventricle**. This small chamber surrounds the intermediate mass of the thalamus. CSF in the third ventricle passes through the **cerebral aqueduct** of the midbrain and enters the **fourth ventricle** posterior to the brain stem between the region of the pons/medulla and the cerebellum. In the fourth ventricle, two **lateral apertures** and a single **median aperture** direct CSF laterally to the exterior of the brain and into the subarachnoid space. CSF then circulates around the brain and spinal cord, and then is reabsorbed from an **arachnoid villus** that projects into the veins of the dural sinuses.

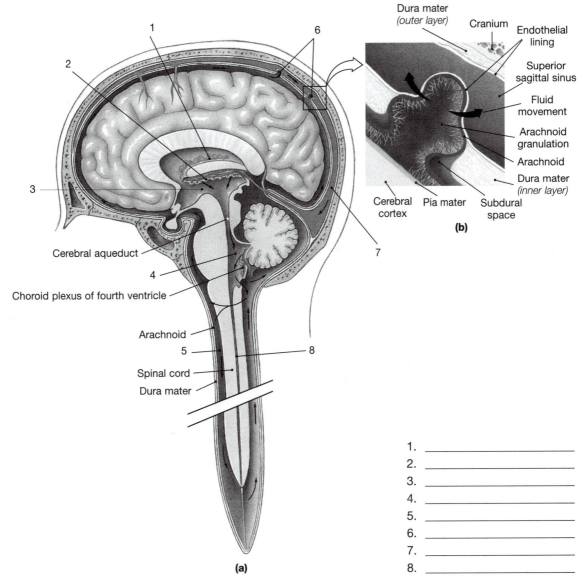

Dura mater
(outer layer)   Cranium   Endothelial
                                    lining

                                    Superior
                                    sagittal sinus

                                    Fluid
                                    movement

                                    Arachnoid
                                    granulation

                                    Arachnoid

                                    Dura mater
                                    (inner layer)

Cerebral   Pia mater   Subdural
cortex   space

(b)

Cerebral aqueduct

Choroid plexus of fourth ventricle

Arachnoid

Spinal cord

Dura mater

(a)

1. _____
2. _____
3. _____
4. _____
5. _____
6. _____
7. _____
8. _____

●**Figure 9.7**
Circulation of Cerebrospinal Fluid

(a) Sagittal section indicating the routes of formation and circulation of cerebrospinal
fluid. **(b)** Orientation of the arachnoid villi.

Laboratory Activity   OBSERVATION OF THE HUMAN BRAIN _____

MATERIALS

human brain models and charts   preserved and sectioned human brain
ventricular system model   (if available)

PROCEDURES

1. Label and review the anatomy presented in Figures 9.3 through 9.7.
2. Identify the structures of the cerebrum, diencephalon, midbrain, pons, medulla,
   and cerebellum.
3. Identify the components of the ventricular system, and trace the circulation of
   cerebrospinal fluid on a brain model.

PART THREE   Sheep Brain Dissection

The sheep brain, like all mammalian brains, is similar in structure and function to the human brain. One major difference between the human brain and that of other animals is the orientation of the brain stem. Humans are vertical animals and walk on two legs. The spinal cord is perpendicular to the ground, so the brain stem must also be vertical. In four-legged animals, the spinal cord and brain stem are parallel to the ground.

Dissecting a sheep brain will enhance your studies of models and charts of the human brain. Take your time while performing the dissection, and follow the directions carefully. Refer to the manual and figures often during the procedures.

Laboratory Activity   SHEEP BRAIN DISSECTION

MATERIALS

gloves, safety glasses
butcher knife (if available)

sheep brain (preferably with dura intact)
dissection pan, scissors, scalpel, forceps, probe

PROCEDURES

Put on your gloves and safety glasses before you open a storage container of brains or handle a brain. A variety of preservatives are used on biological specimens, some of which will cause irritation to your skin and mucous membranes. Your lab instructor will inform you how to handle, dissect, and finally dispose of the sheep brain.

If your sheep brain does not have the outer dura mater, proceed to section B of the dissection, and identify the external anatomy of the brain.

A.   Removal and Examination of the Meninges

1. Examine the intact dura mater on the sheep brain (see Figure 9.8). Locate the **falx cerebri** and the **tentorium cerebelli** from the dorsal surface of the dura mater. How does the intact tissue of the falx cerebri compare to the dura covering a hemisphere?

2. Determine if your brain specimen still has the **ethmoid bone**, a mass of bone on the anterior frontal lobe. If the bone is present, slip your probe between the bone and the dura. Carefully pull the bone off the specimen, using your scissors to snip any attached dura. Examine the removed ethmoid, and identify the **crista galli**, the crest of bone where the meninges attach.

3. Insert a probe between the dura and the brain to gently separate the two. Use your scissors sparingly to cut the dura away from the base of the brain. Take care not to cut or remove a cranial nerve.

4. The dura is strongly anchored at the tentorium cerebelli. To detach the dura in one piece, first loosen the dura from the rest of the brain and gather it at the tentorium. Grasp the dura at the tentorium deep between the cerebellum and cerebrum, and pull it straight out.

5. Open the detached dura, and identify the falx cerebri and tentorium cerebelli.

B.   External Structures of the Sheep Brain

1. Examine the cerebrum, and identify the frontal, parietal, occipital, and temporal lobes. The insula is a deep lobe and is not visible externally. How deep is the longitudinal fissure separating the right and left cerebral hemispheres? Observe the gyri and sulci on the cortical surface. Search the surface between sulci for the arachnoid and pia mater.

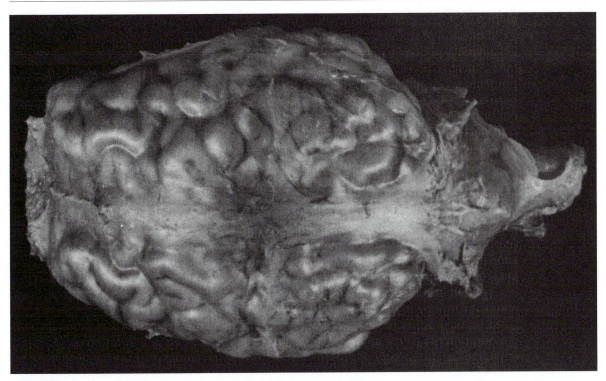

●Figure 9.8
Superficial View of Sheep Brain Showing Dura Mater

2. View the dorsal surface of the cerebellum. Compare the size of the folia with the gyri of the cerebrum. Unlike the human brain, the sheep cerebellum is not divided medially to form two lateral hemispheres.

3. To examine the dorsal anatomy of the midbrain, hold the sheep brain as in Figure 9.9 and gently depress the cerebellum. The midbrain is now visible between the cerebrum and cerebellum. Identify the medial pineal gland of the diencephalon and the four elevated masses of the corpora quadrigemina. Distinguish between the superior colliculi and the inferior colliculi. What is the function of these masses?

4. Turn the brain to view the ventral surface as in Figure 9.10. Note how the spinal cord joins the medulla. Identify the pons and the cerebral peduncles of the midbrain. Locate the single mamillary body on the hypothalamus. The mamillary body of the human brain is a paired mass. The pituitary gland has most likely been removed from your specimen; however, you may still identify the stub of the infundibulum that attaches the pituitary to the hypothalamus.

5. Identify as many cranial nerves on your sheep brain as possible. Nerves I, II, III, and V are usually intact and easy to identify. Your laboratory instructor may ask you to observe several sheep brains to study all the cranial nerves.

C. Internal Structure of the Sheep Brain

1. Lay the sheep brain in the dissecting pan. Place the blade of a large butcher knife into the anterior region of the longitudinal fissure. Point the tip of the knife blade down. With a single smooth motion, move the knife through the brain from anterior to posterior to separate the right and left halves along the midsagittal plane. If only a scalpel is available, attempt to dissect the brain with as few long, smooth cuts as possible.

2. Using Figure 9.11 as a guide, examine a section of the sheep brain, and locate the following structures:

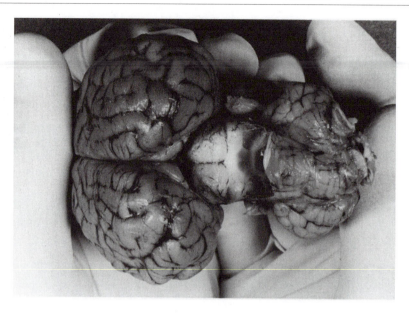

●Figure 9.9
The Midbrain

CEREBRUM

Locate the **corpus callosum**, a white tract interconnecting the cerebral hemispheres. Gently slide a blunt probe inside the **lateral ventricle**. How deep does the ventricle extend into the cerebrum? Look inside the lateral ventricle, and locate the **choroid plexus**, which appears as a granular mass of tissue.

DIENCEPHALON

Identify the **intermediate mass** surrounded by the **third ventricle**. The lateral walls of the third ventricle are the medial margins of the **thalamus**. Inferior to the intermediate mass is the **hypothalamus**. Identify the **infundibulum** and **pituitary gland** if still present on the sheep brain. The posterior mass of the hypothalamus is the **mamillary body**. Anterior to the midbrain is the **pineal gland**.

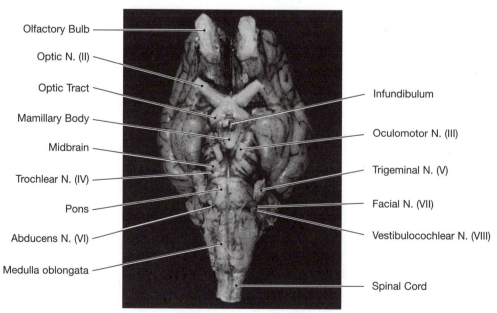

●Figure 9.10
Ventral View of the Sheep Brain

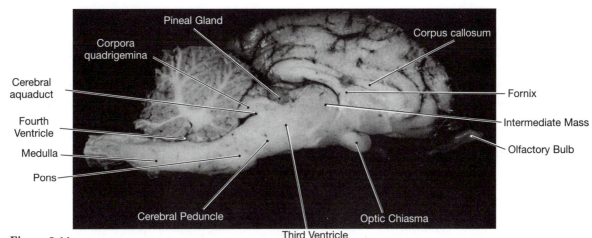

●**Figure 9.11**
Midsagittal Section of the Sheep Brain

MIDBRAIN

The **cerebral aqueduct** transverses the midbrain and separates the inferior **cerebral peduncles** from the upper midbrain. On the superior surface of the midbrain, locate the **corpora quadrigemina** posterior to the pineal gland. Distinguish between the **superior colliculus** and the **inferior colliculus** of the corpora quadrigemina.

PONS AND MEDULLA

Posterior to the midbrain is the **pons**, which may be identified as the swollen mass posterior to the cerebral peduncles. The pons narrows and joins the **medulla**, which is connected to the **spinal cord**.

CEREBELLUM

The cerebellum of the sheep brain is located superior to the medulla and pons. In sagittal section the white matter of the **arbor vitae** is clearly visible. Between the cerebellum and brain stem is the **fourth ventricle**.

3. Discard or store your sheep brain as explained by your laboratory instructor. Proper disposal of all preserved material protects your local environment and is mandated by state and federal regulations.

PART FOUR   THE SPINAL CORD, SPINAL NERVES, AND SPINAL REFLEXES

WORD POWER	MATERIALS
ganglion (ganglia—a knot on a string, a swelling)	spinal cord models and charts
commissure (commis—united)	reflex hammer
	compound microscope
	prepared slides:
	spinal cord cross section
	peripheral nerve cross section

OBJECTIVES

On completion of this part of the exercise, you should be able to:

• Identify the major surface features of the spinal cord, including the spinal meninges.
• Identify the sectional anatomy of the spinal cord.

- Describe the organization and distribution of spinal nerves.
- List the events of a typical reflex arc.

The spinal cord connects the peripheral nervous system (PNS) with the brain. In the PNS, peripheral nerves merge into spinal nerves and enter the spinal cord. Sensory neurons ascend the spinal cord and pass sensory information into the brain for interpretation. Motor neurons descend the spinal cord and carry motor commands from the brain to effectors, the muscles and glands of the body.

A. Gross Anatomy of the Spinal Cord

The spinal cord is continuous with the medulla of the brain. The spinal cord descends approximately 45 cm (18 inches) in the spinal canal of the vertebral column and terminates between the lumbar vertebrae L_1 and L_2. In young children, the spinal cord extends through most of the spine. After the age of four, the spinal cord stops growing in length, yet the vertebral column continues to grow. By adulthood, the spinal cord is shorter than the spine and descends only to the level of the upper lumbar vertebrae.

Figure 9.12 presents the gross anatomy of the spinal cord. The diameter of the spinal cord is not constant along its length. Two swollen regions occur where spinal nerves of the limbs join the spinal cord. The **cervical enlargement** in the neck supplies nerves to the shoulders and arms. The **lumbar enlargement** occurs near the distal end of the cord where nerves supply the pelvis and lower limbs. Inferior to the lumbar enlargement, the spinal cord narrows and stops at the **conus medullaris**. Spinal nerves fan out from the conus medullaris in a group called the **cauda equina**, the horse's tail. A thin thread of tissue from the pia mater called the **filum terminale** extends past the conus medullaris to anchor the spinal cord in the sacrum.

The spinal cord is organized into 31 segments, as shown in Figure 9.12. Each segment has a paired **spinal nerve** formed by the joining of two lateral extensions, the **dorsal** and **ventral roots**. The **dorsal root** contains sensory neurons entering the spinal cord from sensory receptors; the **ventral root** consists of motor neurons exiting the CNS and leading to effectors. Each spinal nerve is therefore a **mixed nerve** and carries both sensory and motor information. The dorsal root expands at the **dorsal root ganglion**, where the cell bodies of sensory neurons mass.

Figure 9.13 illustrates the anatomy of the spinal cord in transverse section. The spinal cord is divided anteriorly by the deep **anterior median fissure** and posteriorly by the shallow **posterior median sulcus**. An H-shaped area called the **gray horns** contains many glial cells and neuron cell bodies. Axons may cross to the opposite side of the spinal cord at the cross bars of the gray horns, called the **anterior** and **posterior gray commissures**. Between the commissures is a hole, the **central canal**. This canal contains cerebrospinal fluid and is continuous with the fluid-filled ventricles of the brain. Surrounding the gray horns are three masses of white matter: the **posterior**, **lateral**, and **anterior white columns**.

Figure 9.14 details the three layers of the **spinal meninges**, protective coverings similar to those of the brain. Cerebrospinal fluid from the ventricular system of the brain also circulates in the subarachnoid space of the spinal meninges.

Laboratory Activity   ANATOMY OF THE SPINAL CORD _____

MATERIALS

dissecting and compound microscopes
prepared slide: transverse (cross) section of spinal cord

PROCEDURES

1. Label and review the anatomy in Figures 9.12, 9.13, and 9.14.
2. Locate each surface feature of the spinal cord on available lab models and charts.
3. Locate the internal anatomy of the spinal cord on a sectional spinal cord model.

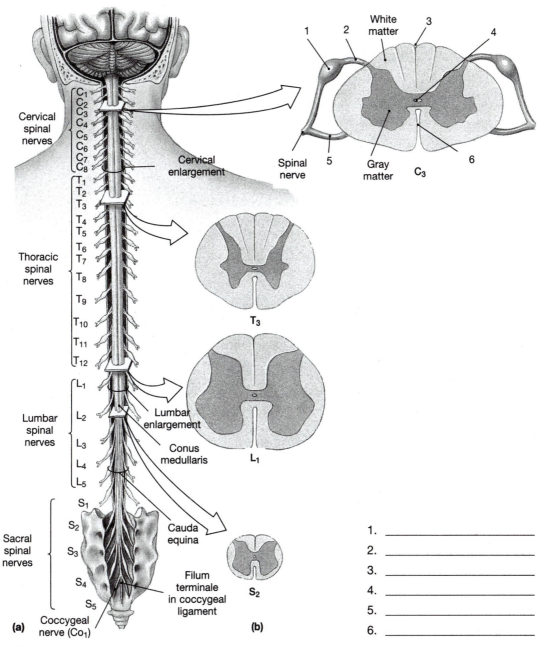

●Figure 9.12
Gross Anatomy of the Spinal Cord

(a) Superficial anatomy and orientation of the adult spinal cord. The numbers to the left identify the spinal nerves and indicate where the nerve roots leave the vertebral canal. Note, however, that the adult spinal cord extends only to the level of vertebrae L_1–L_2. **(b)** Inferior views of cross sections through representative segments of the spinal cord, showing the arrangement of gray and white matter.

4. Examine the microscopic features of the spinal cord in transverse section.

   a. For an orientation to the spinal cord slide, first view the slide at low magnification with a dissection microscope. Which features of the spinal cord enable you to distinguish anterior and posterior regions?

   _____

   _____

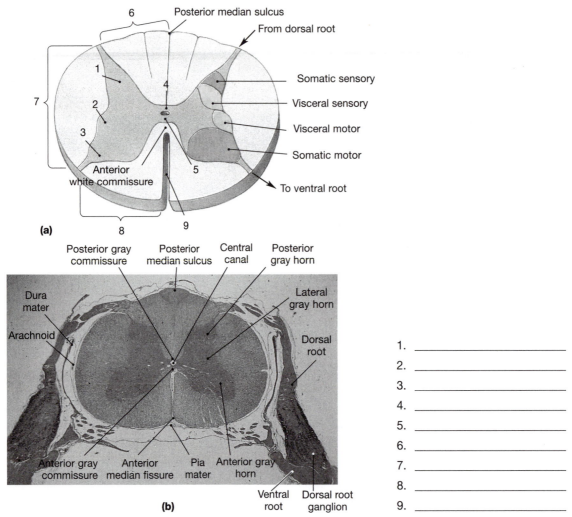

●Figure 9.13
Sectional Organization of the Spinal Cord
(a) The left half of this sectional view shows important anatomical landmarks and major regions of white matter in the posterior white column. The right half indicates the functional organization of the gray matter in the anterior, lateral, and posterior gray horns. (b) Micrograph of a section through the spinal cord, showing major landmarks; compare with part (a).

1. _____
2. _____
3. _____
4. _____
5. _____
6. _____
7. _____
8. _____
9. _____

b. Transfer your slide to a compound microscope. Move the slide around to survey the preparation at low magnification. What structures can you use as a landmark to distinguish between the posterior and anterior aspects of the spinal cord?

_____

_____

c. Examine the central canal and the gray horns. Can you distinguish between the posterior, ventral, and anterior horns? Where are the gray commissures?

_____

_____

d. Examine the white columns. What is the difference between gray and white matter in the central nervous system?

_____

_____

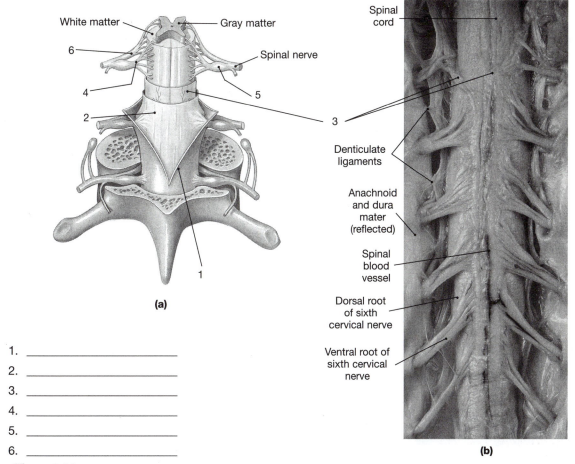

White matter

6

4

2

1

Gray matter

Spinal nerve

5

(a)

Spinal cord

3

Denticulate ligaments

Anachnoid and dura mater (reflected)

Spinal blood vessel

Dorsal root of sixth cervical nerve

Ventral root of sixth cervical nerve

(b)

1. _____

2. _____

3. _____

4. _____

5. _____

6. _____

●Figure 9.14
The Spinal Cord and Spinal Meninges
(a) Posterior view of the spinal cord showing the spinal meninges. **(b)** Posterior view
of spinal cord within the vertebral cord showing the roots of the spinal nerves.

e. Examine the outer layers surrounding the cross-sectioned spinal cord. Which
of the meningeal layers can you locate?

_____

_____

f. Draw and label the spinal cord in cross section in the space provided.

B. **Spinal Nerves**

Two types of nerves connect peripheral sensory receptors and effectors to the central
nervous system: 12 pairs of **cranial nerves** and 31 pairs of **spinal nerves**. Spinal nerves
branch into **peripheral nerves** made up of axons of sensory and motor neurons. Figure

9.12 shows the spinal nerves emerging from the spinal cord. Each nerve passes through an intervertebral foramen between two adjacent vertebrae. There are 8 **cervical nerves** (C_1–C_8), 12 **thoracic nerves** (T_1–T_{12}), 5 **lumbar nerves** (L_1–L_5), 5 **sacral nerves** (S_1–S_5), and a single **coccygeal nerve**. Although there are only 7 cervical vertebrae, there are 8 cervical spinal nerves. The eighth cervical nerve is located between vertebrae C_7 and T_1. Spinal nerves are mixed nerves and contain sensory, visceral motor, and somatic motor neurons. Use your textbook for information regarding individual spinal nerves.

Spinal nerves branch into peripheral nerves. These nerves are compartmentalized by connective tissue much in the same way skeletal muscles are organized. The peripheral nerve is wrapped in an outer covering called the **epineurium** (see Figure 9.15). Beneath this layer is the **perineurium** that separates the axons into bundles called **fascicles**. Inside a fascicle, the **endoneurium** surrounds each axon and isolates it from neighboring axons.

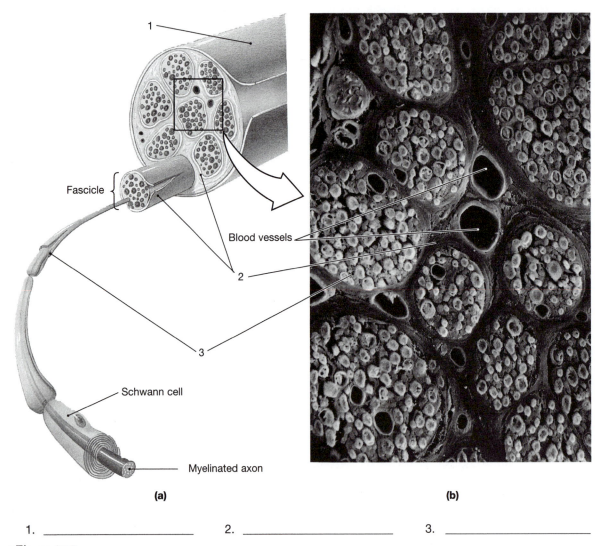

(a)   (b)

1. _____   2. _____   3. _____

●**Figure 9.15**
Organization of Peripheral Nerves

(a) A typical peripheral nerve and its connective tissue wrappings. **(b)** A scanning electron micrograph showing the perineurium and endoneurium in great detail.

Laboratory Activity   SPINAL NERVES _____

MATERIALS

compound microscope
spinal nerve model
prepared slide:
   peripheral nerve

PROCEDURES

1. Label and review the anatomy in Figures 9.14 and 9.15.

2. Locate the spinal nerves assigned by your laboratory instructor on the various lab models.

3. Examine the microscopic preparation of the **peripheral nerve**:

   a. Move the slide around to survey the preparation. Locate the epineurium. Is it continuous around the nerve?

      _____

      _____

   b. Examine a single fascicle. Can you distinguish the perineurium from the epineurium?

      _____

      _____

   c. Locate the individual axons inside a fascicle. Can you distinguish between the myelin sheath of the axon from the endoneurium?

      _____

      _____

   d. Draw and label the peripheral nerve in the space provided.

C. Spinal Reflexes

Reflexes are automatic responses by the nervous system to specific stimuli. Most reflexes have a protective function and cause rapid adjustments to maintain home-ostasis. Touch something hot, and the withdrawal reflex removes your hand to prevent tissue damage. Shine a bright light into someone's eyes, and his or her pupils constrict to protect the retina from overstimulation. Reflexes are used as a diagnostic tool to evaluate the function of specific regions of the brain and spinal cord. The sensory and motor components of a reflex are "prewired" and will initiate the reflex on stimulation. Figure 9.16 details the components of a reflex arc.

You are probably familiar with the "knee jerk" or **stretch reflex**, which occurs by stimulating the tendon over the patella by hitting it with a rubber percussion hammer

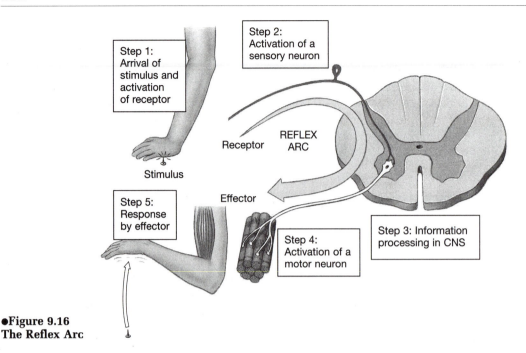

●**Figure 9.16**
The Reflex Arc

(see Figure 9.17). Although it is the tendon that is initially stimulated, the sensory receptors involved in this reflex are specialized skeletal muscle fibers called **muscle spindles**. Sensory neurons wrap around the muscle spindles and detect changes in their length. Stretching the muscle pulls on the fibers, and the muscle spindles send more signals to the spinal cord to promote an increase in muscle tension. As tension increases, the muscle length decreases and the muscle spindles are inhibited. Tapping on the patellar tendon stretches the muscle spindles in the quadriceps muscle group of the anterior thigh. This stimulus evokes a rapid motor reflex to contract the quadriceps and shorten the muscles.

●**Figure 9.17**
The Patellar Reflex

This stretch reflex is controlled by muscle spindles in the muscles that straighten the knee. **(a)** Step 1: A reflex hammer strikes the muscle tendon, stretching the spindle fibers. This stretching results in a sudden increase in the activity of the sensory neurons. These neurons synapse on spinal motor neurons. Step 2: The activation of extrafusal motor units produces an immediate increase in muscle tone and a reflexive kick. **(b)** The changes in muscle tension can be graphically recorded. Note the small secondary kick that may occur if the leg rebounds past its resting position.

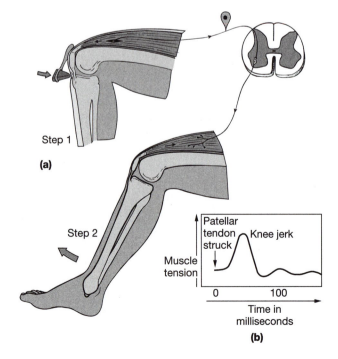

Laboratory Activity   STRETCH REFLEXES _____

MATERIALS

percussion hammer (with rubber head)
lab partner

PROCEDURES

1. Have your lab partner sit and cross his or her legs. Gently tap below the patella with the percussion hammer to cause the muscle spindles of the rectus femoris muscle to suddenly stretch. What is the response? How might this patellar reflex help maintain upright posture?

   _____

   _____

2. The calcaneal reflex tests the response of the gastrocnemius muscle. Have your lab partner hang his or her foot off a chair. Gently tap the calcaneal tendon with the percussion hammer. What is the response?

   _____

   _____

THE NERVOUS SYSTEM
Exercise 9 Laboratory Report

A. Matching

Match the neuron structure on the left with the correct description on the right.

1. _____	dendrite	A. myelinates axons in PNS
2. _____	axon	B. main branches of an axon
3. _____	collateral branches	C. vacuole of neurotransmitters
4. _____	synaptic knob	D. fine branches of axon
5. _____	axon hillock	E. forms blood–brain barrier
6. _____	telodendria	F. directs impulses to cell body
7. _____	astrocyte	G. swollen end of axon
8. _____	synaptic vesicles	H. connects soma to axon
9. _____	axolemma	I membrane of axon
10. _____	Schwann cell	J. single extension from cell body

B. Matching

Match the brain structure on the left with the correct description on the right.

1. _____	folia	A. upper part of the corpora quadrigemina
2. _____	cerebrum	B. white tract between cerebral hemispheres
3. _____	longitudinal fissure	C. forms floor of the diencephalon
4. _____	gyrus	D. duct through the midbrain
5. _____	inferior colliculus	E. white matter of cerebellum
6. _____	arbor vitae	F. site of cerebrospinal fluid reabsorption
7. _____	falx cerebri	G. furrow between gyri
8. _____	hypothalamus	H. chamber of the diencephalon
9. _____	central sulcus	I. endocrine gland of diencephalon
10. _____	dura mater	J. forms the lateral walls of third ventricle
11. _____	sulcus	K. ridge of cerebral cortex
12. _____	vermis	L. lower part of the corpora quadrigemina
13. _____	subarachnoid space	M. cleft between the cerebral hemispheres
14. _____	pons	N. outer meningeal layer
15. _____	tentorium cerebelli	O. site of cerebrospinal fluid circulation
16. _____	insula	P. small folds on the cerebellum
17. _____	third ventricle	Q. narrow central region of cerebellum
18. _____	thalamus	R. separates cerebellum and cerebrum
19. _____	corpus callosum	S. area of brain superior to medulla
20. _____	pineal gland	T. lobe deep to the temporal lobe
21. _____	superior colliculus	U. divides motor and sensory cortical regions
22. _____	arachnoid villus	V. separates cerebral hemispheres
23. _____	cerebral aqueduct	W. contains five lobes

C. Completion

1. Imagine watching a bird fly across your line of vision. What part of the brain is active in keeping the moving object on the retina of the eyeball?

2. You have just eaten a medium-sized pepperoni pizza and lie down to digest. What cranial nerve stimulates the muscular activity of your digestive tract?

3. You walk into an old building and immediately smell a dusty odor. After 20 minutes of exploring, you realize that you no longer notice the smell. What part of the brain is responsible for this apparent loss of sensation?

4. A child is preoccupied with a large cherry lollipop. What part of the child's brain is responsible for the licking and eating reflexes?

D. Matching

Match the spinal cord structure on the left with the correct description on the right.

1. _____	ventral gray horn	A.	site of cerebrospinal fluid circulation
2. _____	posterior median sulcus	B.	sensory branch entering cord
3. _____	subarachnoid space	C.	surrounds axons of peripheral nerve
4. _____	ventral root	D.	contains somatic motor somae
5. _____	posterior gray horn	E.	tapered end of spinal cord
6. _____	conus medullaris	F.	shallow groove
7. _____	endoneurium	G.	motor branch exiting cord
8. _____	dorsal root	H.	contains sensory somae

E. Labeling and Completion

1. Label Figure 9.18, the inferior surface of the brain.
2. Label the anatomy of the spinal cord in Figure 9.19.
3. Label the division of the nervous system in Concept Map I.

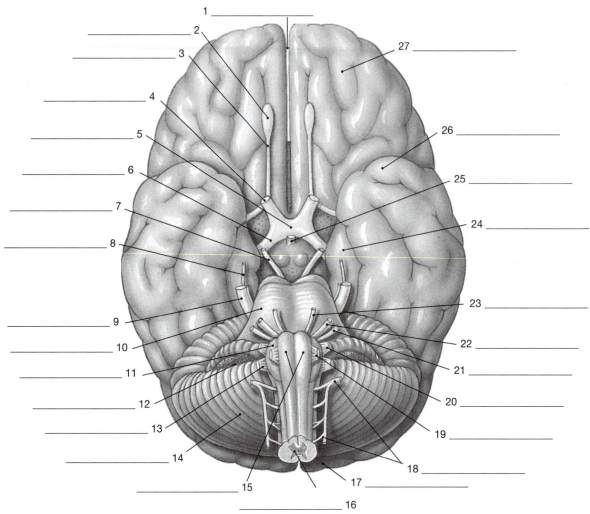

●Figure 9.18
The Inferior Aspect of the Brain

External anatomy of the inferior surface, diagrammatic view.

1. _____
2. _____
3. _____
4. _____
5. _____
6. _____
7. _____
8. _____
9. _____
10. _____

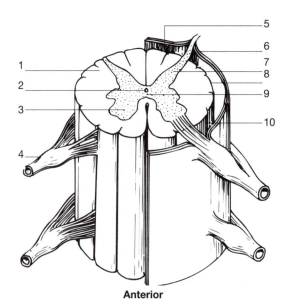

●Figure 9.19
Organization of the Spinal Cord

Anterior

Concept Map I

Using the following terms, fill in the circled, numbered, blank spaces to complete the concept map. Follow the numbers that comply with the organization of the map.

Brain   Peripheral nervous system
Afferent division   Smooth muscle
Sympathetic N.S.   Somatic Nervous System
Motor System

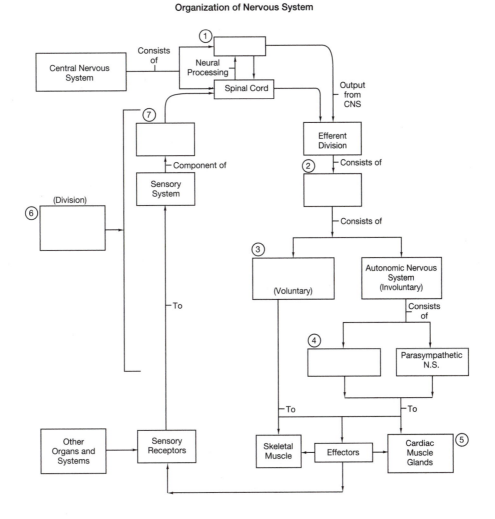

Organization of Nervous System

10 THE ENDOCRINE SYSTEM

WORD POWER	MATERIALS
adenohypophysis (adeno—a gland)	torso models
follicle (follicul—small bag)	endocrine charts
parathyroid glands (para—near)	dissecting and compound microscopes
cortex (cort—bark or shell)	prepared slides
medulla (medull—pith)	hypophysis
acinar (acin—a berry)	thyroid, parathyroid
corpus luteum (corpus—body; luteum—yellow)	thymus
	adrenal
	pancreas
	testis, ovary

OBJECTIVES

On completion of this exercise, you should be able to:

- Compare the two regulatory systems of the body, the nervous and endocrine systems.
- Identify each endocrine gland on laboratory models.
- Describe the histological appearance of each endocrine gland.
- Identify each endocrine gland when viewed microscopically.

Two systems regulate homeostasis: the nervous system and the endocrine system. These systems must coordinate their activities to maintain control of internal functions. The nervous system responds rapidly to environmental changes, sending electrical commands to the body that cause an immediate response. The duration of each electrical impulse is brief, measured in milliseconds. The endocrine system maintains long-term control. In response to stimuli, **endocrine glands** produce regulatory molecules called **hormones** that slowly cause a change in the metabolic activities of target cells, cells with membrane receptors for the hormone. Hormones are secreted into interstitial spaces, and then absorbed and transported by the circulatory system. The effect of the hormone is usually long lasting, over many hours.

The secretion of many hormones is regulated by negative feedback mechanisms. In **negative feedback**, a stimulus causes a response that reduces or removes the stimulus. An example that you are familiar with is the operation of a central air conditioning (A/C) unit. When a room heats up, the warm temperature activates the thermostat on the wall to turn on the compressor of the A/C unit. Air flows in and cools the room and removes the stimulus, the warm temperature. Once the stimulus is completely removed, the A/C shuts off. Negative feedback is therefore a self-limiting mechanism.

An example of negative feedback control of hormonal secretion is the regulation of insulin, a hormone from the pancreas that lowers blood sugar concentration. When blood sugar levels are high, as they are after a meal, the pancreas secretes insulin. Insulin stimulates cells to increase their sugar consumption and storage, thus lowering blood sugar concentration. When sugar levels return to normal, secretion of insulin stops.

In this exercise, you will study the pituitary gland, thyroid and parathyroid glands, thymus gland, adrenal glands, pancreas, and gonads. Figure 10.1 shows the location of and hormones secreted by each endocrine gland.

A. Pituitary Gland

ANATOMY

The **pituitary gland** or **hypophysis** produces hormones that control the activity of many other endocrine glands and is therefore called the "master gland." It is a small gland located in the sella turcica (Turk's saddle) of the sphenoid bone of the skull (see Figure 10.2). A stalk or **infundibulum** attaches the gland to the hypothalamus. Hormones and regulatory molecules from the hypothalamus travel down a plexus of blood vessels in the infundibulum to reach the hypophysis.

The pituitary gland contains two lobes, the anterior and posterior lobes. The anterior lobe, the **adenohypophysis**, produces many hormones that target other endocrine glands, causing them to produce and secrete their hormones.

The posterior lobe, the **neurohypophysis**, does not produce hormones; it secretes only the hormones produced by the hypothalamus. The hypothalamus produces two

●**Figure 10.1**
The Endocrine System

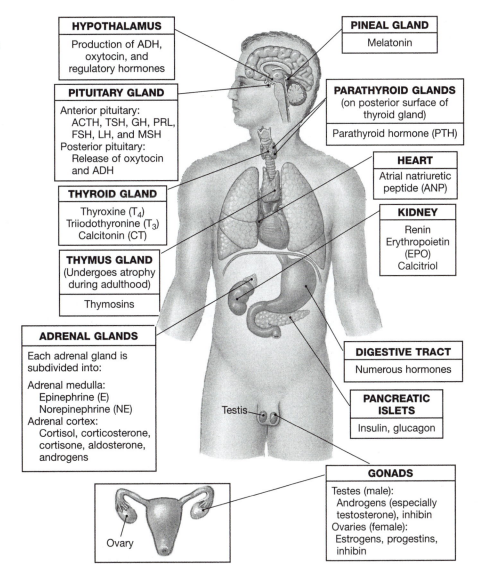

hormones, antidiuretic hormone (ADH) and oxytocin (OT), which travel down the infundibulum to enter the neurohypophysis.

HISTOLOGY

The adenohypophysis is populated by a variety of cell types, each secreting a different hormone. Compared to the neurohypophysis, the adenohypophysis contains more cells and looks denser. Most of the cells in the adenohypophysis stain a reddish color.

Laboratory Activity   MICROSCOPIC OBSERVATIONS OF THE PITUITARY GLAND

MATERIALS

> dissecting and compound microscope
> prepared slide
>   pituitary gland

PROCEDURES

1. Obtain a slide of the pituitary gland. The slide will usually have both lobes of the gland.
2. Use the dissecting microscope to survey the pituitary gland slide at low magnification. Can you distinguish between the two lobes of the gland?
3. Examine the slide at low and medium powers with the compound microscope. Identify the adenohypophysis and neurohypophysis, noting the different cell arrangement of each.

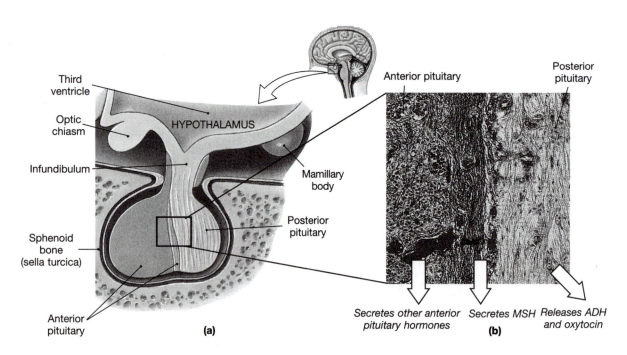

●Figure 10.2
Anatomy and Orientation of the Pituitary Gland
(LM × 62)

B. Thyroid Gland

ANATOMY

The thyroid gland is located in the anterior aspect of the neck, directly inferior to the **thyroid cartilage** (Adam's apple) of the larynx and just superior to the trachea. The gland consists of two lateral lobes that are connected by a central mass, the **isthmus** (IS-mus). Locate the thyroid gland in Figures 10.1 and 10.3a. This gland produces two groups of hormones associated with regulation of cellular metabolism and calcium homeostasis.

HORMONES

Follicular cells produce the hormones **thyroxine (T_4)** and **triiodothyronine (T_3)**, hormones that in part regulate metabolic rate. These hormones are secreted into the lumen of the follicles as a glycoprotein, **thyroglobulin**, which stores thyroid hormones. Thyroglobulin is later reabsorbed by the follicular cells and released into the bloodstream, as needed.

Peripheral to the follicles are larger **C cells (parafollicular cells)**. These cells produce the hormone **calcitonin (CT)**, which decreases blood calcium levels. Calcitonin inhibits osteoclasts in bone tissue from dissolving bone matrix and releasing calcium from the skeleton into the blood.

Laboratory Activity   MICROSCOPIC OBSERVATIONS OF THE THYROID

MATERIALS

> dissecting and compound microscope
> prepared slide
>   thyroid gland

PROCEDURES

1. Label the features of the thyroid gland in Figure 10.3.
2. Scan the thyroid slide with the dissecting microscope. What structures are visible at this magnification?
3. Use a compound microscope to view the thyroid slide at low and medium powers. Locate the following structures: thyroid follicle, follicular cells, thyroglobulin, and C cells.
4. Sketch several thyroid follicles as observed at medium magnification in the space provided.

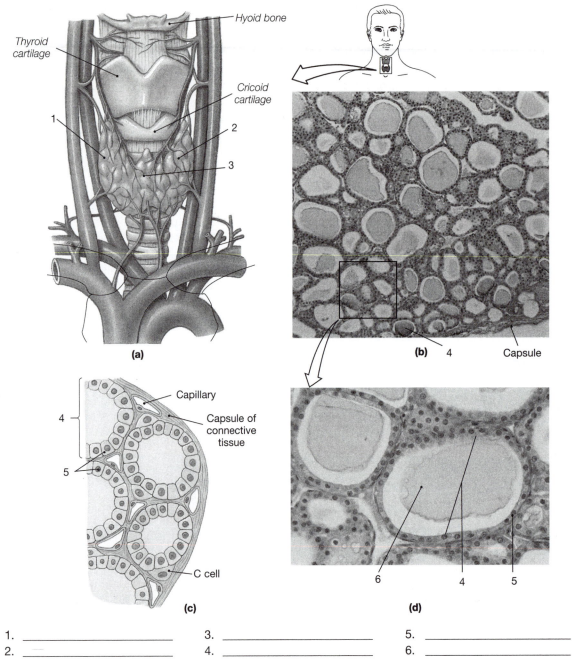

(a)

(b) 4   Capsule

(c)

(d)

1. _____   3. _____   5. _____
2. _____   4. _____   6. _____

●**Figure 10.3**
The Thyroid Gland

(a) Location and gross anatomy of thyroid. (b) History of thyroid. (LM, 108×)
(c) Detail of follicles.

C. Parathyroid Glands

ANATOMY

The parathyroid glands consist of two pairs of oval masses on the posterior surface of
the thyroid gland. Each parathyroid gland is isolated from the underlying thyroid tissue
by parathyroid **capsule**. Locate and label the parathyroid glands in Figures 10.1 and
10.4 and on the laboratory models.

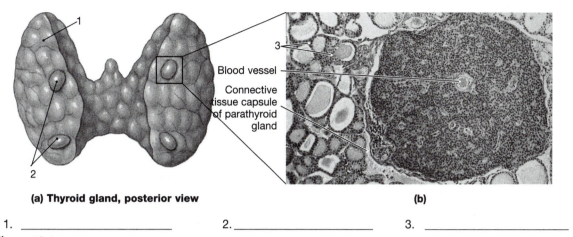

(a) Thyroid gland, posterior view

3 ←— Blood vessel

Connective
tissue capsule
of parathyroid
gland

(b)

1. _____   2. _____   3. _____

●Figure 10.4
The Parathyroid Glands
(a) Location of parathyroids on posterior surface of thyroid lobes. (b) Photomicrograph

HISTOLOGY

The parathyroid glands are mostly made up of **chief cells**. The chief cells are numerous and cluster together in strings or cords. See Figure 10.4 for details on the histology of the parathyroid glands.

HORMONES

Parathyroid glands produce **parathormone** (**PTH**), also called parathyroid hormone, which is antagonistic to the thyroid hormone calcitonin. Parathormone increases blood calcium levels by stimulating osteoclasts in bone to dissolve small areas of bone matrix and release calcium ions into the blood. PTH also stimulates calcium uptake in the digestive system and reabsorption of calcium from the filtrate in the kidneys during urine formation.

Laboratory Activity   MICROSCOPIC OBSERVATIONS OF THE PARATHYROID GLANDS_____

MATERIALS

    dissecting and compound microscope
    prepared slide
       parathyroid gland

PROCEDURES

   1. Label the parathyroid tissue in Figure 10.4.
   2. Examine the parathyroid slide with the dissecting microscope. Scan the gland for thyroid follicles that may be on the slide near the parathyroid tissue.
   3. Observe the parathyroid slide at low and medium powers with the compound microscope. Locate the darkly stained chief cells.
   4. Sketch the parathyroid gland as observed at medium magnification in the space provided.

D. Thymus Gland

ANATOMY

The thymus gland is located inferior to the thyroid gland, in the thoracic cavity posterior to the sternum. Because hormones secreted by the thymus gland facilitate development of the immune system, the gland is larger and more active in infants than adults.

HISTOLOGY

The thymus gland is organized into many lobules, each consisting of a dense outer **cortex** and a light-staining central **medulla**. The cortex is populated by **reticular epithelial cells** that secrete the thymic hormones. The medulla is characterized by the presence of oval **thymic corpuscles** (Hassall's corpuscles). The cells surrounding the thymic corpuscles are lymphocytes. Adipose and other connective tissues are abundant in an adult thymus since the function and size of the gland decrease after puberty. Label the thymus gland in Figures 10.1 and 10.5.

HORMONE

The epithelial cells of the thymus gland produce several hormones; however, the function of only one, **thymosin**, is understood. Thymosin is essential in the development and maturation of the immune system. Removal of the gland during early childhood will usually result in a greater susceptibility to acute infections.

Laboratory Activity   MICROSCOPIC OBSERVATION OF THE THYMUS GLAND_____

MATERIALS
        dissecting and compound microscope
        prepared slide
            thymus gland

PROCEDURES
   1. Label the thymus gland features in Figure 10.5.
   2. Scan a slide of the thymus gland with the dissecting microscope, and distinguish between the cortex and the medulla.
   3. Examine the thymus slide with the compound microscope at low magnification to locate a stained thymic corpuscle. Increase magnification and examine the corpuscle. The cells surrounding the corpuscles are lymphocytes.
   4. Sketch the thymus gland as observed at medium magnification in the space provided.

E. Adrenal Glands

ANATOMY

Above each kidney is an **adrenal gland**, also called the **suprarenal gland** because of its location superior to the kidney (see Figure 10.1). Each gland is encased in an **adrenal capsule**. Locate and label the capsule in Figure 10.6. The gland is organized into two

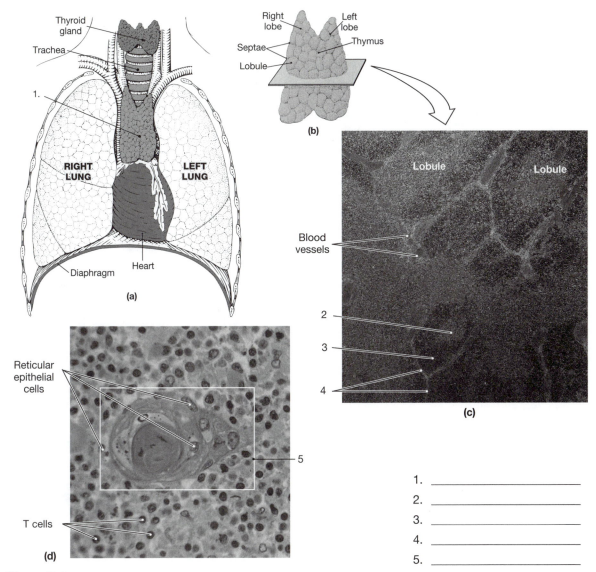

●Figure 10.5
The Thymus

(a) Appearance and position of the thymus on gross dissection; note its relationship to other organs in the chest. **(b)** Anatomical landmarks on the thymus. **(c)** A low-power light micrograph of the thymus. (LM × 40) **(d)** At higher magnification the unusual structure of Hassall's corpuscles can be examined. The small cells in view are lymphocytes in various stages of development. (LM × 532)

1. _____
2. _____
3. _____
4. _____
5. _____

major regions, the outer **cortex** and the inner **medulla**. The cortex is differentiated into three distinct regions, each producing specific hormones.

HISTOLOGY AND HORMONES

The adrenal **cortex** comprises three distinct layers or zones: the zona glomerulosa, zona fasciculatus, and zona reticularis. The **zona glomerulosa** is the outermost cortical region. Cells in this area are stained dark and arranged in oval clusters. The zona glomerulosa secretes a group of hormones called **mineralocorticoids** that regulate, as their name implies, mineral or electrolyte concentrations of body fluids. A good example is **aldosterone**, which stimulates the kidney to reabsorb more sodium from the fluid being processed into urine.

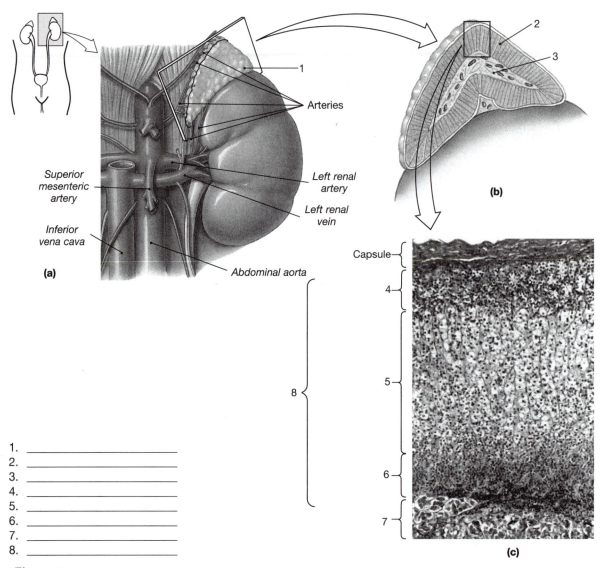

Superior mesenteric artery

Inferior vena cava

Arteries

Left renal artery

Left renal vein

Abdominal aorta

(a)

(b)

Capsule

(c)

1. _____
2. _____
3. _____
4. _____
5. _____
6. _____
7. _____
8. _____

●**Figure 10.6**
The Adrenal Gland

(a) Superficial view of the kidney and adrenal gland. **(b)** An adrenal gland, showing the sectional plane for part (c). **(c)** Light micrograph, with the major hormones identified.

The next layer, the **zona fasciculatus**, is made up of larger cells organized into tight columns. The cells contain large amounts of lipid and thus appear lighter than the surrounding cortical layers, evident in the micrograph in Figure 10.6. The zona fasciculatus produces a group of hormones called **glucocorticoids** that are involved in fighting stress, increasing glucose metabolism, and preventing inflammation. **Cortisol** and **cortisone** are commonly used in creams to control irritating rashes and allergic responses of the skin.

The deepest layer of the cortex, next to the medulla, is the **zona reticularis**. Cells in this area are small and loosely linked into chainlike structures. Interestingly, the zona reticularis produces **androgens**, male sex hormones. Both males and females produce small quantities of androgens in the zona reticularis.

The adrenal **medulla** is the innermost area of the gland. The cells of the medulla are modified sympathetic neurons that secrete **epinephrine** and **norepinephrine**, hormones involved in the sympathetic "fight-or-flight" response. Many blood vessels in this part of the medulla give the tissue a dark red coloration.

Laboratory Activity   MICROSCOPIC OBSERVATIONS OF THE ADRENAL GLANDS _____

MATERIALS

dissecting and compound microscopes
prepared slide
  adrenal gland

PROCEDURES

1. Label the components of the adrenal gland in Figure 10.6.
2. Examine a slide of the adrenal gland with the dissecting microscope, and distinguish between the capsule, adrenal cortex, and medulla.
3. Observe the adrenal gland with the compound microscope and differentiate among the three layers of the adrenal cortex.
4. Sketch the adrenal gland in the space provided to show the detail of the three cortical layers and the medulla.

F.   Pancreas

ANATOMY

The **pancreas** lies between the stomach and the duodenum of the small intestine. The gland is a "double gland," performing both exocrine and endocrine functions. Exocrine glands have ducts or tubules that transport secretions to the body surface or to an exposed internal surface. Exocrine cells of the pancreas secrete enzymes, buffers, and other molecules into pancreatic ducts for the digestive process. Endocrine portions of the pancreas produce hormones that regulate blood sugar metabolism.

HISTOLOGY

Figure 10.7 details the histology of the pancreas, which comprises mainly exocrine cells called **acinar cells** that secrete pancreatic juice for digestive processes. These cells are dark stained and grouped in small ovals or columns. Connective tissues and pancreatic ducts are dispersed in the tissue. Embedded in the acini cells are isolated clusters of **pancreatic islets** (islets of Langerhans). Each islet houses three types of endocrine cells, **alpha cells, beta cells**, and **delta cells**. These cells are difficult to distinguish with routine staining techniques and will not be individually examined here.

HORMONES

Pancreatic hormones regulate carbohydrate metabolism. Alpha cells in pancreatic islets secrete the hormone **glucagon**, which raises blood sugar levels by catabolizing glycogen, the stored form of carbohydrate, into glucose for cellular respiration. This process is called glycogenolysis. Beta cells secrete **insulin**, which accelerates glucose transport into cells and glycogenesis, the formation of glycogen. Insulin lowers blood sugar levels by promoting the removal of sugar from circulation.

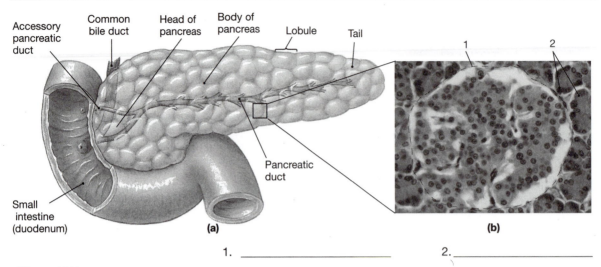

(a)   **(b)**

1. _____   2._____

●**Figure 10.7**
The Endocrine Pancreas
(a) Gross anatomy of the pancreas. **(b)** A pancreatic islet surrounded by exocrine-secreting

Laboratory Activity   Microscopic Observations of the Pancreas _____

MATERIALS
compound microscope
prepared slide
pancreas

PROCEDURES
1. Label the histology of the pancreas in Figure 10.7.
2. Use a compound slide and locate the dark-stained acinar cells and the oval pancreatic ducts of the pancreas. Identify the clusters of pancreatic islets, the endocrine portion of the gland.
3. Sketch the pancreas in the space provided, and label the pancreatic islets and the acinar cells.

G. Gonads

Gonads are male and female reproductive organs that produce gametes, spermatozoa, and ova. The testes and ovaries secrete hormones that regulate reproductive physiology and maturation and maintenance of the sexually mature adult. Locate the gonads in Figure 10.1.

ANATOMY AND HISTOLOGY

Testes are the male gonads, located outside the body in the pouchlike scrotum. A testis or testicle is made up of coiled **seminiferous tubules** that produce spermatozoa. In

Figure 10.8, a micrograph of a testis cross section, the tubule is the large round structure. **Interstitial cells** are endocrine cells between the tubules. They secrete the male sex hormone, **testosterone**, which produces and maintains the secondary male sex characteristics.

The **ovaries** are the female gonads, located in the pelvic cavity. During each ovarian cycle, a small group of immature eggs or oocytes begins to develop an outer capsule of follicular cells. Eventually, one follicle becomes a **Graafian follicle**, a fluid-filled bag containing the oocyte for release during ovulation (see Figure 10.9). The Graafian follicle is a temporary endocrine structure and secretes the hormone **estrogen**, which promotes thickening of the uterine wall. After ovulation, the ruptured Graafian follicle becomes the **corpus luteum**, another temporary endocrine structure, which secretes **progesterone**, the hormone that prepares the uterus for the implantation of a fertilized egg.

Laboratory Activity   MICROSCOPIC OBSERVATIONS OF THE GONADS _____

MATERIALS

> compound microscope
> prepared slides
> > testis
> > ovary

PROCEDURES

I. Testis

1. Label the structures of the testis in Figure 10.8.
2. Scan a slide of the testis at low magnification, and identify the many seminiferous tubules. Increase magnification to locate interstitial cells between the seminiferous tubules.
3. Sketch a cross section of a testis detailing the seminiferous tubules and interstitial cells in the space provided.

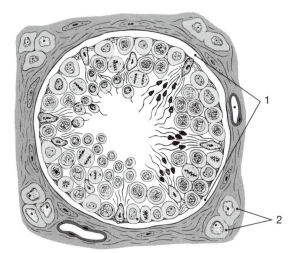

●**Figure 10.8**
Portion of Testis

Portion of testis in cross-section detailing seminiferous tubules and interstitial cells. (LM × 756)

1. _____

2. _____

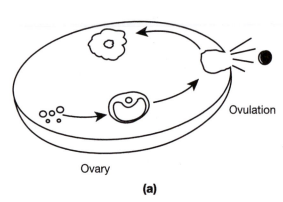

Ovulation

Ovary

(a)

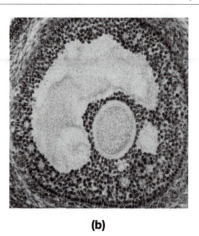

(b)

●**Figure 10.9**
The Ovary

(a) Overview of ovarian cycle highlighting the Graafian follicle, ovulation and corpus luteum. **(b)** Micrograph of Graafian follicle.

II. Ovary

1. Label the ovarian structures in Figure 10.9.
2. Scan the ovary slide at low power to locate the large Graafian follicle. Identify the developing ovum inside the follicle.
3. Sketch the Graafian follicle in the space provided.

ENDOCRINE SYSTEM
Exercise 10 Laboratory Report

A. Matching
Match each endocrine structure on the left with the correct description on the right.

1. _____ thyroid follicle
2. _____ adrenal medulla
3. _____ thymic corpuscle
4. _____ seminiferous tubules
5. _____ zona glomerulosa
6. _____ parathyroid gland
7. _____ acinar cells
8. _____ Graafian follicle
9. _____ adenohypophysis
10. _____ master gland
11. _____ target cell
12. _____ pancreatic islets
13. _____ interstitial cells
14. _____ zona reticularis
15. _____ C cells
16. _____ infundibulum
17. _____ corpus luteum

A. contains an ovum
B. four oval masses on posterior thyroid gland
C. produces testosterone
D. produces insulin
E. cells between the thyroid follicles
F. inner cortical layer of adrenal gland
G. contains hormones in a colloid
H. pituitary gland
I. develops from ruptured Graafian follicle
J. innermost part of gland above kidney
K. stalk of pituitary gland
L. outer cortical layer of adrenal gland
M. produce spermatozoa
N. exocrine cells of pancreas
O. anterior pituitary gland
P. found in thymus gland
Q. cell that will respond to a specific hormone

B. Short Answer Questions

1. Describe how negative feedback regulates the secretion of most hormones.

2. Why is the pituitary gland called the master gland of your body?

3. Describe the hormones produced by the three layers of the adrenal cortex.

4. Describe what structures are involved in the ovulation of an egg.

5. What are the endocrine functions of the pancreas?

6. Compare the histology of an adult thymus with one of an infant.

7. How is blood calcium regulated by the endocrine system?

8. A patient complains of sharp pains in the heel of her foot. You palpate her heel and discover a bone spur, an abnormal deposit of bone on the skeleton. Further examination reveals several other bone spurs. What endocrine imbalance may have caused this excessive calcification of the skeleton?

9. A woman has completed menopause and no longer produces female sex hormones. Over the next few years, she notices growth of coarse facial hair and other hair on her body. How can you explain this phenomenon?

10. Although advertisements on television tell us to eat a candy bar for quick energy, some individuals actually feel depressed after eating chocolate and other high-sugar snacks. What is the cause of their "sugar depression"?

C. Completion

1. Label each endocrine organ in Figure 10.10.

●**Figure 10.10**
The Endocrine System

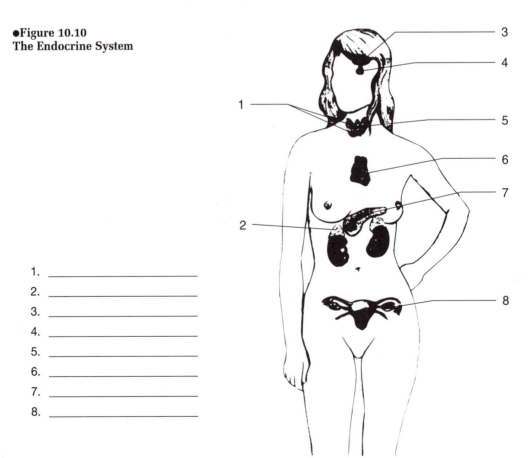

1. _____
2. _____
3. _____
4. _____
5. _____
6. _____
7. _____
8. _____

11 Anatomy and Physiology of the Sensory Organs

Changes in the internal and external environments are detected by special cells called sensory receptors. Each type of receptor is sensitive to a specific stimulus. The taste buds of the tongue, for example, are stimulated by dissolved chemicals and not by sound waves or light rays. Receptors may be grouped into two broad sensory categories, the general senses and the special senses. The **general senses** have simple neural pathways and include the following sensations: touch, temperature, pain, chemical and pressure detection, and body position or proprioception. The **special senses** have complex pathways, and the receptors are housed in specialized organs. The special senses include the sense of gustation (taste), olfaction (smell), vision, auditory (hearing), and equilibrium.

PART ONE   THE GENERAL SENSES

WORD POWER

proprioceptor (propri—one's own)
nociceptor (noci—to injure)

MATERIALS

compass with millimeter scale

OBJECTIVES

On completion of this part of the exercise, you should be able to:

- List the receptors for the general senses and the special senses.
- Discuss the distribution of cutaneous receptors.
- Describe the two point discrimination test.

A variety of senses is included as general senses because of the simple structure of their receptors. A sensory neuron, which may be wrapped in a connective tissue sheath, monitors a specific region called a **receptive field**. Overlap in these fields enables the brain to detect where a stimulus was applied to the body.

Receptors for body position are called **proprioceptors**. **Muscle spindle** receptors in muscles and **golgi tendon organs** in tendons near joints inform the brain about muscle tension and joint position. **Thermoreceptors** have a wide distribution in the body, including the dermis, skeletal muscles, and the hypothalamus of the brain, our internal thermostat. Pressure receptors or **baroreceptors** convey signals about fluid and gas pressures. These receptors are usually simple nerve endings in blood vessels and the lungs that monitor pressure changes. **Chemoreceptors** monitor changes in chemical concentrations in body fluids.

The skin has a variety of touch receptors, as illustrated in Figure 11.1. **Meissner's corpuscles** are nerve endings sensitive to touch and are located in the dermal papillae of the skin. **Merkel's disks** are embedded in the epidermis and are very responsive to touch. **Free nerve endings** are dendrites of **nociceptors** or pain receptors in the epidermis. The **root hair plexuses** are dendrites of sensory neurons wrapped around hair roots. These receptors are stimulated when an insect, for example, touches and moves a hair on the skin. Pressure-sensitive receptors include **Pacinian corpuscles** and **Ruffini corpuscles**. These receptors are located deep in the dermis.

Cutaneous receptors are not evenly distributed in the body. Some areas are densely populated with a particular receptor, whereas other areas may have a few or none of that receptor. This explains why your fingertips, for example, are more sensitive to touch than your scalp. The **two-point discrimination test** is used to map the distribution of touch receptors on the skin. A compass with two points is used to determine the distance between cutaneous receptors. The compass points are gently pressed into the skin, and the subject decides if one or two points are felt. If the sensation is a single compass point, then only one receptor has been stimulated by the compass. By gradually increasing the distance between the points until two distinct sensations are felt, the density of the receptor population in that region may be measured.

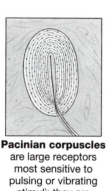

Pacinian corpuscles are large receptors most sensitive to pulsing or vibrating stimuli; they are common in the skin of the fingers, breasts, and external genitalia and in joint capsules, mesenteries, and the wall of the urinary bladder.

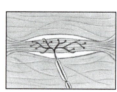

Ruffini corpuscles detect pressure and distortion of the dermis.

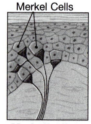

Merkel's discs are nerve endings that contact specialized epithelial cells *(Merkel cells)* sensitive to fine touch.

Merkel Cells

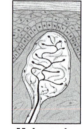

Meissner's corpuscles provide touch sensations from the eyelids, lips, fingertips, nipples, and external genitalia.

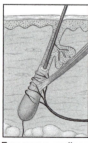

Free nerve endings of the **root hair plexus** are distorted by movement of the hairs.

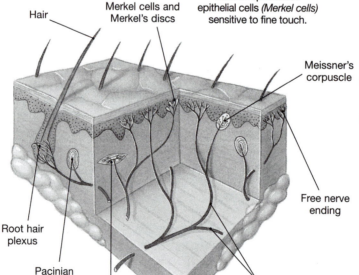

Hair

Merkel cells and Merkel's discs

Meissner's corpuscle

Free nerve ending

Root hair plexus

Pacinian corpuscle

Ruffini corpuscles

Sensory nerves

Free nerve endings are found between epidermal cells and are sensitive to a variety of stimuli.

●**Figure 11.1**
Tactile Receptors in the Skin

Laboratory Activity   TWO-POINT DISCRIMINATION TEST

MATERIALS

compass with millimeter scale

PROCEDURES

1. Gently place the compass on the tip of your laboratory partner's index finger. Slightly spread the compass points and replace them on the same area of the fingertip. Repeat this procedure until the subject feels two distinct points. Record the results in the area provided in Table 11.1.

2. Reset the compass, and test the back of the hand, the back of the neck, and the sides of the nose. Record the data in Table 11.1.

TABLE 11.1	Two-Point Discrimination Test

Area of the Skin	*Minimum Distance for Two-Point Detection (mm)*
Back of neck	
Back of hand	
Tip of index finger	
Side of nose	

PART TWO   THE SPECIAL SENSES: GUSTATION AND OLFACTION

WORD POWER

gustatory cells (gusto—taste)
papillae (papilla—a nipple)
circumvallate papillae (vallate—to surround)
fungiform (fungi—fungus, form—shaped)
filiform (filiform—thread)

MATERIALS

compound microscope
prepared slides of olfactory epithelium and taste buds
phenylthiocarbamide (PTC) papers
small cubes of apple and onion

OBJECTIVES

On completion of this part of the exercise, you should be able to:

- Describe the location and structure of taste buds and papillae.
- Describe the location and structure of the olfactory receptors.
- Identify the microscopic features of taste buds and olfactory epithelium.
- Explain why olfaction accentuates gustation.

A. Gustation

The receptors for **gustation,** or taste, are located in **taste buds** that cover the surface of the tongue, the pharynx, and the soft palate. Taste buds are inside elevations called **papillae**, detailed in Figure 11.2. The base of the tongue has several circular-shaped papillae forming an inverted V called **circumvallate** papillae. These papillae contain

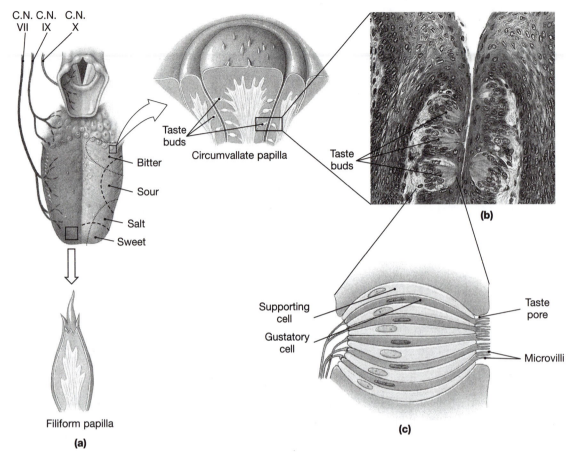

●Figure 11.2
Gustatory Reception

(a) Gustatory receptors are found in taste buds that form pockets in the epithelium of
the tongue. (b) Photomicrograph of taste buds in circumvallate papilla. (LM × 280)
(c) Diagrammatic view of a taste bud.

taste buds that are sensitive to bitter substances. The tip and sides of the tongue contain buttonlike papillae called **fungiform** papillae, which are stimulated by salty,
sweet, and sour substances. Approximately two-thirds of the anterior portion of the
tongue is covered with **filiform** papillae, which provide a rough surface for the movement of food. Filiform papillae do not contain taste buds.

A taste bud contains up to 20 gustatory cells. Each cell has a small hair, or
microvillus, that projects through a small **taste pore**. Contact with dissolved food stimulates the microvilli to produce gustatory impulses. Sensory information from the taste
buds is carried to the brain by parts of three cranial nerves: the vagus nerve (X), which
serves the pharynx; the facial nerve (VII), which serves the anterior two-thirds of the
tongue; and the glossopharyngeal nerve (IX), which serves the posterior third of the
tongue.

Laboratory Activity   GUSTATORY RECEPTORS

MATERIALS

compound microscope
prepared slide of the tongue
PTC paper

PROCEDURES

1. Obtain a slide of the tongue showing a taste bud in cross section. Using Figure 11.2, identify the papillae and taste buds.

2. The sense of taste is a genetically inherited trait. Not all individuals can perceive the same substance. The chemical phenylthiocarbamide (PTC) tastes bitter to some individuals and sweet to others. Others, however, cannot taste the chemical. Approximately 30% of the population are nontasters of PTC.

   a. Take a strip of PTC paper, place it on your tongue, and chew it several times. Are you a taster or a nontaster?

   b. Students will record their results on the blackboard. Calculate the percentage of tasters and nontasters in the class, and then complete Table 11.2.

TABLE 11.2	PTC Taste Experiment

% Tasters	*% Nontasters*

B. Olfaction

Olfactory receptors are located in the roof of the nasal cavity. Most of the air we inhale passes straight through the nasal cavity and into the pharynx. Sniffing increases our sense of smell by pulling air higher into the nasal cavity where the olfactory receptors are found. Locate the olfactory receptors in Figure 11.3.

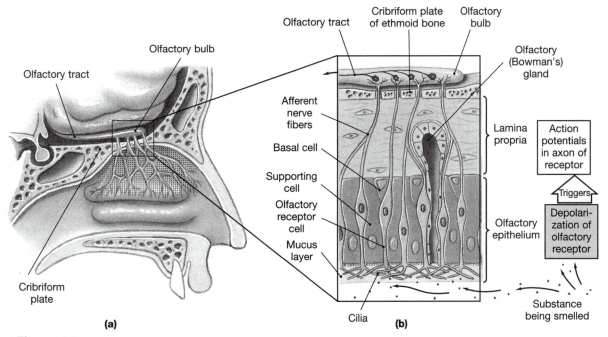

●**Figure 11.3**
The Olfactory Organs

(**a**) The structure of the olfactory organ on the right side of the nasal septum. (**b**) An olfactory receptor is a modified neuron with multiple cilia extending from its free surface.

Olfactory cells have many cilia that are sensitive to airborne molecules. **Goblet cells** interspersed among the olfactory cells secrete a protective coating of mucus over the cilia. To smell a substance, it must diffuse through the mucus before it can stimulate the cilia of the olfactory receptors. Our sense of smell is drastically reduced by colds and allergies because mucus production increases in the nasal cavity and blocks the diffusion of molecules to the olfactory receptors.

Laboratory Activity   OLFACTORY RECEPTORS

MATERIALS

compound microscope   paper towels
prepared slide   apple cubes
  olfactory epithelium   onion cubes

PROCEDURES

1. Obtain a slide of olfactory receptors. Identify the supporting and olfactory cells as seen in Figure 11.3.

2. The sense of taste is thousands of times more sensitive when gustatory and olfactory receptors are stimulated simultaneously. This experiment demonstrates the effect of smell on the sense of taste. You will place small cubes of apple or onion on your laboratory partner's dry tongue and ask if he or she can detect which type of cube is present.

   a. Dry the surface of your partner's tongue with a clean paper towel.

   b. Have your partner hold his or her nose and close his or her eyes.

   c. Place a cube on your partner's dry tongue.

   d. Your partner is to identify the cube as apple or onion when it is placed on his or her tongue, immediately after chewing, and on opening his or her nostrils. Responses to the different food cubes should be recorded in Table 11.3 as yes or no.

TABLE 11.3	**Gustatory and Olfactory Sensations**		
Food	*Dry Tongue*	*After Chewing*	*Open Nostrils*
Apple			
Onion			

PART THREE   SPECIAL SENSES: ANATOMY OF THE EYE

WORD POWER	MATERIALS
canthus (kanthos—corner of the eye)	eyeball in orbit model
lacrimal (lacrima—a tear)	compound microscope
palpebral (palpebra—eyelid)	prepared slide of the retina
caruncle (caruncula—a small, fleshy mass)	sectioned eye model
sclera (scler—hard)	preserved cow eye
fovea (fovea—a cuplike depression or pit)	dissection equipment
	safety glasses, gloves

OBJECTIVES

On completion of this part of the exercise, you should be able to:

- Identify and describe the accessory structures of the eyeball.
- Identify and explain the actions of the six extrinsic eye muscles.
- Describe and identify the external and internal components of the eye.
- Describe the three tunics of the eyeball.
- Identify the microscopic organization of the retina.
- Identify the structures of a dissected cow eye.

A. External Eye Anatomy

The human eye is a spherical organ measuring about 2.5 cm (1 inch) in diameter. Only about one-sixth of the eyeball is visible between the eyelids; the rest of the eyeball is recessed in the bony orbit of the skull. The corners of the eye are called the **lateral canthus** and **medial canthus**. A red, fleshy structure in the medial canthus of the eye, the **lacrimal caruncle**, contains modified sebaceous and sweat glands. Secretions from the caruncle accumulate in the medial canthus during long periods of sleep. Locate these structures in Figure 11.4.

The **accessory structures** of the eye are the eyelids, eyebrows, eyelashes, lacrimal apparatus, and six extrinsic eye muscles. The eyelids or **palpebrae** (PAL-pe-bre) are skin-covered muscles that protect the surface of the eye. Blinking the eyelid keeps the surface of the eye lubricated and clean. A thin mucous membrane called the **conjunctiva** (kon-junk-TI-va) covers the underside of the eyelids and the surface of the eye. The **eyelashes** are short hairs that project from the border of each eyelid. Eyelashes and **eyebrows** protect the eyeball from foreign objects such as perspiration or dust and partially shade the eyeball from the sun.

The **lacrimal glands** are superior and lateral to each eyeball. Each gland contains many **lacrimal ducts** that deliver an antibacterial fluid or **tears** onto the surface of the eye. Tears flow medially across the eye and enter a pair of **lacrimal canals** near the lacrimal caruncle. The canals drain into the **nasolacrimal duct**, which empties into the nasal cavity.

Six extrinsic (external) eye muscles control the movements of the eye (see Figure 11.5). The **superior rectus**, **inferior rectus**, **medial rectus**, and **lateral rectus** are straight muscles that move the eyeball up and down and side to side. The **superior** and **inferior oblique** muscles attach diagonally on the eyeball. The superior oblique has a tendon passing through a trochlea (pulley) located on the upper orbit. The superior oblique muscle rolls the eye down, and the inferior oblique rolls the eye up.

B. Internal Eye Anatomy

The eyeball is divided anatomically into three tunics or layers, each detailed in Figure 11.4c. The outermost layer is called the **fibrous tunic** because of the abundance of dense connective tissue. The **sclera** is the white part of the fibrous tunic that covers the eyeball except where light enters at the transparent **cornea**. The cornea lacks blood vessels and bends light rays as they enter the eye. The border of the cornea and sclera is called the **limbus**.

The **vascular tunic** or **uvea** is the middle layer of the eye. The posteriormost portion, the **choroid**, is highly vascularized and contains a dark pigment (melanin) that absorbs light to prevent reflection. Anteriorly, the uvea is modified into the **ciliary body** and the pigmented iris. The ciliary body contains two structures, the **ciliary process** and the **ciliary muscle**. The ciliary process is a series of folds with thin **suspensory ligaments** extending to the lens. The ciliary muscle adjusts the shape of the lens for near and far vision. The front of the iris is pigmented and has a central aperture called the **pupil**.

●Figure 11.4
External Features and Accessory Structures of the Eye

(a) Gross and superficial anatomy of the accessory structures. **(b)** Details of the organization of the lacrimal apparatus. **(c)** Internal anatomy.

1. ____________________
2. ____________________
3. ____________________
4. ____________________
5. ____________________
6. ____________________
7. ____________________
8. ____________________
9. ____________________
10. ____________________
11. ____________________
12. ____________________
13. ____________________
14. ____________________
15. ____________________
16. ____________________
17. ____________________
18. ____________________
19. ____________________
20. ____________________
21. ____________________

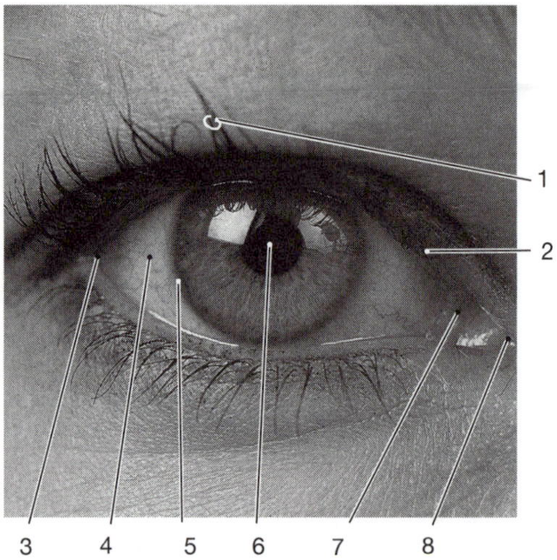
1
2
3
4
5
6
7
8

(a)

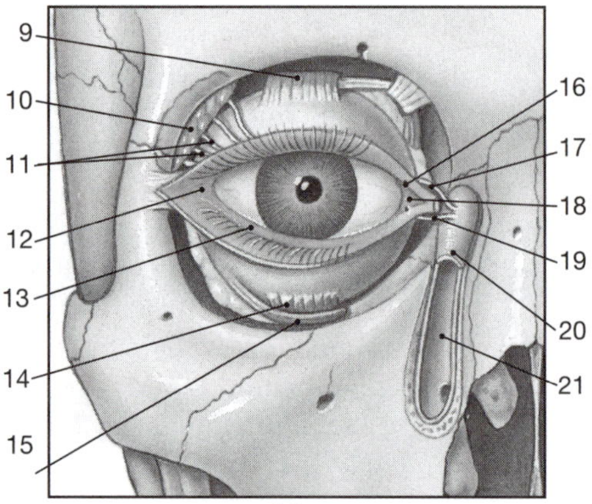
9
10
11
12
13
14
15
16
17
18
19
20
21

(b)

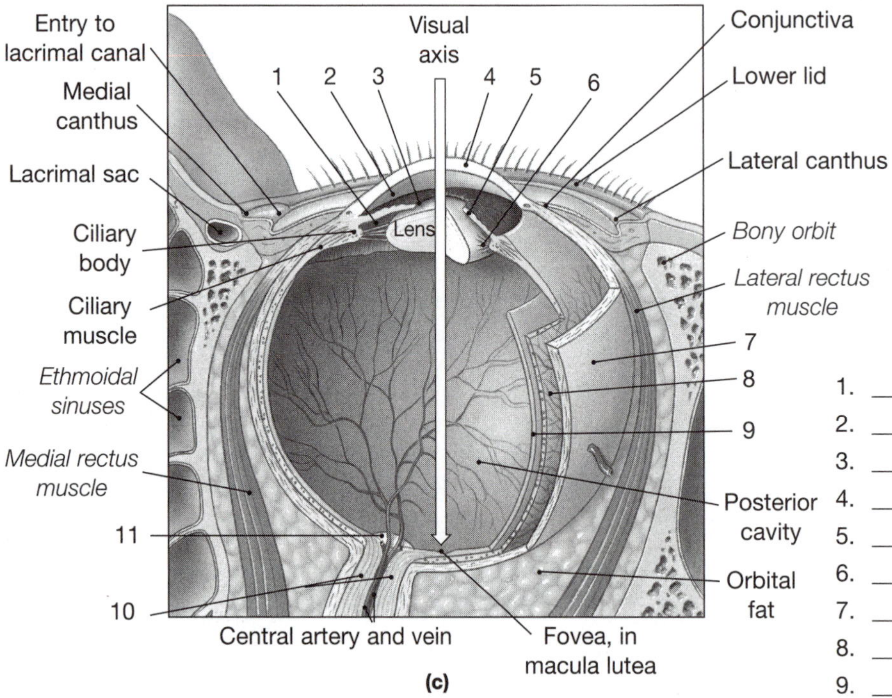
Entry to lacrimal canal
Medial canthus
Lacrimal sac
Ciliary body
Ciliary muscle
Ethmoidal sinuses
Medial rectus muscle
11
10
Visual axis
1
2
3
4
5
6
Lens
Conjunctiva
Lower lid
Lateral canthus
Bony orbit
Lateral rectus muscle
7
8
9
Posterior cavity
Orbital fat
Central artery and vein
Fovea, in macula lutea

(c)

1. ____________________
2. ____________________
3. ____________________
4. ____________________
5. ____________________
6. ____________________
7. ____________________
8. ____________________
9. ____________________

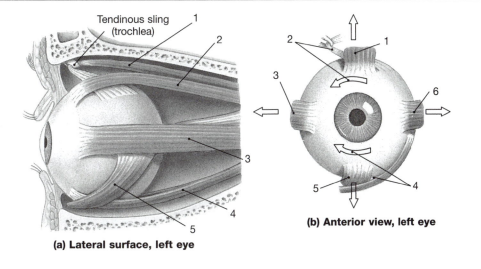

(a) Lateral surface, left eye

(b) Anterior view, left eye

1. _____   3. _____   5. _____

2. _____   4. _____

●Figure 11.5
The Extrinsic Eye Muscles

Circular and radial muscles of the iris change the diameter of the pupil to regulate the amount of light striking the retina.

The **retina**, Figure 11.6, is also called the **neural tunic** because it contains receptors and two layers of sensory neurons. Receptors of the eye are called photoreceptors and are stimulated by rays of light made up of photons. Two types of receptors occur, cones and rods. **Cones** are stimulated by moderate or bright illumination and respond to different wavelengths or colors of light. Our sharpest vision or visual acuity is attributed to cones. **Rods** are sensitive to low illumination and to motion. These receptors are not stimulated by different wavelengths (colors), and our night vision is in black and white. The rods and cones face a **pigmented epithelium** on the posterior of the eyeball. Light passes through the retina, reflects off the pigmented epithelium, and then strikes the photoreceptors. The photoreceptors communicate with a layer of sensory neurons called **bipolar cells**. These cells pass the visual signal on to a layer of **ganglion cells** whose axons converge to an area called the **optic disk** or blind spot where the optic nerve exits the eyeball. The optic disk lacks photoreceptors and is a "blind spot" in your field of vision. Because the visual fields of the eyes overlap, the blind spot is "filled in" and is not noticeable. Lateral to the optic disc is an area of high cone density called the **macula lutea** (yellow spot). In the center of the macula lutea is a small depression called the **fovea**. The fovea is the area of sharpest vision due to the high density of cones. There are no rods in the macula or the fovea, but outside these areas their numbers increase toward the periphery of the retina.

The **lens** divides the eyeball into an **anterior cavity**, an area between the lens and the cornea, and a **vitreous cavity**, an area between the lens and the retina. The anterior cavity is further subdivided into an **anterior chamber** between the iris and cornea and a **posterior chamber** between the iris and lens. Capillaries of the ciliary processes form a watery fluid called **aqueous humor**, which is secreted into the posterior chamber. This fluid circulates through the pupil and into the anterior chamber where it is reabsorbed into the blood. The aqueous humor helps maintain the intraocular pressure of the eyeball and supply nutrients to the avascular lens and cornea. The second, larger cavity of the eyeball is the **vitreous cavity**. This cavity contains the **vitreous body**, a clear jellylike substance that holds the retina against the choroid layer and prevents the eyeball from collapsing.

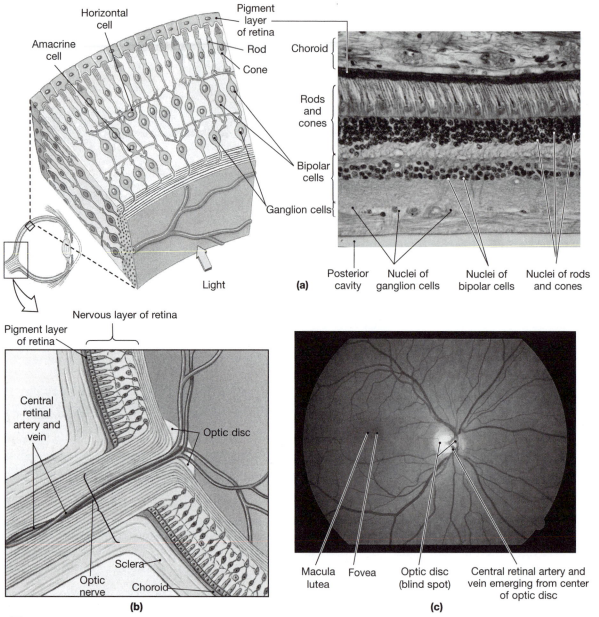

● **Figure 11.6**
Retinal Organization

(a) Cellular organization of the retina. Note that the photoreceptors are located closest to the choroid rather than near the vitreous chamber. (LM × 290) **(b)** The optic disc in diagrammatic horizontal section. **(c)** A photograph of the retina as seen through the pupil of the eye.

Laboratory Activity   ANATOMY OF THE EYE

I. Structure of the Eye

MATERIALS

eyeball models and charts
compound microscope
prepared slide of retina

PROCEDURES

1. Examine the eye models and charts available in the laboratory, and locate each structure discussed in the manual.

2. Study the prepared histology slide of the retina. Using Figure 11.6 as reference, identify the following:

   a. the pigment epithelium

   b. photoreceptor (rods and cones) layer

   c. bipolar cells

   d. ganglion cells

II. Dissection of the Cow Eye

MATERIALS

fresh or preserved cow eyes
dissecting pan, dissecting scissors
blunt probe, scalpel, gloves, safety glasses

PROCEDURES

1. Examine the external features of the eyeball. Note the amount of adipose tissue that serves to cushion the eye. Identify the optic nerve II as it leaves the eyeball, the remnants of the extrinsic muscles of the eye, the conjunctiva, sclera, and cornea. The cornea, which is normally transparent, will be opaque if the eye has been preserved.

2. Securely hold the eyeball and, with the scissors, remove the adipose and extrinsic muscles from the surface of the eyeball taking care not to remove the optic nerve (II).

3. Lay the eyeball on the dissecting pan, and with a sharp scalpel, make an incision about 0.6 cm (1/4 inch) back from the cornea. Continue the cut around the circumference of the eyeball with the scissors, being sure to maintain the 1/4-inch distance back from the cornea.

4. Carefully separate the anterior and posterior portions of the eyeball. The vitreous humor should stay with the posterior portion of the eyeball.

5. Examine the anterior portion of the eyeball. Place the blunt probe between the lens and the ciliary processes, and carefully lift the lens up a little. The halo of delicate transparent filaments between the lens and the ciliary processes are the suspensory ligaments. Notice the ciliary body where the suspensory ligaments originated and the heavily pigmented iris with the pupil in its center.

6. Locate the vitreous body in the posterior portion of the eyeball. The retina is the tan-colored membrane that is easily separated from the heavily pigmented choroid. Examine the optic disk where the retina attaches to the posterior of the eyeball. The choroid has a greenish-blue membrane, the **tapetum lucidum**, which improves night vision in animals such as sheep and cows; humans do not have this structure.

7. After completing the dissection, clean your work area and dispose of the cow eye as indicated by your laboratory instructor.

PART FOUR   SPECIAL SENSES: PHYSIOLOGY OF THE EYE

WORD POWER	MATERIALS
emmetropic (emmetros—according to measure)	Snellen eye chart
astigmatism (stigma—a point)	Astigmatism eye chart
presbyopia (presby—old man)	pen or pencil
myopic (myo—to shut)	
hyperopia (hyper—above, over)	

OBJECTIVES

On completion of this part of the exercise, you should be able to:

- Explain why a blind spot exists in the eye, and describe how it is mapped.
- Demonstrate the use of a Snellen eye chart and the Astigmatism eye chart.
- Explain the terms *myopia*, *hyperopia*, *presbyopia*, and *astigmatism*.
- Describe how to measure accommodation.
- Discuss the role of convergence in near vision.

A. Visual Acuity

The sharpness of vision or **visual acuity** is tested with a Snellen eye chart. This chart contains black letters of various sizes printed on white cardboard. Letters of a certain size are seen clearly by a person with normal vision at a given distance. A person with a visual acuity of 20/20 is considered to have normal or **emmetropic** vision. The lines of letters are marked with numerical values such as 20/20 and 20/30. A reading of 20/30, for example, indicates that you can see at 20 feet what an emmetropic eye can see at 30 feet; 20/30 vision is not as sharp as 20/20 vision. An eye that focuses an image in front of the retina is **myopic** or nearsighted and can clearly see close objects. If an image is focused behind the retina, the eye is **hyperopic** or farsighted and can see only distant objects clearly. Corrective lenses are used to adjust for both conditions.

Astigmatism is a reduction in sharpness of vision due to an irregular cornea or lens. These uneven surfaces bend or **refract** light rays incorrectly, resulting in blurred vision. The astigmatism chart has 12 sets of 3 lines laid out in a circular arrangement resembling a clock. Remove your glasses or contacts before conducting this test.

Laboratory Activity   VISUAL TESTS

I. Visual Acuity Test

The Snellen eye chart should be mounted on a wall at eye level. A strip of masking tape should be placed on the floor to mark a distance of 20 feet from the chart. You will perform this test in groups of two. One person is to stand next to the eye chart to act as the administrator of which line is to be read. The other person will stand at the 20-foot mark and test his or her vision.

MATERIALS

Snellen eye chart
20' mark from chart

PROCEDURES

1. Stand at the 20-foot mark and cover one of your eyes with either a cupped hand or an index card. Read a line that you can easily focus on all the letters. Continue to view progressively smaller lines until you cannot accurately read the letters. Record in Table 11.4 the value of the smallest line that you read without errors. This line is your visual acuity for that eye, 20/20, for example.
2. Repeat this process with your other eye. Record your data in Table 11.4.
3. Now, using both eyes, read the smallest line that you can see clearly. Record your data in Table 11.4.
4. If you wear glasses or contacts, repeat the test while wearing your corrective lenses. Record your data in Table 11.4.

TABLE 11.4	**Visual Acuity**	
	Acuity (without glasses)	*Acuity (with glasses)*
Left eye		
Right eye		
Both eyes		

II. Astigmatism Test

The Astigmatism eye chart should be mounted on a wall at eye level. You will do this test in groups of two. If you wear glasses, perform the test with the glasses both on and off.

MATERIALS

Astigmatism chart

PROCEDURES

1. Look at the white circle in the center of the chart.
2. If all the radiating black lines appear equally sharp and equally black, you have no astigmatism.
3. If some lines appear blurred or are not consistently dark, you have astigmatism.

III. Blind Spot Mapping

The optic disk or blind spot is an area of the retina lacking photoreceptors. Normally, the blind spot is not seen because the visual fields of the eyes overlap and "fill in" the missing information from the blind spot.

MATERIALS

Figure 11.7

PROCEDURES

1. Holding the page 2 inches from your face, close your left eye, and stare at the cross in Figure 11.7.
2. Slowly move the page away from your face. The dot disappears when its image falls on your blind spot.

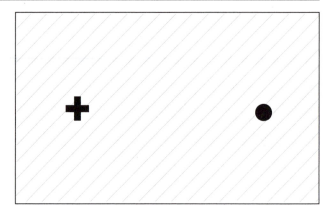

●Figure 11.7
The Optic Disc
Close your left eye and stare at the cross with your right eye, keeping it in the center of your field of vision. Begin with the page a few inches away, and gradually increase the distance. The dot will disappear when its image falls on the blind spot. To check the blind spot in the left eye, close your right eye and repeat this sequence staring at the dot.

B. Accommodation

For objects closer than 20 feet from the eye, the lens must bend the light rays together so that they focus a clear image on the retina. This process is called **accommodation** and is detailed in Figure 11.8.

Objects 20 feet or farther from the eye have parallel light rays that enter the eye. The ciliary muscle is relaxed, and the ciliary body is behind the lens. This causes the suspensory ligaments to pull the lens flat for proper refraction of the parallel light rays. Objects closer than 20 feet to the eye have divergent or spreading light rays that require more refraction to focus them on the retina. The ciliary muscle contracts and the ciliary body shifts forward, releasing the tension on the suspensory ligaments, which causes the lens to bulge and become round. The spherical lens increases refraction, and the divergent rays are bent into focus on the retina. Reading and other activities requiring near vision cause eye strain and fatigue because of the contraction of the ciliary muscle for accommodation. The lens gradually loses its elasticity as we age and causes a form of farsightedness called **presbyopia**. Many individuals have difficulty reading small type by the age of 40 and may require reading glasses to correct for the reduction in accommodation.

Laboratory Activity   ACCOMMODATION DETERMINATION _____

Accommodation is determined by measuring the closest distance from which one can see an object in sharp focus, the **near point** of vision. A simple test for near point vision involves moving an object toward the eye until it becomes blurred.

MATERIALS

pencil
ruler

PROCEDURES

1. Hold a pencil with the eraser up approximately 2 feet from your eyes, and look at the ribs in the metal eraser casing.
2. Close one eye, and slowly bring the pencil toward your open eye.
3. Measure the distance from the eye just before the metal casing blurs. Record your measurements in Table 11.5.
4. Repeat the procedure with the other eye.

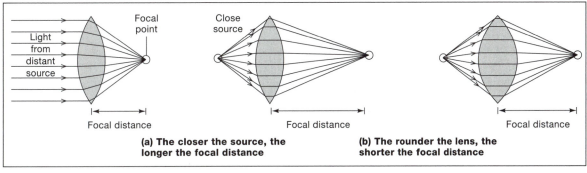

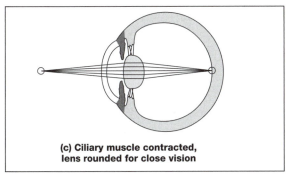

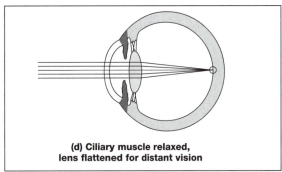

●Figure 11.8
Image Formation and Visual Accommodations

(a) A lens refracts a light toward a specific point. The distance from the center of the lens to that point is the focal distance of the lens. Light from a distant source arrives with all of the light waves traveling parallel to one another. Light from a nearby source, however, will still be diverging or spreading out from its source, when it strikes the lens. Note the difference in focal distance after refraction. **(b)** The rounder the lens, the shorter the focal distance. **(c)** When the ciliary muscle contracts, the suspensory ligaments allow the lens to round up. **(d)** For the eye to form a sharp image, the focal distance must equal the distance between the center of the lens and the retina. The lens compensates for variations in the distance between the eye and the object in view by changing its shape. When the ciliary muscle relaxes, the ligaments pull against the margins of the lens and flatten it.

5. To focus on near objects, the eyes must rotate medially, a process called convergence. To observe convergence, stare at the pencil from arm's length and slowly move it closer to your eyes. Which way did your eyes move?

6. Compare your near point distances with those of classmates of various ages. Record this comparative data in Table 11.5.

7. Why does accommodation change as we age?

TABLE 11.5	Near Point Determination		
Name and Age	*Right Eye*	*Left Eye*	*Both Eyes*

PART FIVE   SPECIAL SENSES: ANATOMY OF THE EAR

WORD POWER	MATERIALS
otolith (oto—ear, lith—stone)	ear models
pinna (pinna—wing)	auditory ossicles model
scala (scala—staircase)	inner ear model
concha (concha—shell)	cochlea model
cupula (cupa—tub)	compound microscope
ossicle (ossicle—bone)	prepared slide of cochlea

OBJECTIVES

On completion of this part of the exercise, you should be able to:

- Identify and describe the components of the external, middle, and inner ear.
- Describe the anatomy of the cochlea.
- Describe the components of the semicircular canals and the vestibule, and identify their roles in static and dynamic equilibrium.

The ear is divided into three regions: the external or outer ear, the middle ear, and the inner ear. The external and middle ear direct sound waves to the inner ear for hearing. The inner ear has three compartments, each containing a specialized population of receptors. The cristae of the semicircular canals and the maculae of the vestibule inform the CNS of our position in the world compared with gravity and our movements. The spiral organ of the cochlea converts vibrations of sound waves into nerve impulses for auditory sensations.

A. The External Ear

The flap or **pinna** of the outer ear funnels sound waves into the **external auditory canal**, a tubular chamber that delivers sounds to the **tympanic membrane** (ear drum). The tympanic membrane is a thin sheet of fibrous connective tissue stretched across the canal. The canal contains wax-secreting cells called **ceruminous glands** and many hairs that prevent dust and debris from entering the ear. Locate the structures of the external ear in Figure 11.9.

B. The Middle Ear

The middle ear (Figure 11.9) is a small space inside the temporal bone. It is connected to the back of the throat (nasopharynx) by the **auditory tube** or **pharyngotympanic tube**. This tube equalizes pressure between the external air and the cavity of the middle ear. Three small bones called **auditory ossicles** transfer vibrations from the external ear to the internal ear. The **malleus** is connected to the tympanic membrane. The **incus** joins the malleus to the third bone, the **stapes**, which connects to the inner ear. Vibrations of the tympanic membrane are transferred to the malleus, which conducts them to the incus and stapes. The stapes, in turn, pushes on the oval window of the cochlea to stimulate the auditory receptors.

C. The Inner Ear

The inner ear consists of three major regions: a helical cochlea, an elongated vestibule, and three semicircular canals. Each region is organized into two fluid-filled chambers:

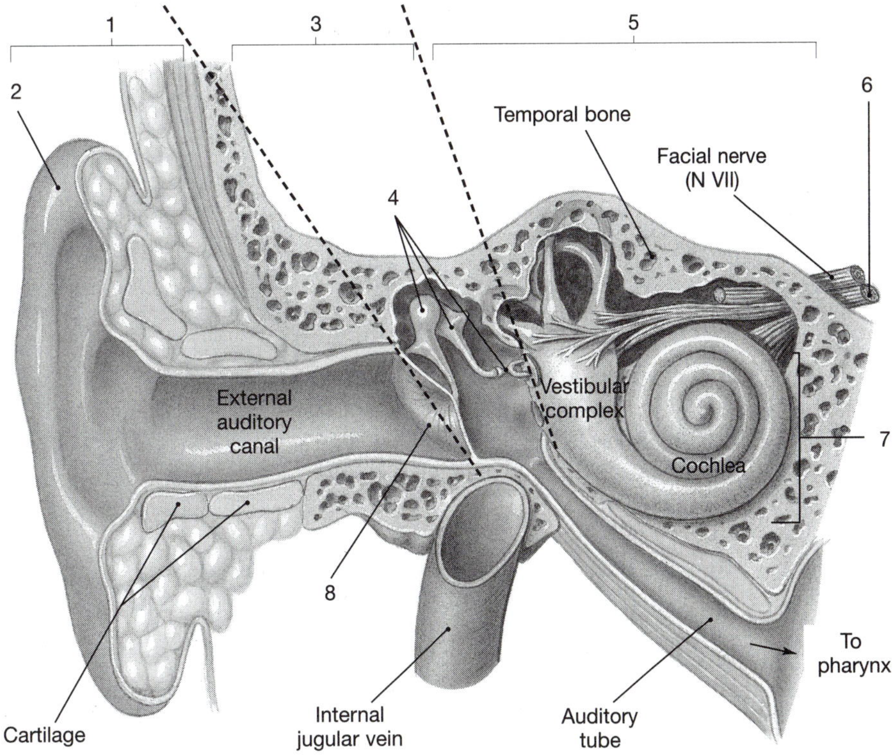

●Figure 11.9
Anatomy of the Ear
The orientation of the external, middle, and inner ear.

1. _____   4. _____   7. _____
2. _____   5. _____   8. _____
3. _____   6. _____

an outer **bony labyrinth** lined by an inner **membranous labyrinth**. The osseous labyrinth is embedded in the temporal bone and contains **perilymph**. The membranous labyrinth is filled with **endolymph**.

The **anterior, lateral,** and **posterior semicircular canals**, shown in Figure 11.10a, are positioned perpendicular to one another. Together they function as a sensory organ of dynamic equilibrium during motion. Inside the membranous labyrinth, the base of each semicircular canal has a swollen **ampulla** that contains a receptor called the **crista**. Each crista has hair cells (receptors) and supporting cells covered by a gelatinous cap called the **cupula**. Movement of the head causes the endolymph to push or pull on the cupula and bend or stretch the embedded hair cells.

The **vestibule** is an area between the semicircular canals and the cochlea. The membranous labyrinth in this area contains two sacs, the **utricle** and the **saccule**, which contain **maculae**, receptors for stationary equilibrium. Like the cristae, the maculae have hair cells and a gelatinous covering. Embedded in the gelatinous membrane are calcium carbonate crystals, called **otoliths**, that shift their position when the head is tilted. This shifting, shown in Figure 11.10d, causes the hairs to bend and trigger sensory impulses. Impulses from both the cristae and the maculae pass into the vestibular branch of the **vestibulocochlear nerve** (cranial nerve VIII).

The coiled **cochlea** contains the sensory receptors for hearing. The membranous labyrinth, located in the center of the cochlea, is called the **cochlear duct** or **scala media**. This duct separates the cochlear cavity into two ducts called the **scala vestibuli**

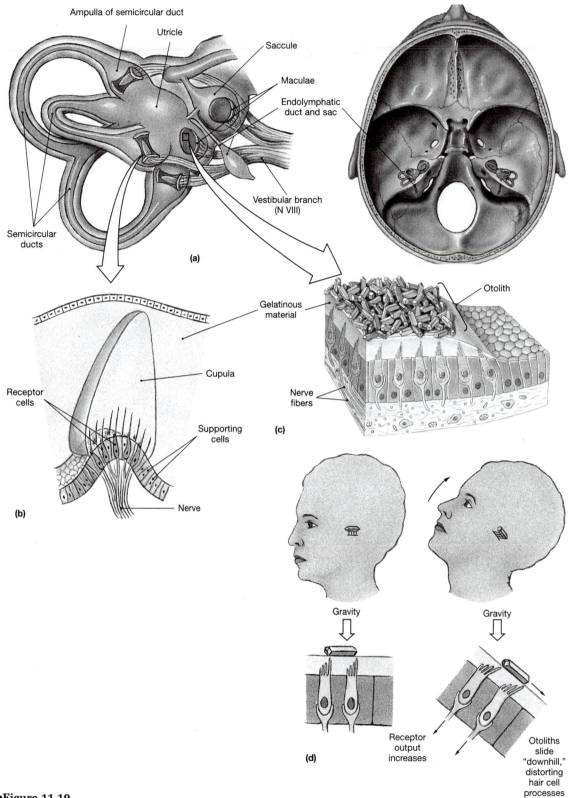

●**Figure 11.10**
The Vestibular Complex

(a) The semicircular ducts, utricle, and saccule, showing the location of sensory
receptors. **(b)** Cross section through the ampulla of a semicircular duct. **(c)** Structure
of a macula. **(d)** Diagrammatic view of macular function when the head is tilted back.

and the **scala tympani** (Figure 11.11). These chambers contain perilymph, whereas the cochlear duct is filled with endolymph. A membrane on the scala vestibuli called the **oval window** is in contact with the stapes of the middle ear. The scala vestibuli follows the twisting of the cochlea and is continuous with the scala tympani. The **round window** of the scala tympani acts as a pressure relief valve when the stapes vibrates and creates pressure waves in the perilymph.

The cochlear duct contains the sensory receptors for hearing called the **spiral organ** (organ of Corti). This organ contains **hair cells**, supporting cells, and a gelatinous membrane called the **tectorial membrane**. The hair cells have long stereocilia that extend into the endolymph and contact the tectorial membrane. The floor of the cochlear duct is made up of a bony shelf and a **basilar membrane**. The **vestibular membrane** (see Figure 11.11) composes the roof of the cochlear duct.

Laboratory Activity   ANATOMY OF THE EAR

MATERIALS

ear models and charts   cochlea model
compound microscope   prepared slide of cochlea

PROCEDURES

1. Identify the structures of the ear on laboratory charts and models. Label the gross anatomy of the ear in Figure 11.9.

2. Complete Figure 11.11 by labeling the parts of the chochlea. Examine a prepared slide of a cross section of the cochlea, and identify the structures shown in Figure 11.11.

PART SIX   SPECIAL SENSES: PHYSIOLOGY OF THE EAR

WORD POWER	MATERIALS
macula (macula—a spot)	Weber's test—tuning forks (100 cps, 1000 cps)
utricle (utriculus—leather bag)	Rinne test—tuning forks (1000 cps, 5000 cps)
saccule (sacculus—a small sac)	
proprioceptors (proprius—one's own)	
cristae (crista—crest)	

OBJECTIVES

On completion of this part of the exercise, you should be able to:

• Explain the difference between static and dynamic equilibrium by performing various comparative tests.

• Test range of hearing by using tuning forks.

• Compare conduction of sound through air versus bone (Rinne test).

A. Equilibrium

Visual awareness of surroundings enhances the sense of equilibrium by comparing your body position to the position of surrounding stationary objects. The maculae in

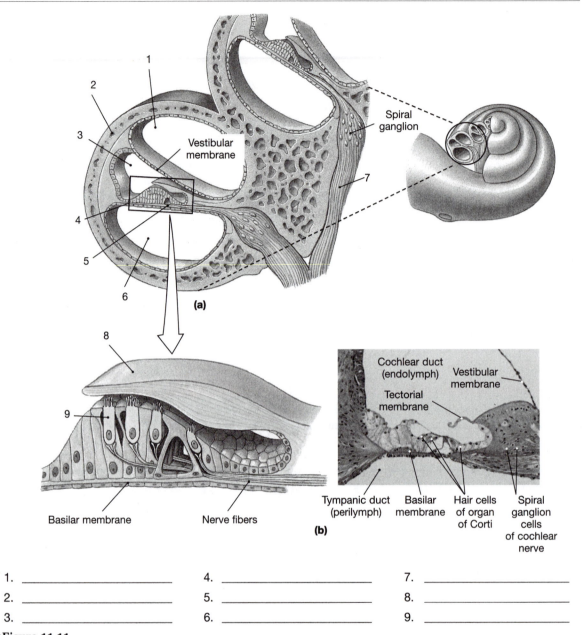

●Figure 11.11
The Cochlea and Organ of Corti

(a) Structure of the cochlea as seen in section. (b) The three-dimensional structure of the tectorial membrane and hair cell complex of the organ of Corti.

the utricle and saccule detect changes in body position in reference to gravity (static equilibrium). They inform the brain if you are leaning in a direction other than perpendicular to the ground. The maculae also determine your orientation compared with gravity. Loss of visual references of the ground and other stationary objects usually results in a loss of balance.

Nystagmus is the trailing of the eyes slowly in one direction followed by their rapid movement in the opposite direction. You will look for this event when a subject is spinning in a swivel chair during the next laboratory activity. As body position changes, nystagmus occurs to produce a brief, crisp image on the retina. Initially, the eyes move slowly in the opposite direction of the rotation, then quickly jump to the other direction. If the body is rotating to the right, the eyes will first slowly track to the left, then rapidly veer to the right. This cycle of slow and fast eye movements produces a brief, stationary image on the retina instead of a blurred image due to the movement.

Laboratory Activity   EQUILIBRIUM TESTS _____

I. Equilibrium and Vision

PROCEDURES

1. Perform the following experiment with your lab partner. Be prepared to help your partner if he or she loses balance and starts to fall.
2. Stand on both feet in a clear area of the room. Raise one foot, and then close your eyes. Record your observations in Table 11.6.

TABLE 11.6 **Equilibrium Tests**	
Standing Position	*Observation*
On one foot, eyes open	
On one foot, one eye closed	
On one foot, both eyes closed	

II. Nystagmus

This experiment involves spinning a volunteer, the subject, in a chair to observe his or her eye movements. Do not use anyone who is prone to motion sickness. The observer will spin the subject and record eye reflexes. To prevent accidents, four other individuals are necessary to hold the base of the chair steady and to serve as spotters to help the subject stay in the chair during the experiment. The subject should remain in the chair for several minutes after the experiment to regain his or her normal equilibrium.

MATERIALS

swivel chair

PROCEDURES

1. Seat the subject in a swivel chair with the spotters supporting the base of the chair with their feet if possible.
2. Instruct the subject to stare straight ahead with both eyes.
3. Spin the chair clockwise for ten revolutions (one rotation per second).
4. Quickly and carefully stop the chair and observe the subject's eye movements. Record your observations in Table 11.7.
5. Keep the subject seated for approximately two minutes so that he or she may regain balance.
6. With other subjects, repeat the experiment with a counterclockwise rotation. Additionally, ask one subject to close his or her eyes during the rotation and then to open them when the chair stops. Record your observations in Table 11.7.

TABLE 11.7	Nystagmus Test	
Rotation and Eyes		*Observation*
Clockwise, eyes open		
Clockwise, eyes closed		
Counterclockwise, eyes open		
Counterclockwise, eyes closed		

B. Hearing

We rely on our sense of hearing for communication and for an awareness of events in our immediate surroundings. Sounds are produced by vibrating objects that cause air to compress and decompress into sound waves. Faster vibrations produce a higher frequency or pitch of sound and cause the tympanic membrane to rapidly vibrate. Similarly, the lower the sound frequency, the slower the tympanic membrane vibrates. Vibrations are conducted to the spiral organ in the cochlea for sound reception. Figure 11.12 and Table 11.8 summarize the physiology of hearing.

C. Middle Ear Conduction Versus Nerve Deafness

Deafness can be the result of many factors, not all of which are permanent. Two categories of deafness are discernable, conduction deafness and nerve deafness. Conduction deafness involves damage to the tympanic membrane or one or more of the auditory ossicles. Proper conduction produces vibrations heard equally in both ears. If a conduction problem exists, sounds are normally heard better in the unaffected ear.

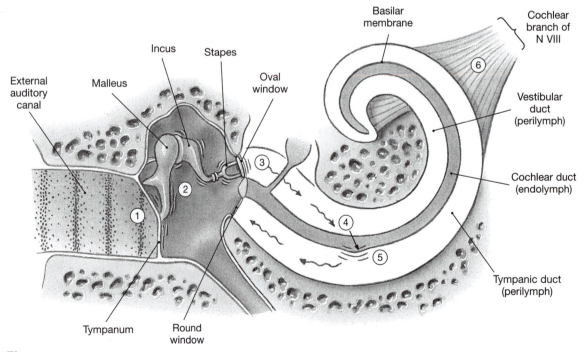

●**Figure 11.12**
The Physiology of Hearing
Steps involved in the reception of sound and the process of hearing. Table 11.8 summarizes each step.

TABLE 11.8	**Steps in the Production of an Auditory Sensation** (See Figure 11.12)

1. Sound waves arrive at the tympanic membrane.
2. Movement of the tympanic membrane causes displacement of the auditory ossicles.
3. Movement of the stapes at the oval window establishes pressure waves in the perilymph of the vestibular duct.
4. The pressure waves distort the basilar membrane on their way to the round window of the tympanic duct.
5. Vibration of the basilar membrane causes vibration of hair cells against the tectorial membrane.
6. Information concerning the region and intensity of stimulation is relayed to the CNS over the cochlear branch of N VIII.

Conduction tests with tuning forks, however, will cause the ear with the deafness to hear the sound *louder* than the normal ear. This is due to an increased sensitivity to sounds in the ear with conduction deafness. Hearing aids are often used to correct for conduction deafness.

Nerve deafness is a result of damage to the cochlea or the cochlear nerve. Repetitive exposure to excessively loud noises such as music and machinery can do damage to the delicate spiral organ. Nerve deafness cannot be corrected and results in a permanent loss of hearing, usually within a specific range of frequencies (tones).

In the following tests, sound vibrations from tuning forks will be conducted through the bones of the skull, bypassing the normal conduction by the external ear and middle ear. Since the inner ear is surrounded by the temporal bone, vibrations will be transmitted directly from the bone into the cochlea.

Laboratory Activity   HEARING TESTS

A quiet environment is necessary for conducting hearing tests. The subject should use his or her index finger to close the ear opposite to the ear being tested. Strike the tuning fork on the heel of your hand, and hold it next to the subject's ear.

I. Weber's Test

MATERIALS

tuning forks (100 cps and 1000 cps)

PROCEDURES

1. Use your index fingers to close both of your ears.
2. Have your partner correctly strike the 100 cps tuning fork and place its base onto your forehead. Did you hear the vibrations from the tuning fork? Was it louder in one ear than the other? _____

   _____

3. Repeat the procedure using the 1000 cps tuning fork. Did you hear one tuning fork better than the other? _____

   _____

II. Rinne Test

MATERIALS

tuning forks (1000 cps and 5000 cps)

PROCEDURES

1. Have your partner sit and find the mastoid process behind his or her right ear. Your partner should close his or her left ear with a finger.
2. Strike the 1000 cps tuning fork, and place the base of its handle against the mastoid process. This bony process will conduct the sound vibrations of the tuning fork to the inner ear.
3. When your partner indicates he or she can no longer hear the tuning fork, quickly hold the fork close to the external ear. In normal hearing, the tuning fork should still be audible. If middle ear deafness exists, no sound will be heard. Repeat this test on the left ear. Record your observations in Table 11.9.
4. Repeat steps 1 through 3 with the 5000 cps tuning fork. Record your observations in Table 11.9.

TABLE 11.9	**Hearing Test Results**	
	Left ear	*Right Ear*
1000 cps		
5000 cps		

ANATOMY AND PHYSIOLOGY OF THE SENSORY ORGANS

Exercise 11 Laboratory Report

A. Matching

Match each general sense on the left with the correct receptor on the right.

1. _____	fine touch of epidermis		A.	Pacinian corpuscle
2. _____	deep pressure		B.	Meissner's corpuscle
3. _____	fine touch of dermis		C.	root hair plexus
4. _____	pain		D.	muscle spindles
5. _____	epidermal contact		E.	Merkel's discs
6. _____	body position		F.	free nerve endings

B. Short Answer Questions

1. Describe the two-point discrimination, and explain how it demonstrates receptor density in the skin.

2. Explain the concept of a receptive field.

C. Matching

Match each special sense on the left with the correct receptor on the right.

1. _____	dynamic equilibrium		A.	taste buds
2. _____	hearing		B.	crista
3. _____	gustation		C.	macula
4. _____	smell		D.	spiral organ
5. _____	vision		E.	olfactory epithelium
6. _____	static equilibrium		F.	retina

D. Short Answer Questions

1. Why does a cold affect your sense of smell?

2. Where are the receptors for taste located?

3. Describe an experiment that demonstrates how the senses of taste and smell are linked.

4. Draw the tongue, and identify where taste buds for sweet, bitter, sour, and salty compounds are concentrated.

E. Matching

Match the description on the left with the correct eye structure on the right.

1. _____	transparent part of fibrous tunic	A.	limbus
2. _____	thin filaments attached to lens	B.	pupil
3. _____	corners of eyes	C.	vitreous body
4. _____	produces tears	D.	optic disc
5. _____	drains tears into nasal cavity	E.	ciliary muscle
6. _____	adjusts shape of lens	F.	caruncle
7. _____	border of sclera and cornea	G.	canthus
8. _____	jelly-like substance of eye	H.	cornea
9. _____	depression in retina	I.	sclera
10. _____	area lacking photoreceptors	J.	aqueous humor
11. _____	white of eyeball	K.	retina
12. _____	watery fluid of anterior eye	L.	lacrimal gland
13. _____	regulates light entering eye	M.	suspensory ligaments
14. _____	contains photoreceptors	N.	fovea
15. _____	red structure in medial eye	O.	nasolacrimal duct

F. Short Answer Questions

1. Explain the difference between viewing a distant and a close object.

2. Why does the accommodation decrease with age?

3. Describe an experiment that demonstrates the presence of a blind spot on the retina.

4. Explain how the Snellen eye chart is used to determine visual acuity.

5. Describe the layers of the retina.

G. Short Answer Questions

1. If conduction deafness exists, why would the sound of a tuning fork be heard best in the deaf ear?

2. List the three bones of the middle ear in order from the tympanic membrane to the oval window of the inner ear.

3. Describe the receptors for dynamic and static equilibrium.

4. Explain the phenomenon of nystagmus.

5. Describe the process of hearing.

H. Matching

Match the description on the left with the correct ear structure on the right.

1. _____ ear drum	A.	otoliths
2. _____ receptors in semicircular canals	B.	basilar membrane
3. _____ coiled region of inner ear	C.	tectorial membrane
4. _____ outer layer of inner ear	D.	membranous labyrinth
5. _____ receptor for hearing	E.	cupula
6. _____ receptors in vestibule	F.	bony labyrinth
7. _____ attachment site for stapes	G.	tympanic membrane
8. _____ jellylike substance of crista	H.	cochlea
9. _____ contains endolymph	I.	semicircular canals
10. _____ membrane over auditory receptors	J.	round window
11. _____ chamber inferior to organ of Corti	K.	crista ampullaris
12. _____ perpendicular loops of inner ear	L.	spiral organ
13. _____ membrane of tympanic duct	M.	oval window
14. _____ membrane supporting spiral organ	N.	macula
15. _____ crystals of maculae	O.	scala tympani

I. Completion

1. Complete Concept Map I of the special senses.

Concept Map I

Using the following terms, fill in the circled, numbered, blank spaces to complete the concept map. Follow the numbers that comply with the organization of the map.

Retina	Audition	Rods and cones	Olfaction
Smell	Ears	Taste buds	Tongue
Balance and hearing			

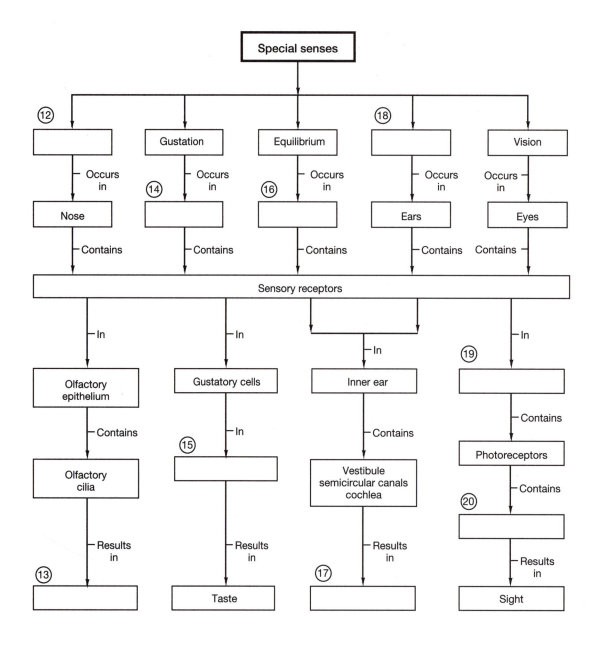

12 THE BLOOD

What would happen if you received a transfusion of blood that did not match your blood type? How can differences in Rh harm a fetus? What procedures are involved in detecting your ABO and Rh blood types? What types of cells occur in the blood, and how do these cells function? In this exercise, you will study the components of blood and perform blood typing and hematocrit tests.

WORD POWER

erythrocyte (erythro—red)
leukocyte (leuko—white)
neutrophil (...phil—loving)
lymphocyte (lympho—watery)
agglutination (agglutinate—to stick together)
hemolytic (hemo—blood)
coagulation (coagul—to curdle)

MATERIALS

compound microscope
prepared slide of blood smear
gloves and safety glasses
bleach solution in spray bottle
paper towels
sterile blood lancets
sterile alcohol prep pads
disposable blood typing plates
  or microscope slides

biohazardous waste container
wax marker
clean toothpicks
anti-A and anti-B typing sera
heparinized capillary tubes
centrifuge
seal ease clay
hematocrit tube reader

OBJECTIVES

On completion of this exercise, you should be able to:

- List the function of the blood.
- Describe each component of the blood.
- Distinguish each type of blood cell on a blood smear slide.
- Describe the antigen-antibody reactions of the ABO and Rh blood groups.
- Safely collect a blood sample using the blood lancet puncture technique.
- Safely type a sample of blood to determine the ABO and Rh blood types.
- Correctly perform a hematocrit test on a sample of blood.
- Discuss how to properly discard of blood contaminated wastes.

Blood is a composite fluid connective tissue that flows through the vessels of the vascular system. In response to injury, blood has the intrinsic ability to change from a liquid to a gel so as to clot and stop bleeding. Blood comprises cells and cell pieces that are collectively called the formed elements. These cells are carried in an extracellular fluid called blood plasma. Blood has many diverse functions, all relating to supplying your cells with essential materials and maintaining the internal environment.

Red blood cells transport respiratory gases to trillions of cells in your body. The blood controls the chemical composition of all interstitial fluid by regulating pH and electrolyte levels. White blood cells are part of the immune system and protect you from microbes by producing antibody molecules and phagocytizing foreign cells.

A. Blood Plasma

Figure 12.1 details the composition of whole blood. A sample of blood is approximately 55% plasma and 45% formed elements. Plasma is approximately 92% water and contains proteins to regulate the osmotic pressure of blood, proteins for the clotting process and antibody proteins that work with the immune system to protect your body from invading pathogens. Electrolytes, hormones, nutrients, and some blood gases are transported in the blood plasma. Later in this exercise, you will do a hematocrit test that separates the blood cells from the plasma.

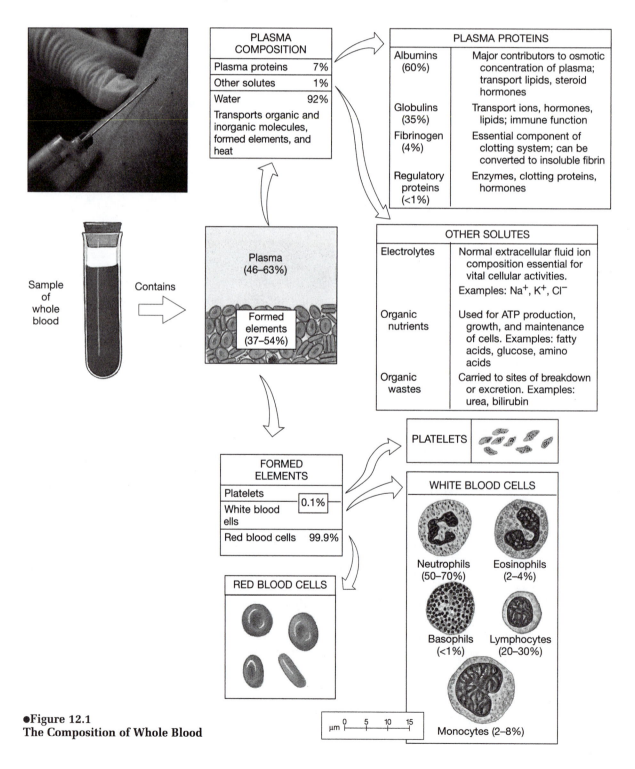

●**Figure 12.1**
The Composition of Whole Blood

B. The Formed Elements

The formed elements may be organized into three groups of cells: the red blood cells or **erythrocytes**, the white blood cells or **leukocytes**, and the **platelets** (see Figure 12.1). When stained, each group is easy to identify with a microscope. The red cells are erythrocytes, the stained cells are leukocytes, and the small cell fragments between the red and white cells are the platelets.

ERYTHROCYTES

Erythrocytes are commonly called red blood cells or RBCs. They are the most abundant of all blood cells. Erythrocytes are biconcave discs and microscopically appear thinner and lighter staining in the middle (see Figure 12.2). The biconcave shape most importantly increases the surface area on the red blood cell for rapid gas exchange between the blood and the tissues of the body. Their shape also allows them to flex and squeeze through narrow capillaries. Erythrocytes are red and lack a nucleus. White blood cells, however, have a nucleus.

The major function of red blood cells is to transport blood gases. Oxygen is picked up in the lungs and carried to the tissue cells of the body. While supplying the cells with oxygen, the blood acquires carbon dioxide from the cells. The plasma and red blood cells convey the carbon dioxide to the lungs for removal during exhalation. To accomplish the task of gas transport, each RBC contains millions of hemoglobin (Hb) molecules. Hemoglobin is a complex protein molecule with four iron atoms that allow oxygen and carbon dioxide molecules to loosely bind to the Hb.

LEUKOCYTES

Leukocytes are white blood cells (WBCs) that contain a darkly stained nucleus. The nucleus often is branched into several lobes, as shown in Figure 12.3. Leukocytes lack hemoglobin and therefore do not transport blood gases. Most leukocytes are phagocytes and are part of the immune system. We will discuss the function of each population of WBCs in the upcoming section.

Two broad classes of leukocytes occur: **granular leukocytes** and **agranular leukocytes**. The granular leukocytes, collectively called **granulocytes**, have granules in their cytoplasm and include the neutrophils, eosinophils, and basophils. Agranular

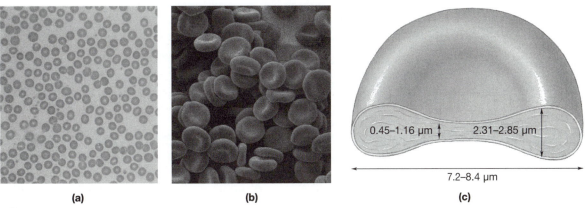

(a)   (b)   (c)

●**Figure 12.2**
Anatomy of Red Blood Cells

(a) When viewed in a standard blood smear, red blood cells appear as two-dimensional objects because they are flattened against the surface of the slide. (LM × 320) **(b)** A scanning electron micrograph of red blood cells reveals their three-dimensional structure quite clearly. (SEM × 1195) **(c)** A sectional view of a mature red blood cell, showing average dimensions.

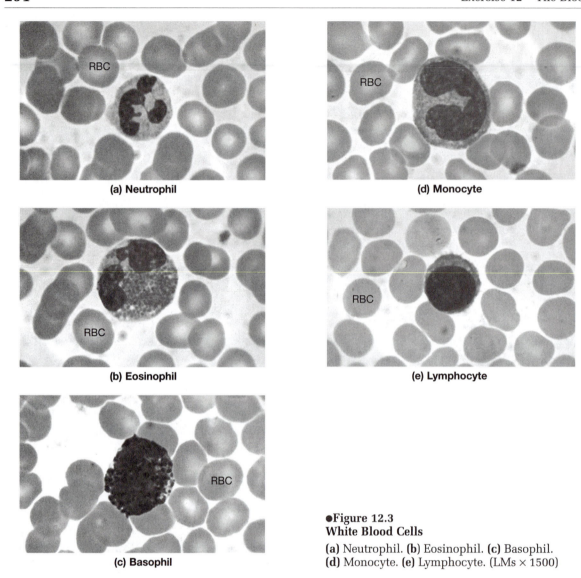

●Figure 12.3
White Blood Cells
(a) Neutrophil. **(b)** Eosinophil. **(c)** Basophil.
(d) Monocyte. **(e)** Lymphocyte. (LMs × 1500)

leukocytes, which include the monocytes and lymphocytes, have few cytoplasmic granules. Figure 12.3 presents a micrograph of each leukocyte.

Neutrophils   The most common leukocytes are the **neutrophils** that compose up to 70% of the entire white blood cell population. These granulocytes are also called **polymorphonuclear** leukocytes because their nucleus is complex and branches into two to five lobes. Neutrophils have many small cytoplasmic granules that stain pale purple, visible in Figure 12.3a.

Neutrophils are the first leukocytes to arrive at a wound site to begin the process of infection control. They phagocytize bacteria and release hormones called **leukotrienes** that attract other phagocytes, such as eosinophils and monocytes, to the site of injury. Neutrophils are short-lived and survive in the blood for up to 10 hours. Active neutrophils in a wound may live only 30 minutes until they succumb to the toxins released by bacteria they have ingested.

Eosinophils   Eosinophils are identified by the presence of medium-sized granules that stain an orange-red color, as shown in Figure 12.3b. Eosinophils are similar in size to neutrophils. The nucleus is conspicuously segmented into two lobes. Eosinophils are phagocytes that engulf bacteria and other microbes that the immune system has coated with antibodies. They also contribute to decrease the inflammatory response at a wound or site of infection. Approximately 3% of the circulating leukocytes are eosinophils.

Basophils   Less than 1% of circulating leukocytes are basophils. These scarce cells are distinguished by large cytoplasmic granules that stain a very dark blue. The granules are so large and numerous that the nucleus is obscured as in the basophil illustrated in Figure 12.3c. Basophils are smaller than neutrophils and eosinophils. They are sometimes difficult to locate on a blood smear because of their low abundance.

Like most leukocytes, basophils can pass between the endothelial cells of capillaries and enter the interstitial spaces of tissues. Once among the tissue cells, basophils release their cytoplasmic granules into the surrounding interstitial fluid. The granules contain heparin that prevents blood clotting and histamines that cause vasodilation and inflammation. These molecules also attract other leukocytes to the site of trauma.

Monocytes   Approximately 2% to 8% of circulating white blood cells are monocytes. On a blood smear slide, monocytes are roundish and may have small extensions, much like an amoeba. These large cells are distinguished by a darkly stained kidney-bean-shaped nucleus surrounded by pale blue cytoplasm (see Figure 12.3d). Monocytes are agranular leukocytes; however, ingested material such as phagocytized bacteria and debris will stain and appear similar to granules.

Monocytes are wanderers. They leave the bloodstream between the capillary endothelium and patrol the body tissues in search of microbes and worn-out tissue cells. They are second to neutrophils in arriving at a wound site. As neutrophils die from phagocytizing bacteria, the monocytes phagocytize the neutrophils.

Lymphocytes   Lymphocytes are the smallest of the leukocytes and are approximately the size of a red blood cell (see Figure 12.3e). The distinguishing feature of lymphocytes is a large nucleus that occupies almost the entire cell, leaving room for only a small halo of pale blue cytoplasm around the edge of the cell. Lymphocytes are abundant in the bloodstream and compose 20% to 30% of all circulating leukocytes. As their name suggests, lymphocytes are the main cells populating the lymphatic system. Many lymphocytes occur in lymph nodes, glands, and other lymphatic structures.

Several types of lymphocytes exist; however, they cannot be individually distinguished with a light microscope. Generally, lymphocytes provide immunity from microbes and defective cells by two methods. **T-cell lymphocytes** attach to and destroy a foreign cell in a cell-mediated response by releasing cytotoxic chemicals to kill the invaders. **B-cell lymphocytes** become sensitized to a specific antigen, then manufacture and pour antibodies into the bloodstream. The antibodies attach to and help destroy foreign cells.

PLATELETS

Platelets are small cellular pieces produced in the red bone marrow when huge megakaryocyte cells fragment. Platelets lack a nucleus and other organelles. They survive in the bloodstream for a brief time and are involved in the coagulation or clotting of blood. Locate the platelet in Figure 12.3e.

Laboratory Activity   OBSERVATIONS OF ERYTHROCYTES, LEUKOCYTES, AND PLATELETS____

MATERIALS

compound microscope
human blood smear slide (Wright's or Giesma stained)

PROCEDURES

1. Obtain a human blood smear slide, and place it on the stage of the microscope. Move the slide so that the blood smear is over the aperture in the stage.

2. Adjust the magnification of the microscope to low power. Slowly turn the coarse focus knob until you can clearly see blood cells. Now use the fine focus adjustment as you examine individual cells on the slide.

3. Scan the slide at low magnification. Notice the abundance of red blood cells. The dark stained cells are the various leukocytes.

4. Increase magnification, and examine each type of formed element. Which cells have cytoplasmic granules?

_____

5. Complete Table 12.1 as you observe each blood cell.

TABLE 12.1	A Summary of the Formed Elements of the Blood		
Cell	*Functions*	*Appearance*	*Drawing*
Erythrocyte			
Neutrophil			
Eosinophil			
Basophil			
Monocyte			
Lymphocyte			
Platelet			

C. ABO and RH Blood Groups

Your blood type is determined by genes you inherited from your parents, and it does not change during your lifetime. Each blood type is due to the presence or absence of specific molecules on the surface of the erythrocytes. These molecules, called **antigens**, are like cellular name tags that inform your immune system that your cells belong to "self" and are not "foreign." The **surface antigens** on erythrocytes are called **agglutinogens**.

Each blood group also has **antibody** molecules, called **agglutinins**, present in the blood plasma. The agglutinins (antibodies) will attach to agglutinogens (antigens) of foreign blood cells and cause hemolysis of the cells.

More than 50 different blood systems occur in the human population. In this section, you will study the two most common blood systems, the ABO and Rh blood groups. Each blood group is controlled by a unique gene so that your ABO blood type does not influence your Rh blood type.

THE ABO BLOOD GROUP

Figure 12.4 illustrates the characteristics of the ABO blood. There are four blood types in the ABO system: A, B, AB and O. Two surface agglutinogens, A and B, occur in different combinations that determine the ABO blood type. Type A blood has the A agglutinogen on the erythrocyte surface, and type B blood has the B agglutinogen. Type AB blood has both A and B agglutinogens, and type O blood has neither A nor B agglutinogens.

Present in the blood plasma are agglutinins. Your plasma agglutinins will not react to your own blood. If a different type of blood is introduced into your bloodstream, your agglutinins will attach to the agglutinogens of the foreign blood cells and cause the cells to clump or **agglutinate**. Hemolysis or bursting occurs, and the intruding cells are destroyed.

Type A blood has A agglutinogens on the membrane surface and type anti-B agglutinins in the plasma. The anti-B antibody will not react with the type A antigen. Type B blood cells are covered with the B agglutinogen, whereas the plasma contains the anti-A agglutinin. AB blood is unique in that each red blood cell contains both the A and B cellular agglutinogens on the membrane surface. The AB blood lacks the anti-A and the anti-B plasma agglutinins. Type O blood cells lack both agglutinogens but has both anti-A and anti-B plasma agglutinins.

To determine an ABO blood type, the presence of agglutinogens is detected by adding drops of anti-sera that contain either the anti-A or the anti-B agglutinins. The serum agglutinins will react with the corresponding agglutinogen on the RBC surface. For example, the anti-A agglutinins will react with the A agglutinogens found in type

TYPE A	TYPE B	TYPE AB	TYPE O
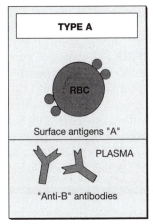	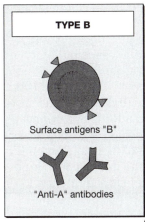	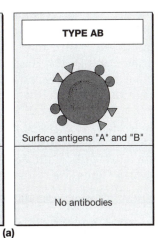	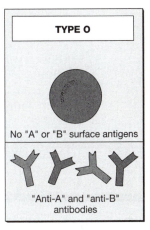
Surface antigens "A"	Surface antigens "B"	Surface antigens "A" and "B"	No "A" or "B" surface antigens
PLASMA "Anti-B" antibodies	"Anti-A" antibodies	No antibodies	"Anti-A" and "anti-B" antibodies

(a)

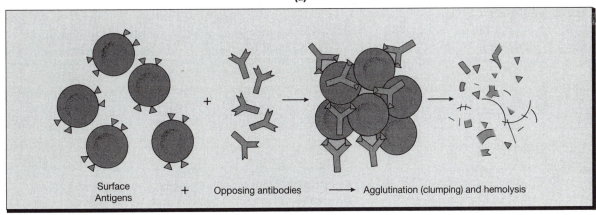

Surface Antigens   +   Opposing antibodies   →   Agglutination (clumping) and hemolysis

(b)

●**Figure 12.4**
Blood Typing

The blood type depends on the presence of agglutinogens on RBC surfaces. The plasma contains agglutinins, antibodies that will react with foreign agglutinogens. The relative frequencies of each blood type in the U.S. population are indicated in Table 12.5.

A blood. The blood will clump or **agglutinate** as more antibodies react with the antigens. Complete Table 12.2 before you type your blood.

TABLE 12.2	A Summary of ABO Blood Characteristics	
Blood Type	*Cellular Agglutinogen*	*Plasma Agglutinin*
Type A		
Type B		
Type AB		
Type O		

RH BLOOD GROUPS

The Rh blood group is named after the Rhesus monkey in which the blood group was first identified. Rh blood groups include Rh positive (+) and Rh negative (−) blood. Although this blood system is separate from the ABO system, the two are usually used together to identify a blood type. For example, a blood sample may be A+ or A-.

Unlike the ABO system with two cellular agglutinogens, the Rh system has only one cellular agglutinogen, called the D agglutinogen. A single Rh surface agglutinin also occurs, designated anti-D. The D agglutinogen is only present on RBCs that are Rh positive; Rh negative individuals lack the D agglutinogens. Rh positive blood cells are covered with the D agglutinogens, and the plasma of this blood lacks the anti-D Rh agglutinins. Interestingly, the Rh negative individual also lacks the anti-D agglutinins. However, if the Rh negative blood is exposed to the Rh D agglutinogens in Rh positive blood, the immune system of the Rh negative person will produce the anti-D agglutinin. This becomes clinically significant in cases of pregnancy with Rh incompatibilities between the mother and her fetus. We discuss this incompatibility in more detail in the Clinical Application box. Complete Table 12.3 before doing the Rh blood test.

TABLE 12.3	A Summary of Rh Blood Characteristics		
Blood Type	*Cellular Agglutinogen*	*Plasma Agglutinin (before sensitization)*	*Plasma Agglutinin (after sensitization)*
Rh positive			
Rh negative			

Clinical Application

If an expectant mother is Rh negative and her baby is Rh positive, a potentially life-threatening Rh incompatibility exists for the baby. Normally, the blood from the fetus does not mix with the mother's bloodstream. The umbilical cord of the baby connects to the placenta where fetal blood capillaries exchange gases, wastes, and nutrients with the mother's blood. If internal bleeding does occur and the mother is exposed to the D agglutinogens in her baby's Rh positive blood, she will produce the anti-D agglutinin. These anti-D antibodies will cross the membranes of the placenta and enter the fetal bloodstream where they will hemolyze the fetal blood cells of this fetus and any other future Rh+ fetuses. This Rh factor is called the **hemolytic disease of the newborn**, or **erythroblastosis fetalis**. A dosage of anti-Rh antibodies, called RhoGam, may be given to the mother during pregnancy and after delivery to destroy any Rh positive fetal cells in her bloodstream. This will prevent her from developing anti-D agglutinins in her blood.

Laboratory Activity   ABO AND RH BLOOD TYPING _____

MATERIALS

    gloves and safety glasses
    biohazardous waste container
    sterile blood lancet (disposable)
    sterile alcohol prep pads (disposable)
    disposable blood typing plates or sterile microscope slides
    wax pencil (if using microscope slides instead of typing plates)
    clean toothpicks
    anti-A, anti-B, and anti-D blood typing sera
    warming box for Rh reaction
    paper towels
    bleach solution in spray bottle

Safety Tips

1. Refer to Exercise 1, "Laboratory Safety," and review the handling and disposal of blood.
2. Some infectious diseases are spread by contact with blood. Follow all instructions carefully, and protect yourself by wearing gloves and working only with your own blood.
3. Materials contaminated with blood must be disposed of properly. Your instructor will inform you of methods to dispose of blood lancets, slides, alcohol prep, and toothpicks.

I. Blood Sample Collection

PROCEDURES

1. Wash both hands thoroughly with soap, then dry hands with a clean paper towel. Obtain an additional paper towel to place blood-contaminated instruments on while collecting a blood sample.

2. Open a sterile alcohol prep pad, and clean the tip of your index finger. Be sure to thoroughly disinfect the entire fingertip, including the sides of the finger. Place used prep pad on the extra paper towel.

3. Open a sterile blood lancet to expose only the sharp tip. Do not use an old blood lancet, even if it was used on one of your own fingers. Use the sterile tip immediately; do not allow time for the sterile tip to inadvertently become contaminated.

4. With a swift motion, jab the point of the lancet into the lateral surface of the finger tip. Place the used blood lancet on the paper towel.

II. Combined ABO and Rh Blood Typing

Your laboratory instructor may ask for a volunteer to "donate" blood to demonstrate the process of blood typing. Alternatively, many biological supply companies sell simulated blood typing kits that contain a bloodlike solution and anti-sera. These kits contain no human or animal blood products and safely show the principles of typing human blood.

PROCEDURES

1. Obtain a blood sample as indicated. Gently squeeze a drop of blood to each individual depression on the blood typing plate or to the circles on the slide. If necessary, slowly "milk" the finger to work more blood out of the puncture site.

2. Add a drop of anti-A serum to the A blood sample. Do not allow blood to touch and contaminate the dropper bottle. Repeat the process by adding a drop of anti-B serum to the B sample and a drop of anti-D to the D sample of blood.

3. Immediately and gently mix each drop of anti-serum into the blood with a clean toothpick. To prevent cross contamination of the blood samples, use a separate clean toothpick to stir each sample. Place all used toothpicks on the paper towel until they can be disposed of in a biohazard container.

4. Place the slide or plate on the warming box, and agitate the blood samples by rocking the box carefully back and forth for two minutes.

5. Examine the blood drops for clumping or agglutination that is visible with the unaided eye, and compare your blood sample to Figure 12.5, which details the agglutinogen–agglutinin reaction for each blood type. Agglutination results when the agglutinins in the anti-serum react with the matching agglutinogen on the red blood cells. Type A blood agglutinates with the anti-A serum, and type B blood agglutinates with the anti-B serum. Type AB blood agglutinates with both anti-A and anti-B sera, and type O blood does not agglutinate with either sera. If the sample agglutinates with the anti-D serum, the blood is Rh positive, whereas no agglutination shows the blood is Rh negative. The anti-D agglutination reaction is often weaker and less easily observed than the A and B agglutination reactions. A microscope may help you observe the anti-D reaction. Record your blood typing results in Table 12.4 by indicating yes or no for the presence of agglutination. Collect blood typing data from several classmates to compare agglutination responses among blood types.

TABLE 12.4	Blood Typing Observations		
Student	*Anti-A Serum*	*Anti-B Serum*	*Anti-D Serum*

6. What is your ABO blood type? What is your Rh blood type? Explain how each drop of blood reacted with the anti-sera.

_____

_____

7. How does your blood type compare to the distribution of human blood types as presented in Table 12.5?

_____

_____

Disposal of Materials and Disinfection of Work Space

1. On completion of the blood typing exercise, dispose of all blood-contaminated materials in the appropriate biohazard box. A box for sharp objects may be available to dispose of blood lancets, toothpicks, and glass microscope slides.

2. Your lab instructor may ask you to disinfect your workstation with a bleach solution. Wear your gloves and safety glasses while wiping the surfaces clean.

3. Lastly, remove your gloves and dispose of them in the biohazard box. Remember to wash your hands after disposing of all materials.

TABLE 12.5	Differences in Blood Group Distribution				
		Percentage with Each Blood Type			
Population	*O*	*A*	*B*	*AB*	*Rh+*
U.S. (average)	46	40	10	4	85
Caucasian	45	40	11	4	85
African-American	49	27	20	4	95
Chinese	42	27	25	6	100
Japanese	31	39	21	10	100
Korean	32	28	30	10	100
Filipino	44	22	29	6	100
Hawaiian	46	46	5	3	100
Native North American	79	16	4	<1	100
Native South American	100	0	0	0	100
Australian Aborigines	44	56	0	0	100

D. Packed Red Cell Volume (Hematocrit)

The **hematocrit** test, or packed cell volume (PCV), measures the volume of packed formed elements, the cells and platelets in the blood. Since red blood cells far outnumber all the other formed elements, the test mainly measures the volume of red blood cells. To perform the test, a drop of blood is collected in a heparinized capillary tube. Heparin is an anticoagulant and prevents blood from clotting inside the capillary tube so that the red blood cells can be separated from the plasma. A microcentrifuge is used to spin the blood sample at a very high speed for approximately four minutes. The gravitational forces produced by the rotation of the centrifuge pack the cells into one end of the capillary tube, therefore the term *packed cell volume*. On completion of the centrifugation, the sample is placed on a tube reader, and the percentage of packed blood cells is measured.

The hematocrit results provide information regarding the oxygen-carrying capacity of the blood. A low hematocrit value indicates that the blood has fewer RBCs to transport oxygen. Average hematocrit values for males range from 40% to 54%; for women the range is lower, from 37% to 47%.

Laboratory Activity   HEMATOCRIT DETERMINATION

MATERIALS

gloves and safety glasses	biohazardous waste disposal container
paper towels	bleach solution in spray bottle
sterile blood lancet (disposable)	sterile alcohol prep pads (disposable)
heparinized capillary tubes	seal-ease clay
microcentrifuge	tube reader

PROCEDURES

Your laboratory instructor may ask for a single volunteer to "donate" blood for demonstration of the hematocrit test.

1. Review the safety tips in the previous section. Obtain a blood sample as previously instructed.

2. Gently squeeze a drop of blood out of your finger. Excess pressure forces interstitial fluid into the blood that may alter your hematocrit reading. If you are

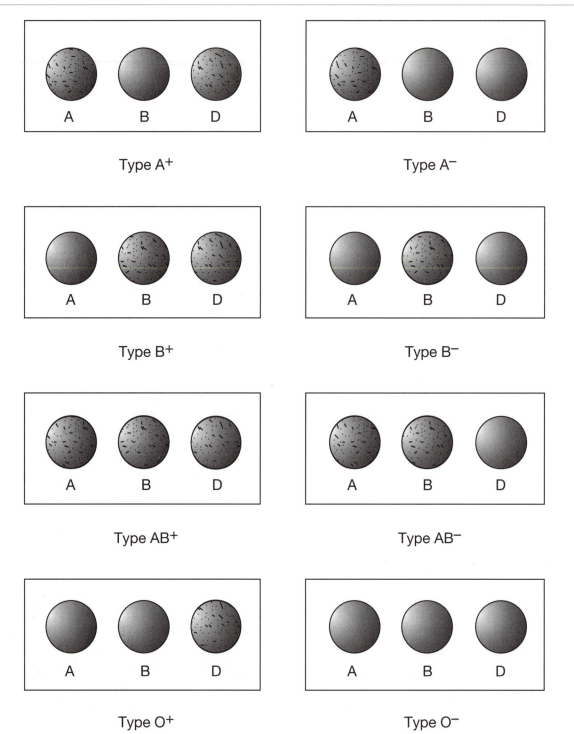

●Figure 12.5
ABO and Rh Blood Typing

having difficulty obtaining a blood sample, lance your finger again with a clean, sterile lancet.

3. Place a sterile heparinized capillary tube on the drop of blood. Orient the open end of the tube downward, as shown in Figure 12.6, to allow the blood to flow into the tube. Fill the tube at least two-thirds full with blood.

●Figure 12.6
Filling Capillary Tube with Blood

4. Carefully seal *one* end of the tube with the seal-ease clay, as shown in Figure 12.7. Do not force the delicate capillary tube into the clay, for it may break and cause you to jam glass into your hand. Clean any blood off the clay container with the bleach solution and a paper towel.

5. Place the tube in the microcentrifuge with the clay end toward the outer margin of the chamber (see Figure 12.8). Because the centrifuge spins at high speeds, the chamber must be balanced by placing tubes evenly in the chamber. Counterbalance your capillary tube by placing another sample directly across from yours. An empty tube with clay may be used if another student blood sample is not available for centrifugation.

6. Screw the inner cover on with the centrifuge wrench. Do not overtighten the lid. Close the outer lid and push the latch in.

7. Set the timer to four to five minutes, and allow the centrifuge to spin. Do not attempt to open or stop the centrifuge while it is turning. Always keep loose hair and clothing away from the centrifuge.

8. After the centrifuge turns off and stops spinning, open the lid and the inner safety cover to remove the capillary tube. Your blood sample should have clear plasma at one end of the tube and packed red cells at the other end.

9. Place the capillary tube in the tube reader. Because there is a variety of tube readers, your instructor will show how to use the reader that is available in your laboratory.

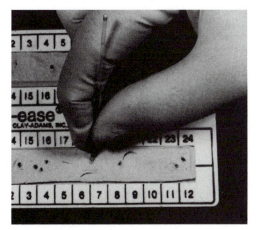

●Figure 12.7
Plugging a Capillary Tube with Clay

●Figure 12.8
Microcentrifuge with Capillary Tubes

10. What is the hematocrit of your blood sample? Is it within the normal hematocrit range?

_____

_____

11. Describe the appearance of your blood plasma.

_____

_____

12. See Disposal of Materials and Disinfection of Work Space on page 210 for instructions on the safe disposal of blood-contaminated materials.

THE BLOOD
Exercise 12 Laboratory Report

A. Matching
Match each term or structure listed on the left with the correct description on the right.

1. _____ erythrocyte
2. _____ polymorphonuclear cell
3. _____ granular leukocytes
4. _____ leukocyte
5. _____ agglutinins
6. _____ type A blood
7. _____ Rh positive
8. _____ red-orange stained blood cell
9. _____ type B blood
10. _____ Rh negative
11. _____ agglutinogens

A. eosinophil
B. molecules on membrane surface
C. has A agglutinogens and anti-B agglutinins
D. has the Rh agglutinogen
E. neutrophil
F. lacks the Rh agglutinogen
G. red blood cell
H. contain cytoplasmic granules
I. reacts with a membrane molecule
J. has B agglutinogens and anti-A agglutinins
K. white blood cell

B. Short Answer Questions

1. Describe what would happen if type A blood were transfused into the bloodstream of someone with type B blood.

2. What happens in the blood of an Rh negative individual who has been exposed to Rh positive blood?

3. What is the main function of red blood cells?

4. List the five types of leukocytes, and describe the function of each.

5. Describe how to do a hematocrit test. What are the average hematocrit values for males and females?

6. Describe how to type blood to detect the ABO and Rh blood groups.

C. Completion

1. Label each blood cell in Figure 12.9.

1. _____

4. _____

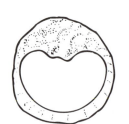

2. _____

5. _____

3. _____

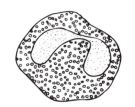

6. _____

●Figure 12.9
Leukocytes

13 ANATOMY AND PHYSIOLOGY OF THE HEART

Your heart beats approximately 100,000 times daily to send blood flowing into miles of blood vessels, providing nutrients, regulating substances and gases, and removing wastes from cells. All organ systems of the body depend on the cardiovascular system, so damage to the heart often results in widespread disruption of homeostasis.

For a drop of blood to complete one circuit through your body, it must be pumped twice by your heart. Each passage through the heart directs blood into a vascular system that loops back to the opposite side of the heart. Deoxygenated blood enters the right side of the heart and is pumped to the lungs to become saturated with oxygen and remove carbon dioxide. This oxygenated blood then returns to the left side of the heart, which must produce enough pressure to force blood to flow to all of your body tissues.

Your anatomical studies in this exercise include the histology of cardiac muscle tissue, external and internal heart structures, blood flow through the heart, and fetal heart structures. Since all mammals have a four-chambered heart, you will dissect a sheep heart to reinforce your observations of the human heart. Further, you will observe the effects of posture and exercise on blood pressure and determine the rate of your pulse from several pressure points.

PART ONE   ANATOMY OF THE HEART

WORD POWER	MATERIALS
pericardium (peri—around)	torso model
myocardium (myo—muscle)	heart models
endocardium (endo—inner)	compound microscope
atrium (atri—a vestibule)	prepared slide of cardiac muscle tissue
ventricle (ventricul—a belly)	sheep heart
septum (septi—a fence)	dissection scissors
pulmonary (pulmo—lung)	blunt probe
systemic (systema—a system)	gloves, safety glasses

OBJECTIVES

On completion of this part of the exercise, you should be able to:

- Identify the gross external and internal anatomy of the heart.
- Identify and discuss the function of the valves of the heart.
- Identify the major blood vessels of the heart.
- Trace a drop of blood through the heart, identifying both circuits.
- Identify the components of the conduction system of the heart.

- Discuss the differences between a prenatal and postnatal heart.
- Identify the anatomy of a dissected sheep heart.

A. Location of Heart and Anatomy of Heart Wall

Figure 13.1 illustrates the location of the heart within the mediastinum of the thoracic cavity. Blood vessels join the heart at the **base** positioned medially in the mediastinum. Because the left side of the heart has more muscle mass, the **apex** at the inferior tip of the heart is more in the left side of the thoracic cavity. A line traced from your right shoulder to your left hip would pass through the axis of your heart, separating the right and left chambers.

Within the mediastinum, the heart lies inside a space, the **pericardial cavity**. This cavity is formed by the **pericardium**, a serous membrane. Recall from your earlier studies that all serous membranes are double membranes with parietal and visceral layers. The pericardial cavity contains **serous fluid** to reduce friction during muscular contraction.

Examine Figure 13.1b, and locate the pericardial cavity. The outer **parietal pericardium** attaches to the heart in the mediastinum. The inner **visceral pericardium** or **epicardium** lines the surface of the heart. Imagine pushing your fist into an inflated balloon. The balloon around your fist represents the visceral pericardium, and the rest of the balloon represents the parietal pericardium. The space inside the balloon is the pericardial cavity.

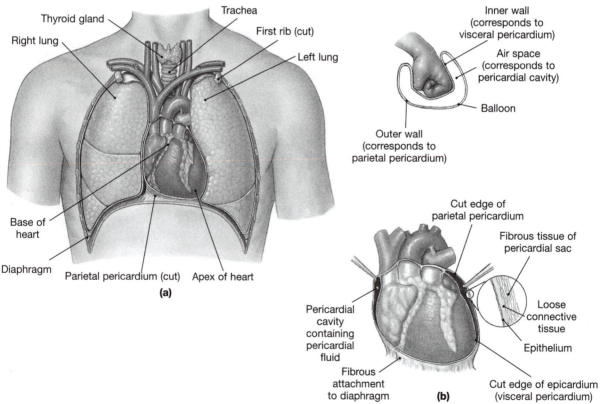

●**Figure 13.1**
Location of the Heart in the Thoracic Cavity
The heart is situated within the anterior portion of the mediastinum, immediately posterior to the sternum. **(a)** Anterior view of the open chest cavity, showing the position of the heart and major vessels relative to the lungs. **(b)** Relationship between the heart and the pericardial cavity: compare with the fist and balloon example.

The heart wall is organized into three layers, the epicardium, myocardium, and endocardium (see Figure 13.2). The **epicardium** is the visceral pericardium. The **myocardium** is made up of cardiac muscle tissue and constitutes most of the heart wall. Cardiac muscle tissue is **striated** and consists of cells called **cardiocytes.** Each cardiocyte is **uninucleated**, containing a single nucleus, and is branched. Cardiocytes interconnect at these branches across **intercalated discs.** Deep to the myocardium is the **endocardium,** a thin layer of endothelial tissue lining the chambers of the heart.

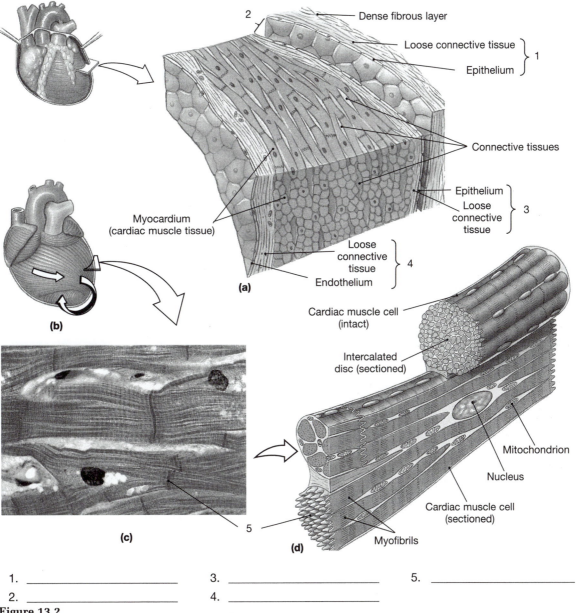

1. _____   3. _____   5. _____

2. _____   4. _____

●**Figure 13.2**
The Heart Wall

(a) A diagrammatic section through the heart wall showing the relative positions of the epicardium, myocardium, and endocardium. **(b)** Cardiac muscle tissue in the heart forms concentric layers that wrap around the atria and spiral within the walls of the ventricles. **(c, d)** Sectional and diagrammatic views of cardiac muscle tissue. Cardiac muscle cells are smaller than skeletal muscle fibers and also have a single, centrally placed nucleus, branching interconnections between cells, and intercalated discs. (LM × 575)

Laboratory Activity   LOCATION AND WALL OF THE HEART _____

MATERIALS

heart models and specimens
compound microscope
prepared slide of cardiac muscle

PRODEDURES

1. Label and review the anatomy presented in Figure 13.1.
2. Identify the layers of the heart wall on laboratory models and specimens.
3. Examine the microscopic structure of cardiac muscle.
   a. Obtain a microscope and a slide of cardiac muscle tissue. Focus the microscope on the heart tissue, starting at low magnification. Refer to Figure 13.2 for the microscopic appearance of cardiac muscle tissue.
   b. Increase the magnification, and locate several cardiocytes. Note the single nucleus in each cell. Intercalated discs are dark-stained lines where cardiocytes connect. Do you see any branched cardiocytes?
   c. Sketch several cardiocytes and intercalated discs in the space provided. Use low and high magnifications for your sketches.

B. Chambers of the Heart

All mammalian hearts have four chambers and are anatomically divided into right and left sides with each side having an upper and a lower chamber. The upper chambers are the **right atrium** and the **left atrium**; the lower chambers are the **right ventricle** and the **left ventricle**. Lining the inside of the right atrium are muscular ridges, the **pectinate muscle**. The wall between the atria is called the **interatrial septum**. The **ventricles** are separated by the **interventricular septum**. Folds of muscle tissue, called **trabeculae carnae**, occur on the inner surface of each ventricle. Locate these structures in Figure 13.3.

The atria are "receiving chambers" and fill with blood that is returning to the heart in veins. Blood in the atria flows into the ventricles, which are "pumping chambers." The ventricles fill with blood, then squeeze their walls together to pressurize the blood and squirt it into two large arteries for distribution to the lungs and body tissues.

Figure 13.4 highlights the external anatomy of the heart. Each atrium is covered externally by a flap, the **auricle**. Fat tissue and blood vessels occur along grooves in the heart wall. The **coronary sulcus** is a deep groove between the right atrium and ventricle, and extends to the posterior surface. The boundary of the right and left ventricles is marked anteriorly and posteriorly by the **anterior interventricular sulcus** and the **posterior interventricular sulcus**. Coronary blood vessels are also shown in Figure 13.4 and will be discussed in another section of this exercise.

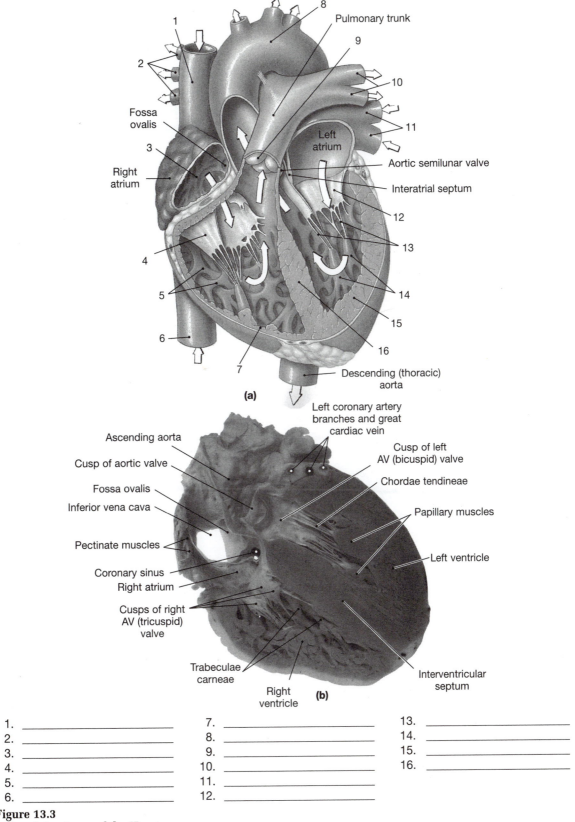

Pulmonary trunk

Fossa ovalis

Right atrium

Left atrium

Aortic semilunar valve

Interatrial septum

Descending (thoracic) aorta

(a)

Left coronary artery branches and great cardiac vein

Ascending aorta

Cusp of aortic valve

Fossa ovalis

Inferior vena cava

Pectinate muscles

Coronary sinus

Right atrium

Cusps of right AV (tricuspid) valve

Trabeculae carneae

Right ventricle

Cusp of left AV (bicuspid) valve

Chordae tendineae

Papillary muscles

Left ventricle

Interventricular septum

(b)

1. _____   7. _____   13. _____
2. _____   8. _____   14. _____
3. _____   9. _____   15. _____
4. _____   10. _____   16. _____
5. _____   11. _____
6. _____   12. _____

●**Figure 13.3**
Sectional Anatomy of the Heart

(a) Diagrammatic frontal view of the internal heart. **(b)** Photograph of a human heart in frontal section.

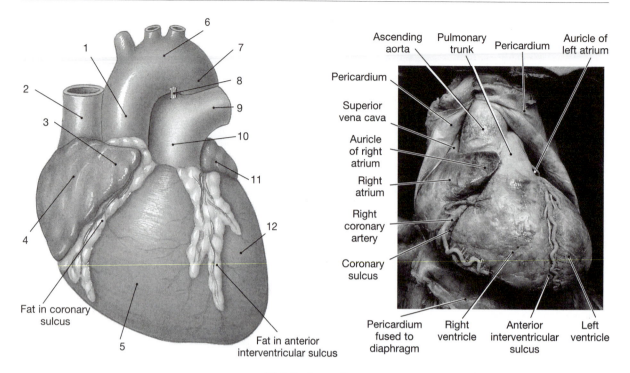

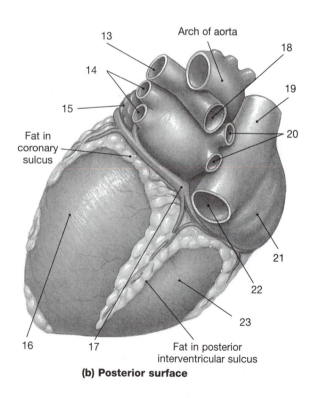

(a) **Anterior surface**

(b) **Posterior surface**

1. _____
2. _____
3. _____
4. _____
5. _____
6. _____
7. _____
8. _____
9. _____
10. _____
11. _____
12. _____
13. _____
14. _____
15. _____
16. _____
17. _____
18. _____
19. _____
20. _____
21. _____
22. _____
23. _____

●**Figure 13.4**
Surface Anatomy of the Heart

(a) Anterior view of the heart, showing major anatomical features.
(b) The posterior surface of the heart.

Laboratory Activity   CHAMBERS OF THE HEART _____

MATERIALS

heart models and specimens

PROCEDURES

1. Label and review the anatomy presented in Figures 13.3 and 13.4.
2. Locate each structure on lab materials.

C. Vessels of the Heart

As a double pump, the heart delivers blood into two circulatory pathways. Each pathway begins in a ventricle that delivers blood into a series of arteries, capillaries, and veins, then concludes with the return of blood to the atrium opposite the initial ventricle. Trace a drop of blood through the heart in Figure 13.5 as each pathway is discussed. The **pulmonary circuit** directs deoxygenated blood to the lungs for conversion into oxygenated blood. The right atrium receives deoxygenated blood from the **superior vena cava** and the **inferior vena cava**. Identify the venae cavae in Figure 13.3. From the right atrium, blood flows into the right ventricle, which pumps it into the **pulmonary trunk**. This large artery branches into the **right** and **left pulmonary arteries** that supply deoxygenated blood to the lungs. In the lungs, the deoxygenated blood is converted into oxygenated blood as gases diffuse between the blood and the lungs. Four **pulmonary veins** drain the lungs and return the oxygenated blood to the left atrium.

●**Figure 13.5**
Overview of Pulmonary and Systemic Circuits

Each circuit starts and ends at the heart and contains a series of arteries, capillaries, and veins.

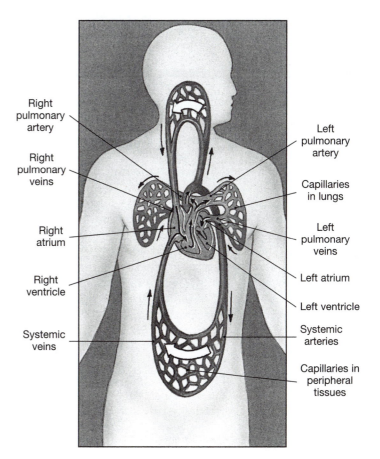

Blood from the left atrium enters the left ventricle for the **systemic circuit**. Blood flows from the left atrium into the left ventricle, which then pumps the oxygenated blood into the **aorta**. The aorta branches into the major systemic arteries, which transport oxygenated blood to all body tissues. In the body tissues, blood is converted into deoxygenated blood and then returns to the right atrium in the vena cavae to repeat the double circuit.

To produce the pressure required to push blood through the vascular system, the heart can never completely rest. **Coronary circulation** supplies the myocardium with oxygenated blood. The right and left **coronary arteries** branch from the base of the aorta and penetrate the myocardium to the outer heart wall. Each coronary artery branches many times to provide an artery for each region of the heart muscle. Locate the coronary arteries in Figure 13.4.

Cardiac veins collect deoxygenated blood from the myocardium. The cardiac veins merge to form the **coronary sinus** situated in the posterior region of the coronary sulcus. The coronary sinus is a large vein that empties into the right atrium. The right atrium also receives deoxygenated blood from the venae cavae. Locate the cardiac veins in Figure 13.4.

Laboratory Activity   VESSELS OF THE HEART

MATERIALS

heart models and specimens

PROCEDURES

1. Trace a drop of blood through the pulmonary and systemic circuits in Figure 13.5.
2. Locate the vessels of the heart on all lab material.
3. Identify the coronary blood vessels on lab materials.
4. Label and review the coronary blood vessels in Figure 13.4.

D. Valves of the Heart

To control and direct blood flow, the heart has two pairs of atrioventricular and semilunar valves. The **atrioventricular (AV) valves** are located between the atria and ventricles on each side of the heart. These valves prevent blood from reentering the atria when the ventricles contract. The right atrioventricular valve has three flaps or cusps and is also called the **tricuspid valve**. The left atrioventricular valve has two cusps and is called the **bicuspid valve** or the **mitral valve**. Observe in Figure 13.3 that the cusps of each AV valve have small cords, the **chordae tendineae**, which are attached to **papillary muscles** at the floor of the ventricles. When the ventricles contract, the AV valves are held closed by the papillary muscles pulling on the chordae tendineae.

The other pair of valves are the **pulmonary** and **aortic semilunar valves** located between the ventricles and the base of the pulmonary trunk and the aorta. Semilunar valves comprise three small cusps and prevent back flow of blood into the ventricles when the ventricles are relaxed. Identify each pair of valves in Figure 13.3.

Figure 13.6a illustrates the function of the AV and the semilunar valves. When a ventricle contracts, a phase called **ventricular systole**, pressure forces the atrioventricular valve closed. To keep this valve from reversing like an umbrella in a strong wind, the papillary muscles contract and pull on the chordae tendineae to secure the cusps. Increased pressure now forces the semilunar valve open, and blood flows into the artery. Figure 13.6b represents valve function during **ventricular diastole**, the relaxation phase of the ventricle. Blood is flowing from the atrium into the relaxed ventricle through the open atrioventricular valve. Note in the figure that the papillary muscles and the chordae tendineae are relaxed. The semilunar valve is now closed to prevent back flow of blood from the artery into the ventricle.

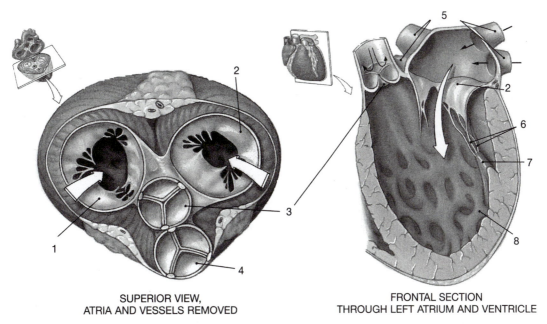

SUPERIOR VIEW,
ATRIA AND VESSELS REMOVED

FRONTAL SECTION
THROUGH LEFT ATRIUM AND VENTRICLE

(a) Ventricular diastole

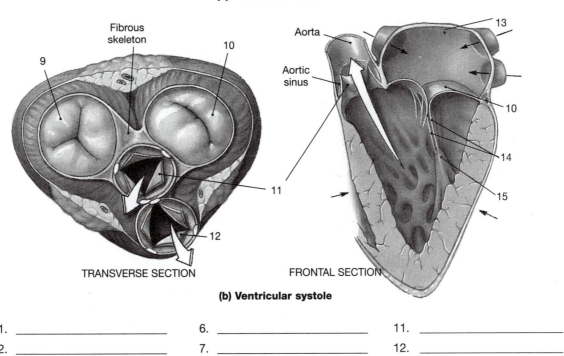

Fibrous
skeleton

Aorta

Aortic
sinus

TRANSVERSE SECTION

FRONTAL SECTION

(b) Ventricular systole

1. _____   6. _____   11. _____
2. _____   7. _____   12. _____
3. _____   8. _____   13. _____
4. _____   9. _____   14. _____
5. _____   10. _____   15. _____

●**Figure 13.6**
Valves of the Heart

(a) Valve position during ventricular diastole (relaxation), when the AV valves are open and the semilunar valves are closed. Note that the chordae tendineae are slack and the papillary muscles are relaxed. **(b)** The appearance of the cardiac valves during ventricular systole (contraction), when the AV valves are closed and the semilunar valves are open. In the frontal section, note the attachment of the bicuspid valve to the chordae tendineae and papillary muscles.

Laboratory Activity   HEART VALVES _____

MATERIALS

heart models

PROCEDURES

1. Label and review valve anatomy in Figures 13.3 and 13.6.
2. Identify each valve on all laboratory material.

E. Conduction System

Cardiac muscle tissue is unique in that it is **autorhythmic**, producing its own contraction and relaxation phases without stimulation from nerves. Nerves may increase or
decrease the heart rate, yet a living heart removed from the body will continue to contract on its own.

Figure 13.7 details the **conduction system** of the heart. Special cells called **nodal
cells** produce and conduct electrical currents to the myocardium, which coordinate the
heart's contraction. The pacemaker of the heart is the **sinoatrial (SA) node**, located
where the superior vena cava empties into the upper right atrium. Nodal cells in the
SA node self-excite faster than other areas of the heart and thus set the pace for the
heart's contraction. The **atrioventricular (AV) node** is located in the lower medial floor
of the right atrium. The SA node stimulates the AV node, which then spreads the
impulse toward the ventricles through the **atrioventricular bundle** (bundle of His). The
atrioventricular bundle passes into the interventricular septum and branches into right
and left **bundle branches**. The bundle branches divide into fine **Purkinje fibers**, which
distribute the electrical impulses to the cardiocytes.

●**Figure 13.7**
The Conduction System of the Heart

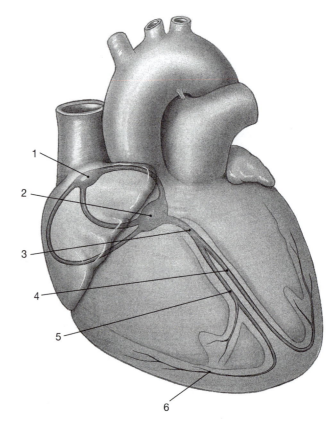

1. _____

2. _____

3. _____

4. _____

5. _____

6. _____

Laboratory Activity   THE CONDUCTION SYSTEM _____

MATERIALS

heart models and charts

PROCEDURES

1. Identify all structures of the conduction system on lab materials.
2. Label and review the conduction system in Figure 13.7.

F. Fetal Heart Structures

A fetus receives oxygen from the mother through the placenta, a vascular organ grown by the fetus into the wall of the mother's uterus. During development, the fetal lungs are filled with amniotic fluid, and for efficiency, some blood is shunted away from the fetal pulmonary circuit by several structures, as shown in Figure 13.8. The **foramen ovale** is a hole in the interatrial wall. Much of the blood entering the right atrium from the inferior vena cava passes through the foramen ovale to the left atrium and avoids the right ventricle and the pulmonary circuit. Some blood that does enter the pulmonary trunk may bypass the lungs through a connection with the aorta, the **ductus arteriosus**. At birth, the foramen ovale closes and becomes a depression on the interatrial wall, the **fossa ovalis**. The ductus arteriosus closes and becomes the **ligamentum arteriosum**.

Laboratory Activity   FETAL HEART STRUCTURES _____

MATERIALS

heart models and charts

PROCEDURES

1. Label and review fetal structures in Figure 13.8.
2. Identify the fossa ovalis and ligamentum arteriosum on lab models.

PART TWO   SHEEP HEART DISSECTION

MATERIALS

gloves	safety glasses
fresh or preserved sheep heart	dissecting pan
dissecting scissors	blunt probe

OBJECTIVES

Upon completion of this part of the exercise, you should be able to:

- Identify the gross external and internal anatomy of a sheep heart.
- Identify the major blood vessels of the sheep heart.

PROCEDURES

1. Put on gloves and safety glasses. Review Exercise 1, "Laboratory Safety."

●**Figure 13.8**
Fetal Circulation

(a) Blood flow to and from
the placenta. **(b)** Blood flow
through the fetal heart.

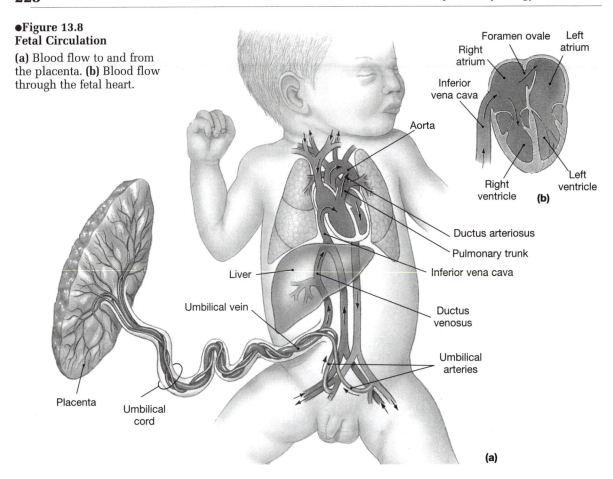

2. Wash sheep heart with cold water to flush out preservatives and blood clots.
   Minimize your skin and mucous membrane exposure to the preservatives.
3. Carefully follow the instructions. Cut into the heart only as instructed.

I. External Anatomy

Figure 13.9 details the external anatomy of a sheep heart. Examine the surface of the
heart for the **pericardium**. Often this serous membrane has been removed in preserved
specimens. Using your scalpel, carefully scrape the outer heart muscle to loosen the
epicardium. Next, locate the anterior surface by orientating the heart so that the **auricles**
face you. Under the auricles are the **right** and **left atria**. Note the **base** of the heart above
the atria where the large blood vessels occur. The **apex** is the inferior tip of the heart.
Squeeze gently above the apex to locate the **right** and **left ventricles**. Locate the **anterior
interventricular sulcus**, the fat-laden groove between the ventricles. Carefully remove
some fat tissue to locate coronary blood vessels. Identify the other grooves, the **coronary
sulcus** between the right atrium and ventricle and the **posterior interventricular sulcus**
between the ventricles on the posterior surface.

Identify the **pulmonary trunk** and the **aorta**. The pulmonary trunk is anterior to
the aorta. If the pulmonary trunk was cut long, you can identify the right and left **pulmonary arteries** branching off the trunk. Posterior to the pulmonary trunk is the aorta.
The **brachiocephalic artery** is the first major branch of the aorta and is often intact on
preserved material.

Follow along the inferior margin of the right auricle to the posterior surface. The
prominent vessel at the end of the auricle is the **superior vena cava**. At the base of this
vessel is the **inferior vena cava**. Next, examine the posterior aspect of the left atrium and

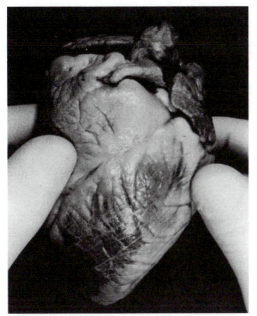
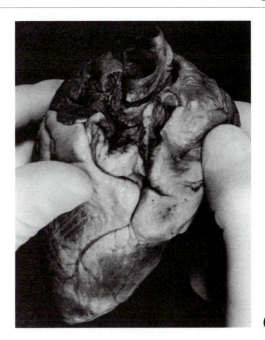

(a)   (b)

●Figure 13.9
External Anatomy of the Sheep Heart
(a) Anterior view (b) posterior view.

find the four **pulmonary veins**. You may need to carefully remove some fatty tissue around the superior region of the left atrium to locate the veins.

II. Internal Anatomy

Follow the instructions carefully to improve your dissection technique and identification of structures. Use Figure 13.10 as a reference to the internal anatomy of the sheep heart.

1. Place a blade of your dissection scissors into the **superior vena cava** and cut down into the upper portion of the **right atrium**. Cut approximately 1/2 inch or until you have exposed the **tricuspid valve**. Do not cut through the tricuspid valve.

2. Examine the right atrium for the comblike **pectinate muscles** lining the inner wall. Identify the opening of the **coronary sinus**, located between the superior and inferior venae cavae.

3. To observe the function of the **tricuspid valve**, slowly fill the right atrium with water. Watch the water drain into the lower right ventricle. Gently squeeze the outer wall of the right ventricle, and observe the cusps of the tricuspid valve closing. Drain the water from the heart to continue the dissection.

4. Use your dissection scissors and continue cutting the right atrium. Cut through the tricuspid valve to the inferior part of the right ventricle. Open the ventricle with your fingers, and identify the **chordae tendineae** and **papillary muscles** of the tricuspid valve. Note the folds of **trabeculae carnae** along the inner ventricular walls. Squeeze the outer wall of the left ventricle to help locate the **interventricular septum**.

5. In the right ventricle, carefully insert a blunt probe into the opening of the **pulmonary trunk**. Slide the probe into the vessel until it exits the pulmonary trunk outside the heart. Remove the probe.

6. To expose the **pulmonary semilunar valve**, use scissors to cut through the right ventricular wall to the pulmonary trunk. Completely open the pulmonary trunk by cutting the entire length of the vessel. Spread the trunk open, and observe the small cusps of the pulmonary semilunar valve at the base of the vessel.

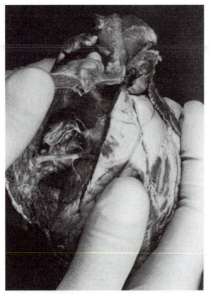
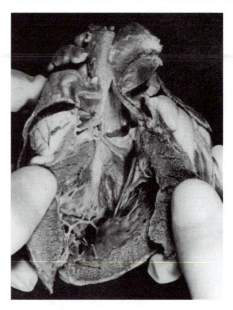

(a)   **(b)**

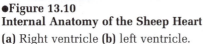

●**Figure 13.10**
Internal Anatomy of the Sheep Heart
(a) Right ventricle **(b)** left ventricle.

7. Next, dissect the left side of the heart. Place the pointed blade of your dissecting scissors into a **pulmonary vein**, and cut into the left atrium to expose the **bicuspid valve**. Fill the left atrium with water, and gently squeeze the outer wall of the left ventricle to observe the cusps of the **bicuspid valve**.

8. Continue cutting through the left atrium past the bicuspid valve and into the **left ventricle**. Review the anatomy of the left atrioventricular (bicuspid) valve.

9. Slip a blunt probe into the **aorta**. Remove the probe, and cut the left ventricular wall toward the aorta, then continue with a longitudinal cut along the length of the aorta. Spread the vessel open, and locate the **aortic semilunar valve** at the aortic base. Examine the base, and identify the right and left **coronary arteries**.

10. On completion of the dissection, dispose of the sheep heart as directed by the lab instructor; wash your hands and dissecting instruments.

PART THREE   CARDIOVASCULAR PHYSIOLOGY

WORD POWER   MATERIALS

systole (systol—contraction)   sphygmomanometer
diastole (diastol—standing apart)   stethoscope
sphygmomanometer (sphygm—the pulse)   alcohol swabs
pulse (puls—a beat)
auscultation (auscult—to listen)

OBJECTIVES

On completion of this part of the exercise, you should be able to:

- Describe the pressure changes that occur during a cardiac cycle.
- Describe and demonstrate the steps involved in blood pressure determination.

- Explain the differences in blood pressure due to body position.
- Take a pulse rate from several locations.

One complete heart beat is called a **cardiac cycle**. During a cardiac cycle, the atria contract and relax, and the ventricles contract and relax. The contraction phase of a chamber is termed **systole**, and the relaxation phase is termed **diastole**. The human heart averages 75 cardiac cycles or heart beats per minute, with each cardiac cycle occurring in 0.8 second. Figure 13.11 illustrates the events of a cardiac cycle. The cycle begins with the brief systole of the atria, lasting 0.1 second (100 milliseconds), to fill the resting ventricles. Next, the ventricles enter their systolic phase and contract for 0.4 second to pump blood out of the heart. For the remaining 0.4 second of the cardiac cycle, all chambers are in diastole and fill with blood in preparation for the next heart beat. Most blood enters the ventricles during this resting period. Atrial systole contributes only 30% of the blood volume in the ventricle before ventricular systole. The remaining 70% of blood gravity and pressure feeds into the ventricles from the atria.

Each cardiac cycle is marked with an increase and a decrease in blood pressure, both within the heart and in the arteries. When a chamber contracts, pressure increases owing to the squeezing together of the chamber walls. The increase in blood pressure forces blood to circulate into the next chamber or out of the ventricle. As a ventricle relaxes, its walls move apart and pressure decreases. This drop in pressure draws blood into the ventricle and refills it for the next systole.

When all chambers of the heart are in diastole, the atrioventricular valves (AV valves) are open, and blood flows from the atria into the ventricles. The semilunar

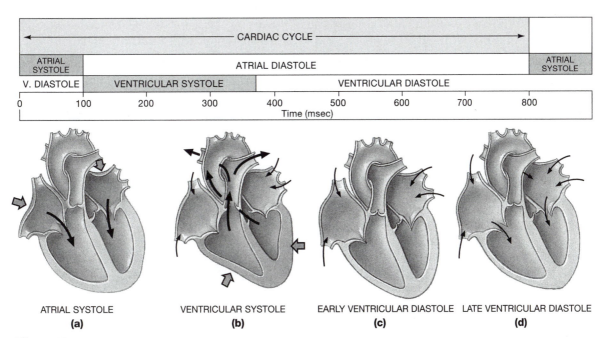

●Figure 13.11
The Cardiac Cycle

The atria and ventricles go through repeated cycles of systole and diastole. The timing of systole and diastole differs between the atria and ventricles. A cardiac cycle consists of one period of systole and diastole; we will consider a cardiac cycle as determined by the state of the atria. **(a)** Atrial systole. During this period the atria contract and the ventricles become filled with blood. **(b)** Ventricular systole. Blood is ejected into the pulmonary and aortic trunks. **(c)** Ventricular diastole. Passive filling of the ventricles occurs for the duration of this cycle and through the period of atrial systole in the next cardiac cycle. (d) Condition of the heart at the end of a cardiac cycle, with both the atria and ventricles in diastole.

valves (SL valves) are closed to prevent back flow of blood from the aorta and pulmonary artery into the left and right ventricles, respectively. When the left ventricle contracts, pressure increases to a point where it exceeds the pressure of the blood in the aorta holding the aortic SL valve shut. The ventricle forces blood through the valve, and arterial blood pressure increases owing to the increase in blood volume. When the ventricle relaxes, aortic and arterial pressures drop, and the aortic SL valve closes. Similar events are occurring on the right side of the heart with the pulmonary SL valve.

In this exercise, you will investigate the physiology of blood pressure and the effect of posture on blood pressure. You will also listen to heart sounds and practice taking a subject's pulse.

A. Heart Sounds

Listening to internal sounds of the body is called **auscultation**. A stethoscope is used to amplify the sounds to an audible level. The heart, lungs, and digestive tract are the most frequently auscultated systems. Auscultation provides the listener valuable information concerning fluid accumulation in the lungs or blockages in the digestive tract. Auscultation of the heart is used as a diagnostic tool to evaluate valve function and heart sounds.

During a single cardiac cycle two distinct heart sounds are produced, a "lubb" and a "dupp" sound. The first sound, the "lubb," is caused by the closure of the atrioventricular valves during the onset of ventricular systole. The second sound, the "dupp," occurs as the semilunar valves close at the start of ventricular diastole. Relate each heart sound with the cardiac cycle, as in Figure 13.12b.

Figure 13.12a illustrates the landmarks for proper placement of the stethoscope to effectively auscultate each heart valve. Notice in the figure that the sites for auscultation do not overlie the anatomical location of the heart valves. Interference of sound conduction by soft tissues and bone deflects the sound waves to locations lateral to the valves.

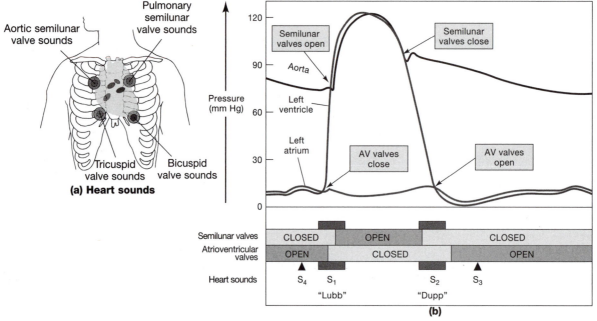

●**Figure 13.12**
Heart Sounds

(a) Placement of a stethoscope for listening to the sounds produced by individual valves. **(b)** Timing of sound production during the cardiac cycle.

Clinical Application

An unusual heart sound is called a **murmur**. Not all murmurs indicate an anatomical or functional anomaly of the heart. The sound may originate from turbulent flow in a heart chamber. Some murmurs, however, are diagnostic for certain heart defects. **Septal defects** are holes in a wall between two chambers. A murmur is heard as blood passes through the hole and into the chamber on the opposite side of the heart. Abnormal operation of a valve also produces a murmur. In **mitral valve prolapse**, the left atrioventricular valve does not seal completely during ventricular systole, and blood **regurgitates** upward into the left atrium.

Laboratory Activity   Auscultation of the Heart _____

MATERIALS

> stethoscope
> alcohol wipes
> laboratory partner

PROCEDURES

1. Auscultation of the heart is best achieved with the shirt open. Your lab partner may feel more comfortable holding the bell of the stethoscope, slipping it inside his or her shirt, and locating each valve site.

2. Clean the ear pieces of the stethoscope with a sterile alcohol wipe. Dispose of the used wipe in the trash can.

3. Wear the stethoscope by placing the angled ear pieces facing forward into the external ear canal.

4. Using Figure 13.12a as a guide, auscultate the AV valves and the SL valves. Can you discriminate between the "lubb" and the "dupp" sounds?

_____

5. What causes the first heart sound? The second heart sound?

_____

B. Determination of Blood Pressure

Blood pressure is a measure of the force exerted by blood on the walls of systemic arteries. Blood pressure in arteries is not constant: pressure increases when the left ventricle contracts and pumps blood into the aorta. When the left ventricle relaxes, less blood flows into the aorta and pressure decreases until the next contraction. Two pressures are therefore used when expressing blood pressure, a systolic pressure and a lower diastolic pressure. Average blood pressure is considered to be 120/80 mm Hg. Do not be surprised when you take your blood pressure in the following exercise and discover it is not "average." Cardiovascular physiology is a dynamic mechanism, and pressures regularly change to adjust to the demands of the body.

Blood pressure is measured using an inflatable cuff called a **sphygmomanometer** or blood pressure cuff. Figure 13.13 demonstrates the proper placement of the sphygmomanometer. The cuff is wrapped around the arm just superior to the elbow and then inflated to approximately 160 mm Hg to compress the brachial artery and block blood flow in the artery. A stethoscope is placed on the antecubital region, and pressure is gradually vented from the cuff. Once pressure in the cuff equals pressure in the brachial artery, blood will spurt through the artery and make a ticking sound, called **Korotkoff's sounds**, which are audible through the stethoscope. Noting the pressure gauge when the first tick is heard records the **systolic pressure**. As more pressure is

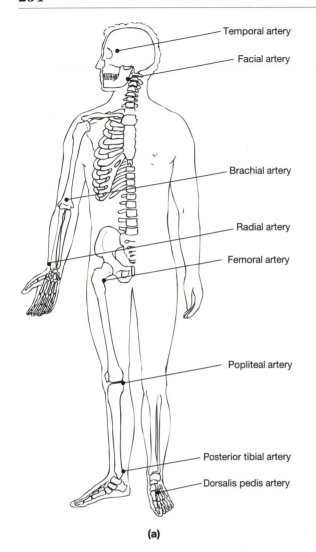

(a)

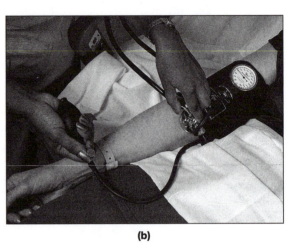

(b)

●**Figure 13.13**
Checking the Pulse and Blood Pressure

(a) Pressure points used to check the presence and strength of the pulse. **(b)** Use of a sphygomanometer to check arterial blood pressure.

relieved from the cuff, blood flow becomes less turbulent and quieter. The last sound occurs when the cuff pressure matches the **diastolic pressure** of the artery.

Laboratory Activity   DETERMINATION OF BLOOD PRESSURE _____

Safety Tip

Measuring blood pressure involves temporarily stopping the flow of blood in an artery by inflating a blood pressure cuff over the artery. Read and understand all the procedures before attempting to measure blood pressure on your laboratory partner. *Never* leave a cuff inflated on an arm for more than one minute. Deflate the cuff if you are unsure of a procedure.

MATERIALS

sphygmomanometer
stethoscope
alcohol wipes
laboratory partner

I. Resting Blood Pressure

PROCEDURES

1. Sit your lab partner comfortably in a chair and ask him or her to relax for several minutes. Roll up the sleeve of your partner's right arm to expose the upper brachium. Clean the stethoscope with a sterile alcohol pad, and then dispose of the used pad in the trash.

2. Force all air out of the sphygmomanometer by compressing the cuff against the laboratory table. Loosely wrap the deflated cuff around the right arm of your partner so that the lower edge of the cuff is just superior to the antecubital region of the elbow, as in Figure 13.13.

3. If the cuff has orientation arrows, line the arrows up with the antecubitis; otherwise, position the rubber tubing over the antecubital region. Tighten the cuff so that it is snug against the arm.

4. Gently close the valve on the cuff, and then inflate the cuff to approximately 160 mm Hg. *Do not* leave the cuff inflated for more than one minute. The inflated cuff is preventing blood flow to the forearm, and the disruption in blood flow could lead to fainting.

5. Wearing the stethoscope, place the bell below the cuff with the diaphragm over the brachial artery at the antecubitis.

6. Carefully open the valve to the sphygmomanometer, and slowly deflate the cuff while listening for the Korotkoff's sounds. When the first sound is heard, note the pressure reading on the gauge.

7. Continue to vent pressure from the cuff while listening with the stethoscope. When you hear the last faint sound, note the pressure on the gauge.

8. Open the pressure valve completely, and quickly finish deflating the cuff and remove it from your partner.

9. Record the observed blood pressures in the space provided. Repeating these procedures and then averaging the readings may yield a more accurate blood pressure determination.

Systolic Pressure_____   Diastolic Pressure_____

II. Effect of Posture on Blood Pressure

PROCEDURES

Changing posture affects blood pressure. This is readily apparent when standing on your head. In this section, your partner will lie in the supine position (on the back) for approximately five minutes to allow for cardiovascular adjustments. Blood pressure will be determined and compared to the pressure measured in Procedure I.

1. Ask your partner to lie on the lab table and relax for five minutes. Have your partner slouch down in a chair and tilt his or her head back if lying on the lab table isn't possible.

2. Wrap the sphygmomanometer around your partner's right arm, and measure his or her supine blood pressure.

3. What was the effect of the supine posture on the blood pressure?

_____

_____

III. Effect of Exercise on Blood Pressure

PROCEDURES

In this investigation, you will examine the effect of mild exercise on blood pressure. Be sure that you have taken the resting blood pressure of your subject as outlined in Procedure I before this section. The blood pressure observed in Procedure I will be used as a baseline to compare with the exercise blood pressure reading.

1. Secure the cuff around your partner's arm loose enough so that it is comfortable for exercise yet is in position for taking pressure readings.

2. Have your partner jog in place for five minutes without stopping if possible. This is not a stress test, so if the subject becomes excessively winded or tired, he or she should stop immediately.

3. On completion of the jogging, take a blood pressure reading. Repeat the readings once every two minutes until the pressure returns to the resting values.

4. Construct a graph showing the effects of exercise on blood pressure using the pressure data you have collected.

C. Determination of Pulse

Heart rate is usually determined by measuring the **pulse** or pressure wave in an artery. During ventricular systole, blood pressure increases and stretches the walls of arteries. When the ventricle is in diastole, blood pressure decreases and the arterial walls rebound to their relaxed diameter. This change in vessel diameter is felt as a throb at a pressure point. The most commonly used pressure point is the radial artery on the lateral forearm just superior to the thumb. Other points include the common carotid artery in the neck and the popliteal artery of the posterior knee (see Figure 13.13). Counting the number of pulses shows the number of cardiac cycles for that interval of time. The **pulse pressure** is the difference between the systolic and diastolic blood pressures, and accounts for the pressure wave.

Laboratory Activity   DETERMINATION OF PULSE _____

MATERIALS

watch or clock with a second hand
laboratory partner

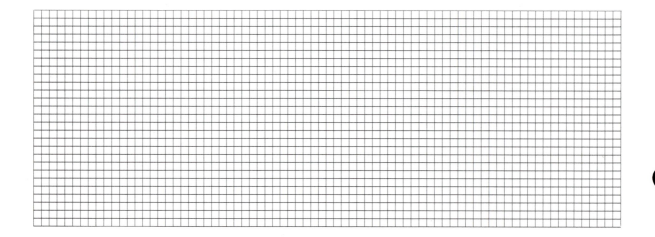

PROCEDURES

1. Have your partner relax in a chair for several minutes.
2. Locate your partner's pulse in the right radial artery located just above the thumb. Use your index finger or index and middle fingers to palpitate and identify the pulse.
3. Apply light pressure to the pulse point, and count the pulse rate for 15 seconds; then multiply the pulse by 4 to obtain the rate per minute. Record your data in Table 13.1.
4. Repeat the pulse determination at the carotid artery and the popliteal artery. Record your data in Table 13.1.
5. Are there differences in the pulse at these various pulse points?

TABLE 13.1	Pulse Measurement
	Pulse Rate
Radial pulse	
Carotid pulse	
Popliteal pulse	

ANATOMY AND PHYSIOLOGY OF THE HEART
Exercise 13 Laboratory Report

A. Matching

Match each heart structure listed on the left with the correct description on the right.

1. _____ tricuspid valve	A. empties into left atrium
2. _____ superior vena cava	B. left atrioventricular valve
3. _____ right ventricle	C. muscle folds of ventricles
4. _____ aorta	D. pumps blood to body tissues
5. _____ interventricular septum	E. opening in interatrial septum
6. _____ left ventricle	F. major systemic artery
7. _____ pulmonary veins	G. muscular ridges of right atrium
8. _____ semilunar valve	H. artery with deoxygenated blood
9. _____ bicuspid valve	I. groove on right side of heart
10. _____ pulmonary trunk	J. visceral pericardium
11. _____ foramen ovale	K. drains coronary veins into heart
12. _____ trabeculae carnae	L. wall between the ventricles
13. _____ pectinate muscle	M. inferior tip of heart
14. _____ coronary sulcus	N. cardiac muscle tissue
15. _____ auricle	O. aortic or pulmonary valve
16. _____ coronary sinus	P. empties into right atrium
17. _____ myocardium	Q. attached to AV valves
18. _____ epicardium	R. right atrioventricular valve
19. _____ chordae tendineae	S. pumps blood to lungs
20. _____ apex	T. external flap of atria

B. Short Answer Questions

1. What type of blood, oxygenated or deoxygenated, does the right ventricle pump?

2. What type of blood does the pulmonary trunk contain? Why is it an artery rather than a vein?

3. Why is the wall of the left ventricle thicker than the wall of the right ventricle?

4. What keeps the atrioventricular valves from reversing?

5. List the layers of the heart wall.

6. List in order the spread of an impulse through the conduction system.

7. Explain the events of the cardiac cycle.

8. Explain how blood pressure is measured.

C. Completion and Labeling

1. Complete Figure 13.14 by labeling the anatomy of the heart.

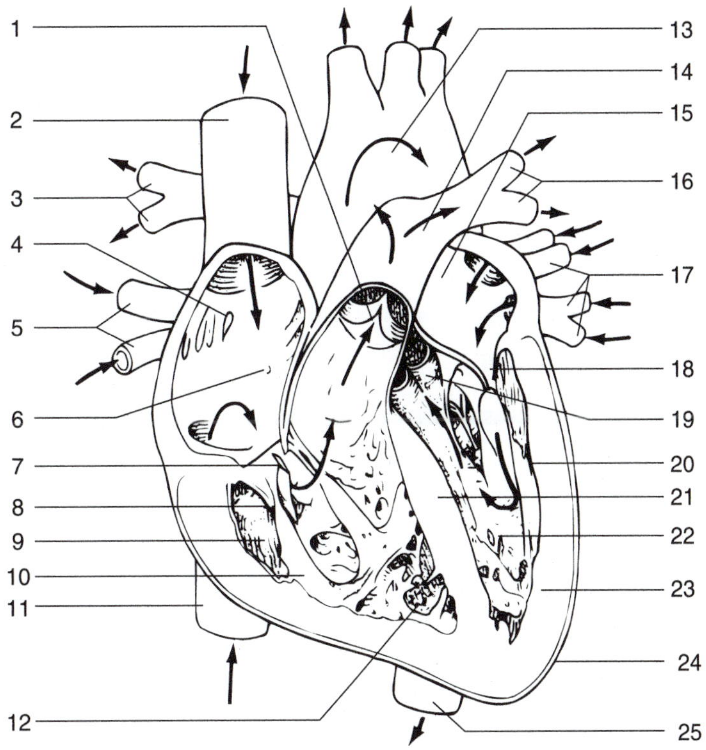

1. _____	10. _____	18. _____
2. _____	11. _____	19. _____
3. _____	12. _____	20. _____
4. _____	13. _____	21. _____
5. _____	14. _____	22. _____
6. _____	15. _____	23. _____
7. _____	16. _____	24. _____
8. _____	17. _____	25. _____
9. _____		

●Figure 13.14
Anatomy of the Heart

14 ANATOMY OF SYSTEMIC CIRCULATION

WORD POWER	MATERIALS
artery (arteri—an artery)	cardiovascular system chart
capillary (capill—a hair)	torso model
vein (ven—a vein)	arm model
tunica (tunic—a covering)	leg model
aorta (aort—the great artery)	abdominal model
anastomosis (anastomos—coming together)	head model
hepatic (hepa—liver)	vascular charts
portal (porta—a gate or door)	compound microscope
	prepared slides:
	artery and vein

OBJECTIVES

On completion of this exercise, you should be able to:

- Compare the histology of an artery, capillary, and vein.
- Compare the circulation of blood to the right and left arms.
- Describe the anatomy and the importance of the circle of Willis.
- Trace a drop of blood from the ascending aorta into each abdominal organ and to the lower limb.
- Trace a drop of blood returning to the heart from the calf.

The body has more than 60,000 miles of blood vessels to transport blood to the trillions of cells in the tissues. The basic circulatory route includes **arteries**, which distribute oxygen and nutrient-rich blood to microscopic networks of vessels called **capillaries**. At the capillaries, diffusion of nutrients, gases, wastes, and cellular products occurs between the blood and the cells. **Veins** drain deoxygenated blood from the capillaries and direct the blood toward the heart, which then pumps it to the lungs to pick up oxygen and release carbon dioxide.

The vascular system is similar to a figure eight. For a drop of blood to circulate through the body, it must be pumped twice. The pulmonary circuit conducts deoxygenated blood to the lungs and returns oxygenated blood to the left side of the heart. The left ventricle pumps the oxygenated blood into the systemic circuit that supplies the organ systems. In Exercise 13, we presented an overview of circulatory pathways and the vessels of the pulmonary circuit. In this exercise, you will study the major arteries and veins of the systemic circuit.

A. Comparison of Arteries and Veins

The walls of arteries and veins have three layers: an outer tunica externa, a middle tunica media, and an inner tunica interna. Locate these layers in Figure 14.1. The **tunica externa** is a connective tissue covering that anchors the vessel to surrounding

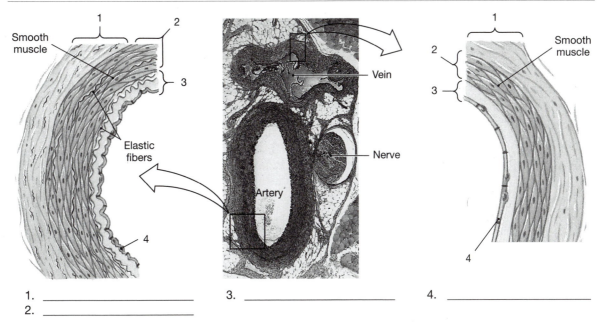

1. _____   3. _____   4. _____
2. _____

●**Figure 14.1**
A Comparison of a Typical Artery and Typical Vein
(LM × 74)

tissues. Collagen and elastic fibers give this layer strength and flexibility. The **tunica media** is a middle layer of smooth muscle tissue. Contraction of the smooth muscle in small arteries called **arterioles** causes the arteriole to narrow or **vasoconstrict**, and results in a decrease in blood flow. Relaxation opens the vessel and promotes blood flow, a process called **vasodilation**. Arteries also have elastic fibers that allow stretching and recoiling of the vessel in response to blood pressure changes; veins have collagen fibers for strength. Lining the vessels is the **tunica interna**, a thin layer of **endothelium**.

Because blood pressure is much higher and fluctuates more in arteries than in veins, the walls of arteries are thicker than the walls of veins. Figure 14.1 compares an artery and a vein in cross section. Notice how the artery is round and has a thick tunica media and a folded tunica interna. When blood pressure increases or vasodilation occurs, the artery expands, and the pleats in the tunica interna flatten as the vessel stretches.

Capillaries lack a tunica externa and tunica media. They consist of a single layer of endothelium that is continuous with the tunica interna of the artery and vein that supply and drain the capillary. Capillaries are so narrow that blood cells must line up in a single file to squeeze through.

Veins have a thinner wall and have a smooth tunica interna. Their walls will collapse if emptied of blood. Blood pressure is low in veins, and to prevent back flow of blood, the peripheral veins have **valves** that function to keep blood flowing in one direction, toward the heart.

Laboratory Activity   ARTERY AND VEIN COMPARISON _____

MATERIALS

compound microscope
prepared slide of an artery and vein

PROCEDURES

1. Label and review the structure of arteries and veins in Figure 14.1.

2. Locate the artery and vein on the slide. Most of these slide preparations have one artery, an adjacent vein, and a nerve. The blood vessels are hollow and most likely have blood cells in the lumen. The nerve appears as a round, solid structure.

3. Identify the tunica externa, tunica media, and tunica interna in the artery and vein. Examine each layer at a high magnification. How does the endothelial layer of the vein differ from that of the artery?

_____

_____

4. Draw and label the artery and vein in the space provided. Include as much detail as possible in your drawing to show the anatomical differences between the vessels.

B. The Arterial System

ARTERIES OF THE HEAD, NECK, AND UPPER LIMB

Figure 14.2 illustrates the major vessels of the systemic arterial system. Oxygenated blood is pumped out of the left ventricle of the heart and into the aorta for distribution to the organ systems. The aorta is curved like a question mark. The **ascending aorta** begins at the base of the heart and curves upward as the **aortic arch** then passes through the thorax as the **thoracic aorta**. The descending aorta pierces through the diaphragm and becomes the **abdominal aorta**. Arteries that branch off the aortic arch serve the head, neck, and upper limb. Branches off the abdominal aorta serve the abdominal organs. The abdominal aorta enters the pelvic cavity and divides to send a branch into each lower limb.

Study Tip

On vascular models, systemic arteries are usually painted red and systemic veins are painted blue to show oxygenated and deoxygenated blood, respectively. Further, when identifying arteries, start at the heart and trace vessels to the periphery. Study veins from the periphery to the heart.

Figure 14.3 illustrates the arteries of the chest and upper limb. The aortic arch has three major arteries that serve the head, neck, and arms. The first branch, the brachio-cephalic or **innominate artery**, is short and divides into the **right common carotid artery** and the **right subclavian artery**. These arteries supply blood to the right side of the head and neck and to the right arm, respectively. The left common carotid artery and left subclavian artery originate directly off the aorta. Only the right common carotid artery and right subclavian artery are derived from a brachiocephalic artery. The **vertebral artery** branches off the subclavian artery and supplies the brain.

Each subclavian artery passes under the clavicle and crosses the armpit as the **axillary artery**, and then continues into the arm as the **brachial artery**. Blood pressure is usually taken at the brachial artery. At the antecubitis, the elbow, the brachial artery divides into the lateral **radial artery** and the medial **ulnar artery**, each named after the bone they follow. In the palm of the hand, these arteries are interconnected by the

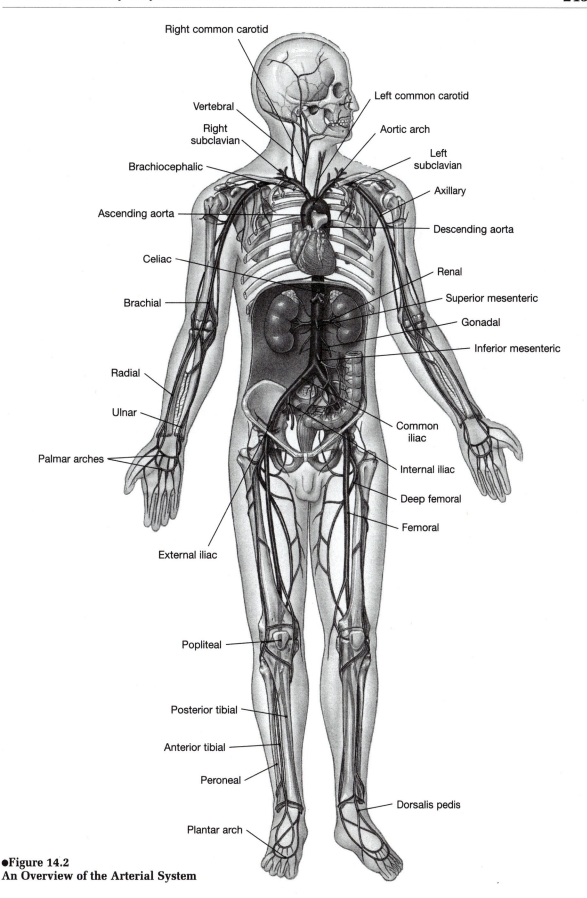

●**Figure 14.2**
An Overview of the Arterial System

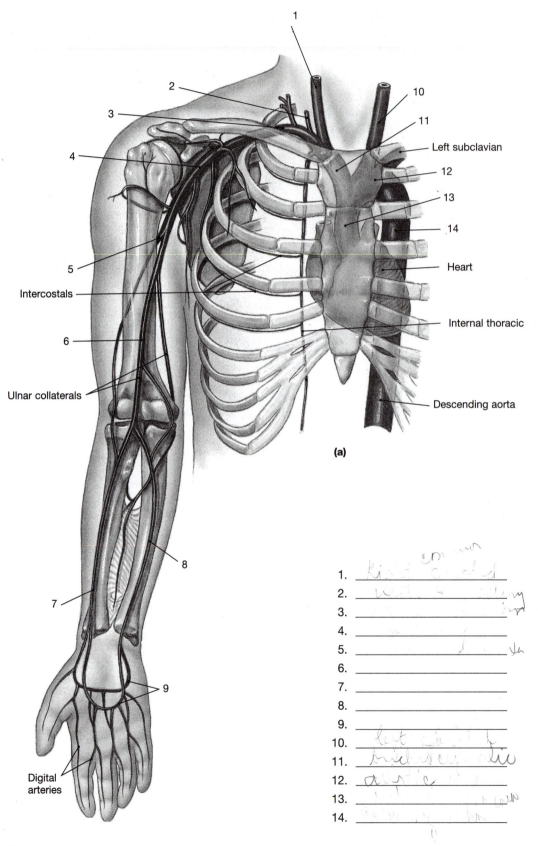

Left subclavian

Heart

Internal thoracic

Intercostals

Descending aorta

Ulnar collaterals

Digital
arteries

(a)

1. _____
2. _____
3. _____
4. _____
5. _____
6. _____
7. _____
8. _____
9. _____
10. _____
11. _____
12. _____
13. _____
14. _____

●Figure 14.3
Arteries of the Chest and Upper Limb

superficial and deep **palmar arches**, which send small **digital arteries** to the fingers. Except for the brachiocephalic artery on the right side, the vascular anatomy is symmetrical between the right and left upper limbs.

The right and left carotid arteries supply blood to the structures of the neck, face, and brain, shown in Figure 14.3. The **right common carotid artery** originates off the brachiocephalic artery, whereas the **left common carotid artery** branches directly off the peak of the aortic arch. The term *common* suggests the vessel will divide into an external and internal vessel. Each common carotid artery ascends deep in the neck and divides at the larynx into an **external carotid** artery and an **internal carotid** artery. The external carotid artery branches to supply blood to the structures of the neck and face. The pulse in the external carotid artery may be felt by placing your fingers lateral to your thyroid cartilage, the Adam's apple. The internal carotid artery ascends to the base of the brain and divides into vessels that supply the eyes and the brain.

The brain has a high metabolic rate and a voracious appetite for oxygen and nutrients. A reduction in blood flow to the brain may result in permanent damage to the inflected area. To ensure the brain has a continuous supply of oxygen and nutrients, branches of the internal carotid artery and other arteries interconnect, or **anastomose**, as the **cerebral arterial circle**, also called the **circle of Willis**. Follow the vascular anatomy in Figure 14.4 as the circle is described. The right and left **vertebral arteries**, which arise off the subclavian arteries, ascend in the transverse foramina of the cervical vertebrae and enter the skull at the foramen magnum. These arteries fuse into a single **basilar artery** on the inferior surface of the brain stem. The basilar artery branches into **posterior cerebral arteries** and **posterior communicating arteries**. The posterior communicating arteries form an anastomosis with the carotid arteries. **Anterior communicating arteries**

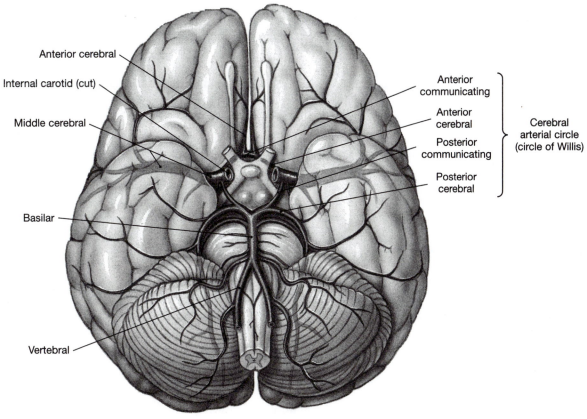

●**Figure 14.4**
Arteries of the Brain
Inferior surface of the brain, showing the arterial distribution.

branch off the internal common carotid arteries, pass anteriorly, and meet to complete the ring of blood vessels at the base of the brain.

Laboratory Activity   ARTERIES OF THE HEAD, NECK, AND ARM

MATERIALS

cardiovascular system chart   torso model
head model   arm model

PROCEDURES

1. Review the arteries presented in Figures 14.3 and 14.4. Label the arteries in Figure 14.3. Locate each artery on the laboratory models and charts.
2. On your body, trace the location of the arteries from the base of the aorta to your right thumb.
3. On your body, trace the location of the arteries from the base of the aorta to the left side of your face.
4. In the space provided, draw and label the first three branches of the aortic arch.

ARTERIES OF THE ABDOMINOPELVIC CAVITY AND LOWER LIMB

The thoracic aorta descends through the diaphragm and becomes the **abdominal aorta**. Many arteries stem from the abdominal aorta (see Figure 14.5). An easy way to identify the branches of the abdominal aorta is to distinguish between paired arteries that have a right and left vessel, and unpaired arteries that have a single branch.

There are three unpaired branches of the abdominal aorta. The celiac trunk is a short artery arising off the abdominal aorta inferior to the diaphragm. It splits into three arteries, the **common hepatic artery**, the **left gastric artery**, and the **splenic artery**. The common hepatic artery divides to supply blood to the liver, gallbladder, and part of the stomach. The left gastric artery also supports blood flow to the stomach. The splenic artery supplies blood to the spleen, stomach, and pancreas. Inferior to the celiac trunk is the next unpaired artery, the **superior mesenteric artery**. This vessel supplies the blood to the large intestine, parts of the small intestine, and other abdominal organs. The third unpaired artery, the **inferior mesenteric artery**, originates before the abdominal aorta divides to enter the pelvic cavity and legs. This artery supplies the parts of the large intestine and the rectum.

Four pairs of major arteries arise off the abdominal aorta. The **suprarenal arteries** arise near the superior mesenteric artery and branch into the adrenal glands located on top of the kidneys. The right and left **renal arteries**, which supply the kidneys, stem off the abdominal aorta just inferior to the suprarenal arteries. A pair of **gonadal arteries** appear near the inferior mesenteric artery. These arteries bring blood to the reproductive organs of the male and female. A pair of **lumbar arteries** originate near the terminus of the abdominal aorta and service the lower body wall.

At the level of the hips, the abdominal aorta divides into the right and left **common iliac arteries** (see Figures 14.5 and 14.6). These arteries descend through the pelvic cavity and branch into **external iliac arteries**, which enter the legs, and **internal iliac arteries**, which supply blood to the organs of the pelvic cavity. When an external iliac artery pierces the abdominal wall, it becomes the **femoral artery** of the thigh. A **deep femoral artery** arises to supply deep thigh muscles. The femoral artery passes though the posterior knee as the **popliteal artery** and then divides into the **posterior tibial artery** and the **anterior tibial artery**, each supplying blood to lower leg structures. The **peroneal artery** stems laterally off the posterior tibial artery. The arteries of the lower leg send branches into the foot, which anastomose at the **dorsal** and **plantar arches**.

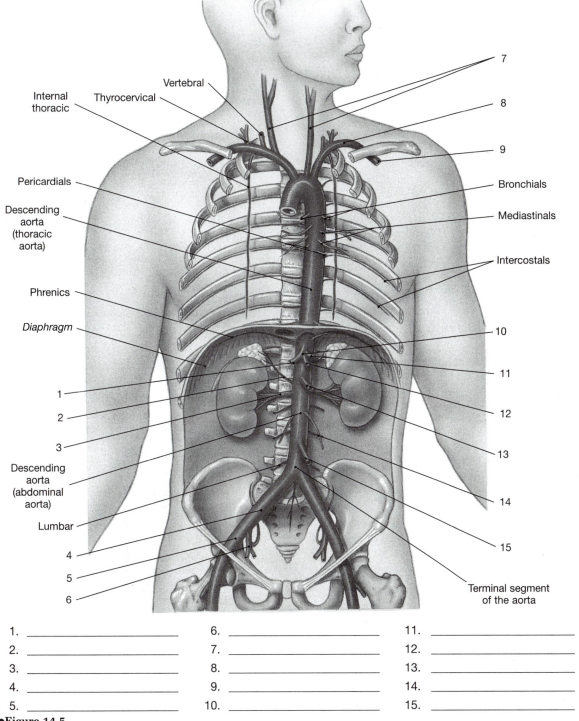

●Figure 14.5
Major Arteries of the Trunk

1. _____   6. _____   11. _____
2. _____   7. _____   12. _____
3. _____   8. _____   13. _____
4. _____   9. _____   14. _____
5. _____   10. _____   15. _____

Laboratory Activity   ARTERIES OF THE ABDOMEN AND LEG _____

MATERIALS

cardiovascular system chart
torso model
leg model

PROCEDURES

1. Label and review the arteries presented in Figures 14.5 and 14.6.
2. Locate each artery on the laboratory models and charts.
3. On your body, trace the location of your abdominal aorta, common iliac artery, external iliac artery, femoral artery, popliteal artery, and posterior tibial artery.

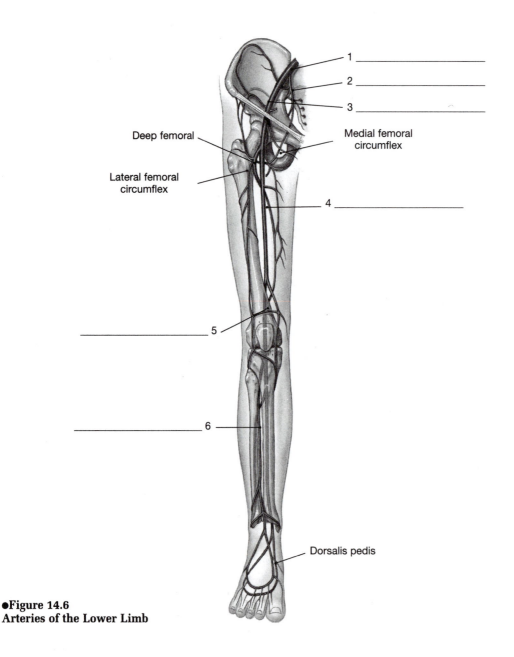

1 _____

2 _____

3 _____

Deep femoral

Medial femoral circumflex

Lateral femoral circumflex

4 _____

_____ 5

_____ 6

Dorsalis pedis

●Figure 14.6
Arteries of the Lower Limb

4. In the space provided, draw and label the abdominal aorta and the major arterial branches.

C. Systemic Veins

Once you have learned the major systemic arteries, the identification of systemic veins is easy because most arteries have a vein that drains the corresponding area that the artery supplied. This similarity in distribution is seen when comparing Figures 14.2 and 14.7. Unlike arteries, many veins are superficial and easily seen under the skin.

VEINS OF THE HEAD, NECK, AND UPPER LIMB

Blood in the brain drains into large veins called sinuses. These large veins drain into the **internal jugular vein**, which exits the skull via the jugular foramen, descends the neck, and empties into the **brachiocephalic vein** (see Figure 14.8). Superficial veins that drain the face and scalp empty into the **external jugular vein**, which also descends the neck to join the brachiocephalic vein. The right and left brachiocephalic veins merge at the **superior vena cava**, which empties deoxygenated blood into the right atrium of the heart. Blood in the right atrium enters the right ventricle, which contracts and pumps the blood to the lungs via the pulmonary circuit.

Study Tip

The venous system has both a right and left brachiocephalic vein, whereas the arterial system has a single brachiocephalic artery that branches into the right common carotid artery and right subclavian arteries. On the left side of the body, the common carotid artery and subclavian artery originate directly off the aortic arch.

Figure 14.8 illustrates the venous drainage of the arm. The hand drains into **digital veins**, which empty into a network of **palmar venous arches**. These vessels drain into the **cephalic vein** and the **basilic vein** that ascend the forearm. The **median cubital vein** joins the basilic vein at the elbow. The median cubital vein is often used to collect blood samples from an individual. The **radial** and **ulnar veins** ascend the forearm and drain into the **brachial vein**. The brachial and basilic veins meet at the armpit as the **axillary vein**, which then joins the cephalic vein at the **subclavian vein**. Subclavian veins and veins from the neck and head drain into the **brachiocephalic vein**, which then empties into the superior vena cava that drains blood into the right atrium.

Laboratory Activity   VEINS OF THE HEAD, NECK, AND ARM_____

MATERIALS

cardiovascular system chart   head model
torso model   arm model

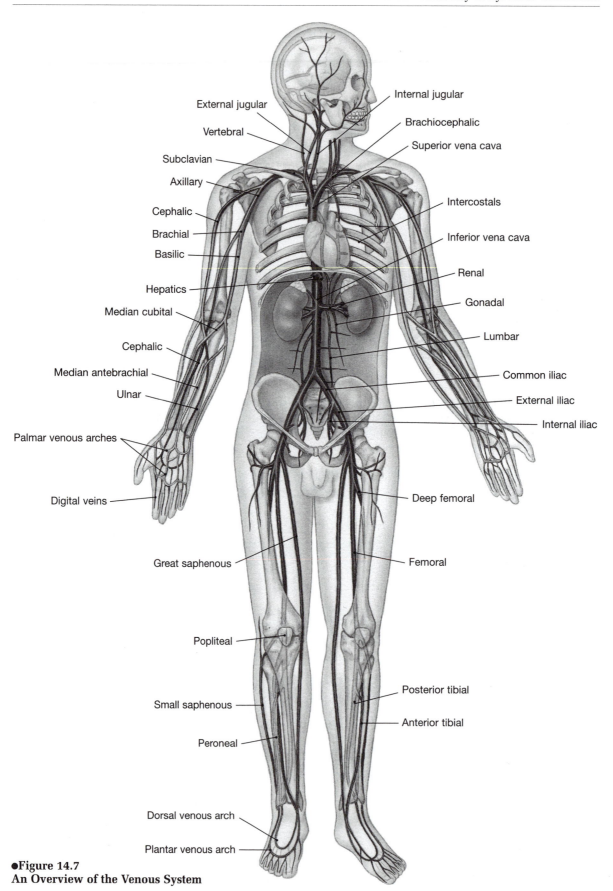

●Figure 14.7
An Overview of the Venous System

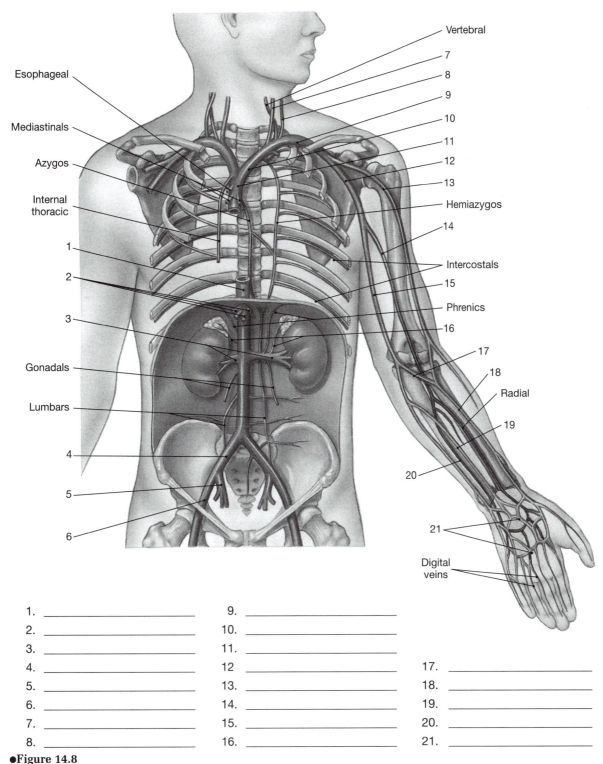

•Figure 14.8
The Venous Drainage of the Abdomen and Chest

1. _____   9. _____
2. _____   10. _____
3. _____   11. _____
4. _____   12. _____   17. _____
5. _____   13. _____   18. _____
6. _____   14. _____   19. _____
7. _____   15. _____   20. _____
8. _____   16. _____   21. _____

PROCEDURES

1. Review the veins presented in Figures 14.7 and 14.8. Label each vein in Figure 14.8.

2. Locate each vein on the laboratory models and charts.

3. On your body, trace your cephalic vein, subclavian vein, brachiocephalic vein, and superior vena cava.

4. Draw and label the venous anatomy near the heart in the space provided.

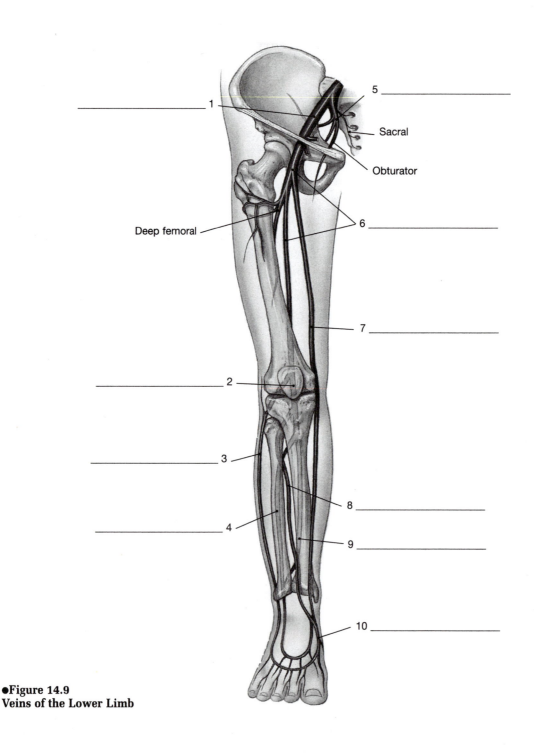

5 _____

Sacral

Obturator

1 _____

Deep femoral

6 _____

7 _____

2 _____

3 _____

8 _____

4 _____

9 _____

10 _____

●Figure 14.9
Veins of the Lower Limb

VEINS OF THE LOWER LIMB AND ABDOMEN

Figure 14.9 illustrates the venous drainage of the lower limb. The foot has many anastomoses that drain into the lateral **peroneal vein** and the **anterior tibial vein**, located on the medial aspect of the anterior shin. These veins, along with the **posterior tibial vein**, merge and become the **popliteal vein** of the posterior knee. The **small (lesser) saphenous vein** ascends laterally from the ankle to the knee, drains blood from superficial veins, and empties into the popliteal vein. Superior to the knee, the popliteal vein becomes the **femoral vein**, which ascends along the femur to the lower pelvic girdle, where it joins with the **deep femoral vein** at the **external iliac vein**. The **great saphenous vein** ascends from the medial side of the ankle to the upper thigh and drains into the external iliac vein. In the pelvic cavity, the external and **internal iliac veins** fuse and form the **common iliac vein**. The right and left common iliac veins merge at the **inferior vena cava**.

Figure 14.10 shows the vascular supply to the liver. The **inferior** and **superior mesenteric veins** drain the nutrient-rich blood of the digestive organs. These veins empty into the **hepatic portal vein**, which passes the blood through the liver for regulation of blood sugar concentration. Phagocytic cells consume microbes that may have entered the bloodstream through the mucous membrane of the digestive system. Blood from the hepatic arteries and hepatic portal vein mixes in the liver and is returned to the inferior vena cava by the **hepatic veins**.

Six major veins from the abdominal organs drain blood into the inferior vena cava (see Figure 14.8). The **lumbar veins** drain the muscles of the lower body wall and the spinal cord, and empty into the inferior vena cava close to the common iliac veins. A pair of **gonadal veins** empty blood from the reproductive organs into the inferior vena cava above the lumbar veins. A pair of **renal** and **suprarenal veins** drain into the inferior vena cava next to their respective organ. Before entering the thoracic cavity to drain blood into the right atrium, the inferior vena cava collects blood from the **hepatic veins** that drain the liver.

Laboratory Activity   VEINS OF THE ABDOMEN AND LEG _____

MATERIALS

    cardiovascular system chart
    torso model
    leg model

PROCEDURES

1. Label and review the veins presented in Figures 14.8, 14.9, and 14.10.
2. Locate each vein on the laboratory models and charts.
3. On your body, trace the location of the veins in your lower limb.
4. Although you have studied the arterial and venous systems separately, they are anatomically interconnected by capillaries. To reinforce this organization, trace blood through the following systemic routes:
   a. Trace a drop of blood from the heart, through the vessels of the left arm, and return to the heart.

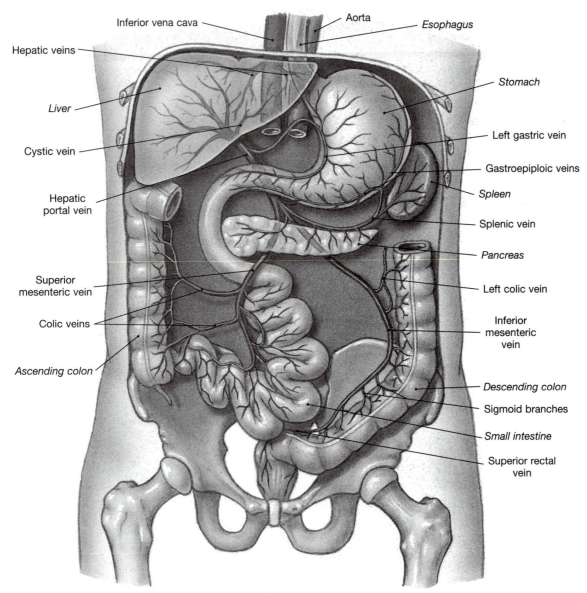

●**Figure 14.10**
The Hepatic Portal System

    b. Trace a drop of blood from the heart, through the vessels of the liver, and return to the heart.

    c. Trace a drop of blood from the heart, through the vessels of the right lower limb, and return to the heart.

ANATOMY OF SYSTEMIC CIRCULATION
Exercise 14 Laboratory Report

Discussion Questions

1. What is the function of valves in the peripheral veins?

2. How is the arterial anatomy to the right arm different from the arterial anatomy to the left arm?

3. Describe the major vessels that return deoxygenated blood to the right atrium of the heart.

4. Compare the wall of an artery to the wall of a vein. Why are these vessels so different?

B. Matching

Match each description on the left with the correct term on the right.

1. _____ artery in armpit	A.	subclavian
2. _____ artery with three branches	B.	superior mesenteric
3. _____ vein used for blood samples	C.	popliteal
4. _____ artery on right side only	D.	cephalic
5. _____ long vein of leg	E.	carotid
6. _____ deoxygenated supply to liver	F.	gonadal
7. _____ artery to large intestine	G.	valves
8. _____ long vein of arm	H.	axillary
9. _____ vein in knee	I.	great saphenous
10. _____ artery to reproductive organ	J.	median cubital
11. _____ major artery in neck	K.	hepatic portal vein
12. _____ vein under clavicle	L.	celiac
13. _____ found only in veins	M.	brachiocephalic artery

C. Completion

Complete the following figures:

1. Figure 14.11, The Venous System.
2. Figure 14.12, The Arterial System.

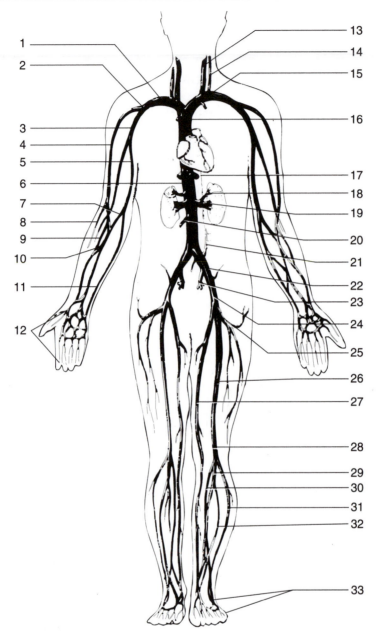

1. _____   12. _____   23. _____
2. _____   13. _____   24. _____
3. _____   14. _____   25. _____
4. _____   15. _____   26. _____
5. _____   16. _____   27. _____
6. _____   17. _____   28. _____
7. _____   18. _____   29. _____
8. _____   19. _____   30. _____
9. _____   20. _____   31. _____
10. _____   21. _____   32. _____
11. _____   22. _____   33. _____

●Figure 14.11
The Venous System

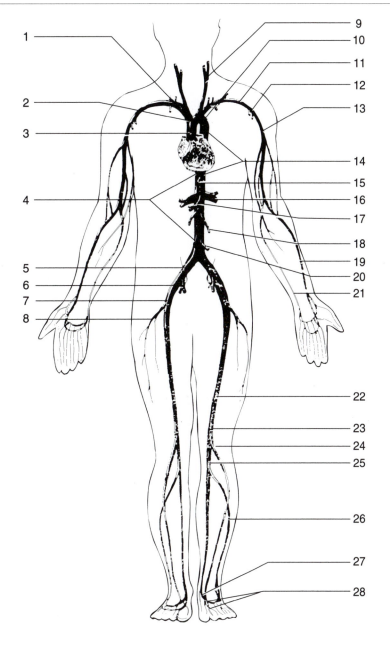

●Figure 14.12
The Arterial System

1. _____   11. _____   21. _____
2. _____   12. _____   22. _____
3. _____   13. _____   23. _____
4. _____   14. _____   24. _____
5. _____   15. _____   25. _____
6. _____   16. _____   26. _____
7. _____   17. _____   27. _____
8. _____   18. _____   28. _____
9. _____   19. _____
10. _____   20. _____

15 LYMPHATIC SYSTEM

WORD POWER

lymph (lymph—water)
cisterna chyli (cistern—storage well; chyl—juice)
node (nod—swelling)
trabeculae (trabeculate—containing cross bars)

MATERIALS

torso model
lymphatic system chart
spleen model
compound microscope
prepared slides
    lymph node
    spleen

OBJECTIVES

On completion of this exercise, you should be able to:

- List the functions of the lymphatic system.
- Discuss exchange of blood plasma, interstitial fluid, and lymph.
- Describe the structure of a lymph node.
- Explain how the lymphatic system drains into the vascular system.
- Describe the gross anatomy and basic histology of the spleen.

The lymphatic system, Figure 15.1, includes lymphatic vessels and lymph nodes, tonsils, the spleen, and the thymus gland. Lymphatic vessels transport a fluid called lymph from the tissue spaces to the veins of the cardiovascular system. Scattered along a lymphatic vessel are lymph nodes that contain lymphocytes and phagocytic macrophages. The macrophages remove invading bacteria and other substances from the lymph before the lymph is returned to the blood. Although lymphocytes are a formed element of the blood, they are the main cells of the lymphatic system and colonize dense populations in lymph nodes and the spleen. **Antigens** are substances capable of causing the production of antibodies. As macrophages capture antigens in the lymph, lymphocytes are exposed to the antigens and activate the immune system to respond to the intruding cells. Some lymphocytes, the B cells, produce **antibodies** that chemically combine with and destroy specific antigens. The thymus gland is involved in the development of the functional immune system in infants. In adults, the thymus gland controls the maturation of lymphocytes. We covered the thymus gland with the endocrine system in Exercise 10 of the lab manual.

Figure 15.2 illustrates fluid circulation in the body. Pressure in blood capillaries forces **plasma** fluids and solutes out of the capillary and into the interstitial spaces. This filtrate, called **interstitial fluid**, bathes the cells with nutrients, dissolved gases, hormones, and other materials. Most of the interstitial fluid reenters the capillary owing to osmotic pressure. The remaining interstitial fluid is returned to the blood by the lymphatic system. Interstitial fluid slowly enters lymphatic vessels and becomes **lymph**. Lymphatic vessels join lymph nodes where phagocytes remove abnormal cells and microbes from the lymph. A pair of lymphatic ducts join with veins near the heart and return the lymph to the blood. Approximately three liters of fluid per day are forced out of the capillaries and flow through lymphatic vessels as lymph.

●Figure 15.1
Components of the Lymphatic System

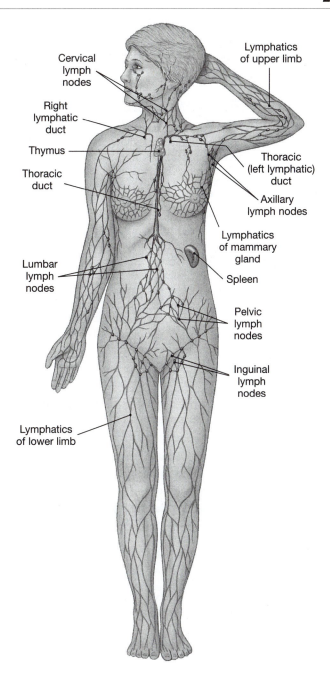

Cervical lymph nodes

Right lymphatic duct

Thymus

Thoracic duct

Lumbar lymph nodes

Lymphatics of lower limb

Lymphatics of upper limb

Thoracic (left lymphatic) duct

Axillary lymph nodes

Lymphatics of mammary gland

Spleen

Pelvic lymph nodes

Inguinal lymph nodes

A. Lymphatic Vessels

Lymphatic vessels occur next to the vessels of the vascular system. The major components of the lymphatic vessels and related structures are illustrated in Figure 15.3. Lymphatic vessels, or simply lymphatics, collect fluid lost from blood capillaries, filter the fluid in lymph nodes, and return it to the blood circulation. Lymphatic vessels are structurally similar to veins. The vessel wall has similar layers and valves to prevent back flow of fluid. Lymphatic pressures are very low, and the lymphatic valves are close together to keep the lymph circulating toward the body trunk. Unlike blood capillaries, lymphatic capillaries are not continuous with other vessels at both ends. Lymphatic capillaries originate as closed tubules near the blood capillaries. The tubules elongate into capillaries that merge into larger lymphatic vessels. These vessels conduct lymph toward the body trunk and into larger lymphatics that empty into veins near the heart.

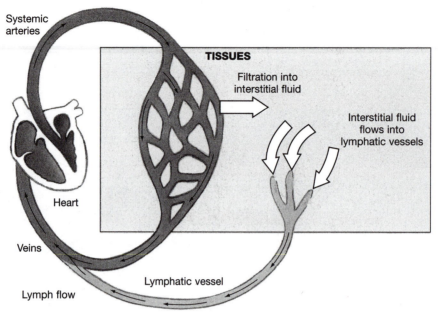

● **Figure 15.2**
Fluid Connective Tissues

The functional relationships between blood and lymph. Blood travels through the
circulatory system, pushed by the contractions of the heart. In capillaries, hydrostatic
(blood) pressure forces fluid and dissolved solutes out of the circulatory system. This
fluid mixes with the interstitial fluid already in the tissue. Interstitial fluid slowly
enters lymphatic vessels; now called lymph, it travels along the lymphatics and
re-enters the circulatory system at one of the veins that returns blood to the heart.

Two large lymphatic vessels, the **thoracic duct** and the **right lymphatic duct**, return
lymph to venous circulation. Use Figures 15.1 and 15.4 to trace the drainage of the
lymphatic system. Most of the lymph is returned to circulation by the thoracic duct.
The thoracic duct commences at the level of the second lumbar vertebra, L_2, on the
posterior abdominal wall behind the abdominal aorta. Lymphatics from the legs,
pelvis, and abdomen drain into an inferior saclike portion of the duct called the **cisterna chyli**. Lymph flows into the thoracic duct that pierces the diaphragm and
ascends to the base of the heart, where it joins with the left subclavian vein to return
the lymph to the blood. The only lymph that does not drain into the thoracic duct is
from lymphatic vessels in the right arm and the right side of the chest, neck, and head.
These areas drain into the **right lymphatic duct** near the right clavicle. This duct empties
lymph into the right subclavian vein located near the base of the heart.

Laboratory Activity   LYMPHATIC VESSELS_____

MATERIALS

torso model or lymphatic system chart

PROCEDURES

1. Locate the thoracic duct and the cisterna chyli on a chart or lab model, and label
them in Figure 15.4. What areas of the body drain lymph into this duct? Where
does this duct return lymph to the blood?

_____

_____

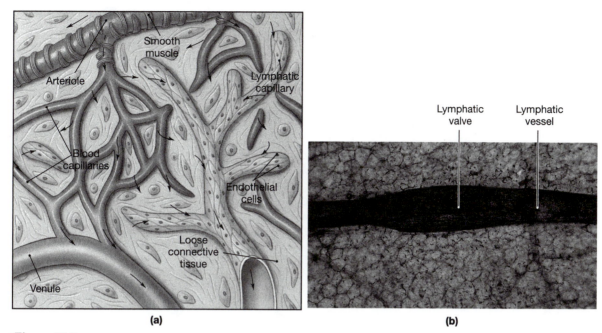

(a)   **(b)**

●**Figure 15.3**
Lymphatic Capillaries

(a) A three-dimensional view of the association of blood capillaries, tissue, interstitial fluid, and lymphatic capillaries. Arrows show the direction of interstitial fluid and lymph movement. **(b)** A valve within a small lymphatic vessel. (LM × 43)

> **2.** Locate the right lymphatic duct on a chart or lab model. Where does this vessel join the vascular system? What regions of the body drain lymph into this duct?
>
> _____
>
> _____

B. Lymphoid Tissues and Lymph Nodes

Two major groups of lymphatic structures occur in connective tissues: **encapsulated lymph organs** and **diffuse lymphoid tissues**. The encapsulated lymph organs include lymph nodes, the thymus gland, and the spleen. Each encapsulated organ is separated from the surrounding connective tissue by a fibrous capsule. Diffuse lymphoid tissue, such as tonsils and lymphoid nodules, do not have a defined boundary from the connective tissue.

A **lymph node** is an oval nodular organ that functions like a filter cartridge. As lymph passes though the node, phagocytes remove microbes, debris, and other antigens from the lymph. Lymph nodes are scattered throughout the lymphatic system, as depicted in Figure 15.1. Lymphatics from the lower limbs pass through a network of **inguinal** lymph nodes. **Pelvic** and **lumbar** nodes filter lymph from the pelvic and abdominal lymphatics. Many lymph nodes occur in the upper limbs and in the **axillary** and **cervical** regions. The breasts in women also have many lymphatic vessels and nodes. Infections often occur in a lymph node before they spread systemically. A swollen or painful lymph node suggests an increase in lymphocyte abundance and general immunological activity in response to antigens in the lymph nodes.

Figure 15.5 details the organization of a lymph node. Each lymph node is encased in a dense connective tissue **capsule**. Collagen fibers from the capsule extend as partitions called **trabeculae** into the interior of the lymph node. The outer **cortex** of the

1. _____

2. _____

3. _____

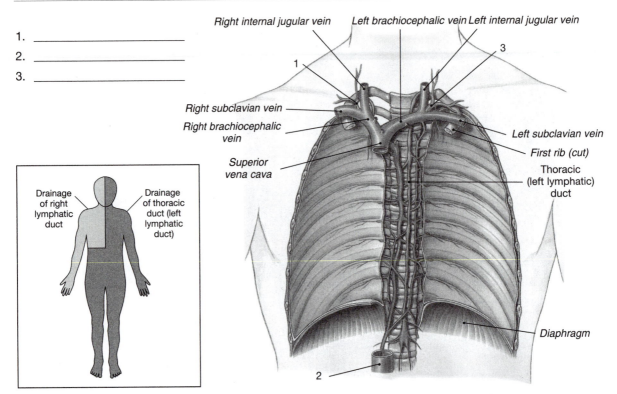

●**Figure 15.4**
The Lymphatic Ducts and the Venous System
The thoracic duct carries lymph originating in tissues inferior to the diaphragm and
from the left side of the upper body. It empties into the left subclavian vein. The right
lymphatic duct drains the right half of the body superior to the diaphragm. It empties
into the right subclavian vein.

node contains sinuses full of lymph fluid. Masses of lymphocytes are produced in
and surround many pale-staining **germinal centers**. Deep to the cortex is the **medulla**,
where **medullary cords** of lymphocytes extend into sinuses. Lymph enters a node
in **afferent lymphatic vessels**. As lymph flows through the cortical and medullary
sinuses, macrophages phagocytize abnormal cells, pathogens, and debris. Draining
the lymph node are **efferent lymphatic vessels** that exit the node at a slit called the
hilus.

Lymphocytes are the most abundant cells in the lymphatic system. This diverse
group of white blood cells protects against infection. Some lymphocytes produce
antibody proteins to destroy microbes. When lymphocytes are exposed to an antigen, the immune system responds with chemical and cellular defenses to eradicate
the antigen.

Lymphoid **nodules** are found in connective tissues under the lining of the digestive, urinary, and respiratory systems. Microbes that penetrate through the exposed
epithelial surface pass through lymphoid nodules where lymphocytes and
macrophages destroy and remove the foreign cells from the lymph. Some nodules have
a germinal center where lymphocytes are produced by cell division.

Tonsils are lymphoid nodules in the pharynx. A pair of **lingual tonsils** occurs at
the posterior base of the tongue. The **palatine tonsils** are easily viewed hanging off the
posterior arches of the oral cavity. A single **pharyngeal tonsil**, or **adenoids**, is located
in the upper pharynx near the opening to the nasal cavity.

Clinical Application

The lymphatic system usually has the "upper hand" in the immunological battle against invading bacteria and viruses. Occasionally, microbes manage to populate a lymphoid nodule. When tonsils are infected, they swell and become irritated. This condition is called tonsillitis and is treated with antibiotics to control the infection. If the problem is recurrent, the tonsils are removed during a surgical procedure called a tonsillectomy. Usually, the palatine tonsils are removed. If the pharyngeal tonsils are also infected or are abnormally large, they are removed, too.

Laboratory Activity   LYMPH NODES

MATERIALS

compound microscope
prepared slide of a lymph node

PROCEDURES

1. Label and review the structure of a lymph node in Figure 15.5.
2. Obtain a prepared slide of a lymph node. Examine the slide at low magnification, and identify the capsule and trabeculae. Within the cortex of the node, examine a germinal center at a higher magnification. What types of cells are produced in this region?

_____

_____

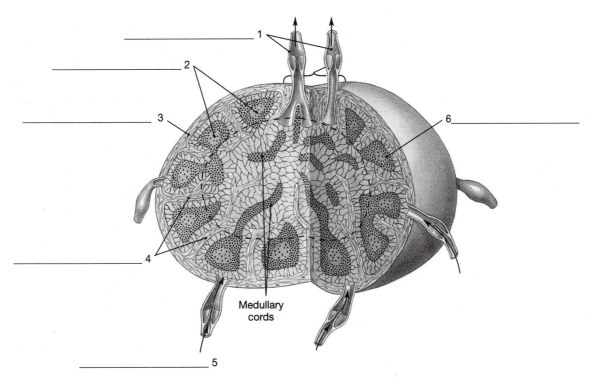

●Figure 15.5
Structure of a Lymph Node

3. Make a sketch of the lymph node and label the following structures: capsule, trabeculae, germinal center, cortex, medulla, and medullary cords.

C. The Spleen

The **spleen** is the largest lymphatic organ in your body. It is located lateral to the stomach along the greater curvature, shown in Figure 15.6.

Histologically, the spleen is surrounded by a capsule that protects the underlying tissue called pulp. The pulp consists of two types of tissues, **red pulp** and **white pulp**. Branches of the splenic artery, called **trabecular arteries**, are surrounded by white pulp that contains large populations of lymphocytes. Capillaries of the trabecular arteries open into the red pulp. As blood flows through the red pulp, free and fixed phagocytes remove abnormal red blood cells and other antigens.

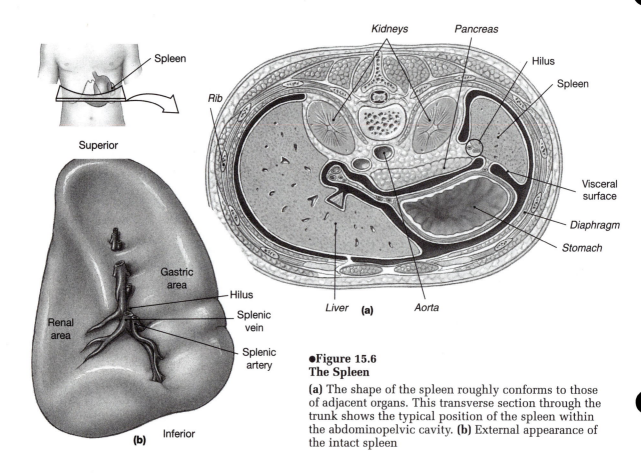

●**Figure 15.6**
The Spleen

(a) The shape of the spleen roughly conforms to those of adjacent organs. This transverse section through the trunk shows the typical position of the spleen within the abdominopelvic cavity. **(b)** External appearance of the intact spleen

Laboratory Activity   THE SPLEEN _____

MATERIALS

torso model or chart with spleen
compound microscope
prepared slide of spleen

PROCEDURES

1. Locate the spleen on laboratory models and charts. Identify the hilus, splenic artery, and splenic vein. On the visceral surface, locate the gastric and renal areas that are in contact with the stomach and kidneys, respectively.
2. Obtain a prepared slide of the spleen. Examine the slide at low magnification, and identify the dark-stained regions of white pulp and the lighter regions of red pulp. Examine several white pulp masses for the presence of an artery. Make a sketch of the white and red pulps in the space provided.

LYMPHATIC SYSTEM

Exercise 15 Laboratory Report

A. Matching

Match each spinal cord structure listed on the left with the correct description on the right.

1.	_____ efferent vessel	A.	empties into right subclavian vein
2.	_____ medullary cords	B.	empties into lymph node
3.	_____ cisterna chyli	C.	splenic tissue with RBCs
4.	_____ right lymphatic duct	D.	fluid in lymphatic vessels
5.	_____ red pulp	E.	lymphocytes deep in node
6.	_____ lymph node	F.	empties into left subclavian vein
7.	_____ thoracic duct	G.	full of macrophages and lymphocytes
8.	_____ white pulp	H.	drains lymph node
9.	_____ lymph	I.	lymphocytes surrounding trabecular artery
10.	_____ afferent vessel	J.	sac of thoracic duct

B. Short Answer Questions

1. Describe the organization of a lymph node.

2. Discuss the major functions of the lymphatic system.

3. How are blood plasma, interstitial fluid, and lymph interrelated?

4. Explain how lymph is returned to the blood.

5. Describe the anatomy of the spleen.

C. Labeling and Completion

1. Complete Figure 15.7 by labeling the organs of the lymphatic system.

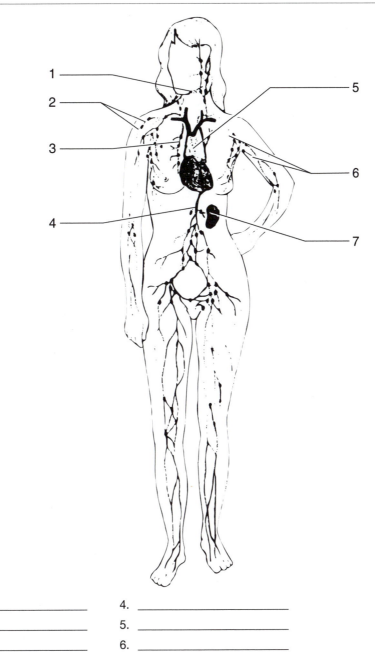

1. _____   4. _____

2. _____   5. _____

3. _____   6. _____

●**Figure 15.7**
The Lymphatic System

16 ANATOMY AND PHYSIOLOGY OF THE RESPIRATORY SYSTEM

All cells require a constant supply of oxygen (O_2) to maintain cellular respiration. A major waste product of cellular metabolism is carbon dioxide (CO_2). The respiratory system is responsible for oxygenating the blood and removing carbon dioxide. In this exercise, you will study the anatomy of the respiratory system and measure respiratory volumes of your lungs.

PART ONE   ANATOMY OF THE RESPIRATORY SYSTEM

WORD POWER

vestibule (vestibulum—a chamber)
septum (septum—a partition)
concha (concha—a shell)
meatus (meatus—a passageway)
pharynx (pharynx—throat)
cricoid (cricoid—a ring)
glottis (glottis—aperture)
pleura (pleura—a rib, the side)

MATERIALS

lung model
head model
torso model
respiratory system chart
hand mirrors
compound microscope

prepared slides
  nasal epithelium
  trachea
  lung tissue

OBJECTIVES

On completion of this exercise, you should be able to:

- Identify and describe the structures of the nasal cavity.
- Identify and describe the three regions of the pharynx.
- Identify and describe the cartilages and folds of the larynx.
- Describe the lining epithelium of the upper respiratory system.
- Identify and describe the structures of the trachea and the branching of the bronchial tree.
- Identify and describe the gross anatomy of the lungs.
- Identify the histological organization of the lung on a prepared slide.

The respiratory system, shown in Figure 16.1, consists of the nose, nasal cavity and sinuses, pharynx, larynx, trachea, bronchi, and lungs. The upper respiratory system includes the nostrils, nasal cavity, paranasal cavities, and pharynx. These structures filter, warm, and moisten air before it enters the lower respiratory system. The larynx, trachea, bronchi, and lungs compose the lower respiratory system. The larynx regulates the opening into the lower respiratory tract and produces speech sounds. The trachea and bronchi maintain an open airway to the lungs. In the lungs, exchange of gases occurs between millions of alveolar sacs and the blood in pulmonary capillaries.

Most of the respiratory system is lined with **pseudostratified ciliated columnar epithelium** that contains many **goblet cells** (Figure 16.2). The goblet cells secrete a

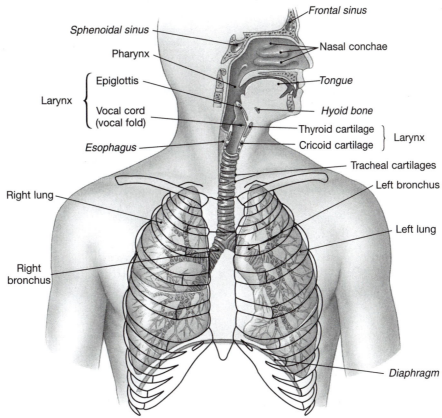

●**Figure 16.1**
Components of the Respiratory System

sticky mucus that covers the ciliated epithelium with a sticky carpet to trap particles in the inspired air. The cilia beat to move the mucus toward the pharynx, where it is swallowed. This process of mucus movement is often called the **mucus escalator**.

A. The Nose and Pharynx

The primary route for air to enter the respiratory system is through two openings called **external nares** or nostrils (Figure 16.3). Just inside each external naris is an expanded **vestibule** that contains coarse hairs. The hairs help prevent large airborne materials like dirt and insects from entering the respiratory system. A midsagittal **nasal septum** divides the **nasal cavity**. This bony septum is formed by the union of the perpendicular plate of the ethmoid bone with the vomer. The external portion of the nose consists of cartilage that forms the bridge and tip of the nose and part of the nasal septum. The nasal cavity is a vibrating or resonating chamber for the voice.

The **superior, middle,** and **inferior nasal conchae** are bony shelves that project from the lateral walls of the nasal cavity. The bone of each concha curls underneath and forms a tube or **meatus** that causes air to swirl in the nasal cavity. This turbulence moves air across the sticky epithelial lining where dust and debris are removed from the inhaled air. The floor of the nasal cavity is the superior portion of the **hard palate**, formed by the maxillae, the palatine bones, and the muscular **soft palate**. Hanging off the posterior edge of the soft palate is the conical **uvula**. Two openings, the **internal nares**, lead from the nasal cavity into the uppermost portion of the throat.

The throat or **pharynx** is divided into three regions: the upper nasopharynx, the middle oropharynx, and the lower laryngopharynx. The **nasopharynx** is located above the soft palate and serves as a passageway for air flow from the nasal cavity. A

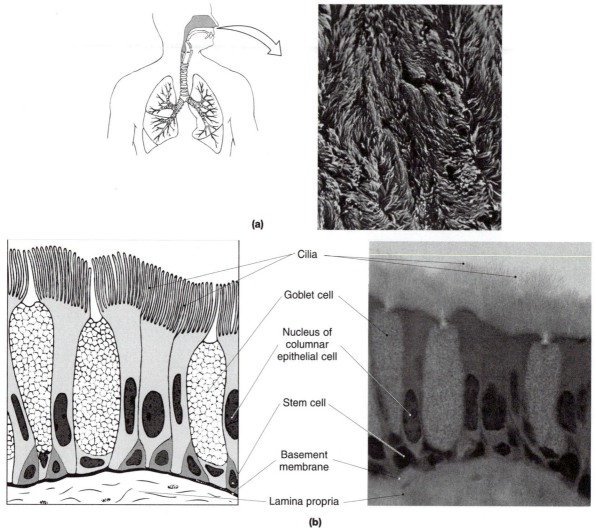

(a)

Cilia

Goblet cell

Nucleus of columnar epithelial cell

Stem cell

Basement membrane

Lamina propria

(b)

●**Figure 16.2**
The Respiratory Epithelium of the Nasal Cavity

(a) A surface view of the epithelium, as seen with the scanning electron microscope. The cilia of the epithelial cells form a dense layer that resembles a shag carpet. The movement of these cilia propels mucus across the epithelial surface. (SEM × 1614)
(b) Sketch and light micrograph showing the sectional appearance of the respiratory epithelium. (LM × 1062)

pseudostratified ciliated columnar epithelium lines the nasopharynx to warm, moisten, and clean inspired air. During swallowing, food pushes past the uvula, and the soft palate raises to prevent food from entering the nasal cavity.

Located on the posterior wall of the nasopharynx is the **pharyngeal tonsil**, commonly known as the adenoid. On the lateral walls are the openings of the **pharyngotympanic tubes**. These tubes allow for the equalization of pressure between the middle chamber of the ears and the atmosphere.

The oral cavity joins the throat at an opening called the **fauces**. The **oropharynx** extends from the soft palate down to the epiglottis. It contains the **palatine** and **lingual** tonsils. The **laryngopharynx** is located between the hyoid bone and the entrance to the esophagus. Because food passes through the oropharynx and laryngopharynx, these areas are lined with a stratified squamous epithelium.

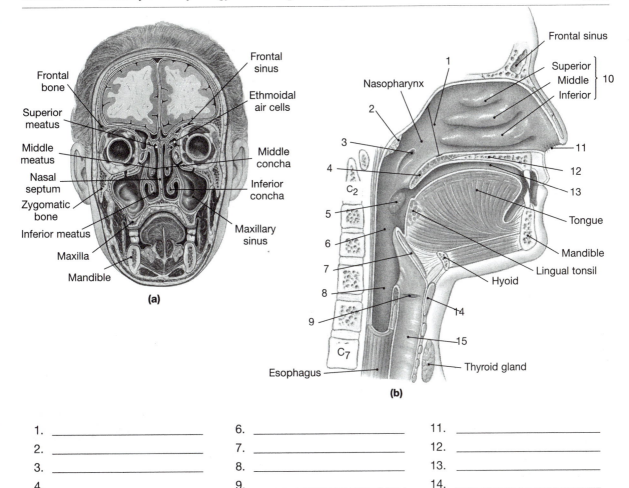

1. _____   6. _____   11. _____
2. _____   7. _____   12. _____
3. _____   8. _____   13. _____
4. _____   9. _____   14. _____
5. _____   10. _____ 15. _____

●**Figure 16.3**
The Nose, Nasal Cavity, and Pharynx

(a) The meatuses and the maxillary and ethmoidal sinuses. (b) The nasal cavity and pharynx as seen in sagittal section, with the nasal septum removed.

Laboratory Activity   THE NOSE AND PHARYNX_____

MATERIALS

head model
respiratory system chart
hand mirror

PROCEDURES

1. Label and review the gross anatomy of the nose in Figure 16.3. Locate these structures on the models and charts available in the laboratory.
2. Label and review the three regions of the pharynx in Figure 16.3. Locate each region on the laboratory models.
3. Using a hand mirror, examine the inside of your mouth. Locate your hard and soft palates, uvula, palatine tonsils, and oropharynx.

B. THE Larynx

The **larynx** or voice box (see Figure 16.4) is a cartilaginous structure that joins the laryngopharynx with the trachea. Three cartilages, the thyroid cartilage, the epiglottis, and the cricoid cartilage, form the body of the larynx. The **thyroid cartilage** or Adam's apple consists of hyaline cartilage. It is visible under the skin on the anterior neck, especially in males. The **cricoid cartilage** is a ring of hyaline cartilage that connects the trachea to the base of the larynx. The **epiglottis** is a tongue-shaped piece of elastic cartilage that falls over the opening or **glottis** during swallowing to prevent ingested food from entering the respiratory tract.

Spanning the glottis from the thyroid cartilage are two pairs of folds. The upper pair of folds are the **false vocal cords**. These folds prevent foreign materials from entering the glottis and tightly close the glottis during coughing and sneezing. The lower pair of folds are the **true vocal cords**. To produce sounds, air is exhaled over the true vocal cords. Laryngeal muscles adjust the tension on the cords to change the pitch or frequency of the sound. Tightly stretched cords produce a high-pitched sound.

Laboratory Activity   THE LARYNX _____

MATERIALS

head model
torso model
respiratory system chart

PROCEDURES

1. Label and review the gross anatomy of the larynx in Figure 16.4. Locate these structures on the models and charts available in the laboratory.

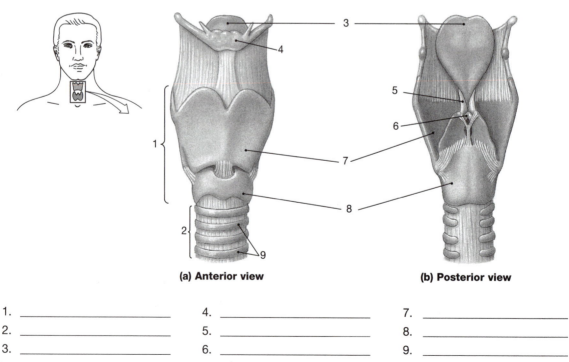

(a) Anterior view   **(b) Posterior view**

1. _____   4. _____   7. _____
2. _____   5. _____   8. _____
3. _____   6. _____   9. _____

●**Figure 16.4**
Anatomy of the Larynx and Vocal Cords

(a) Anterior view of the larynx. **(b)** Posterior view of the larynx.

2. Put your finger on your thyroid cartilage, and swallow. How did your thyroid cartilage move when you swallowed? Is it possible to swallow and make a sound simultaneously? _____

_____

3. While holding your thyroid cartilage, make a high-pitched and then a low-pitched sound. Describe the tension of the true vocal cords for each sound.

_____

_____

C. The Trachea and Bronchial Tree

The **trachea** or windpipe, shown in Figure 16.1, is a tubular structure approximately 11 cm (4.25 inches) in length and 2.5 cm (1 inch) in diameter. It lies anterior to the esophagus and can be felt on the front of the neck below the thyroid cartilage of the larynx. Along the length of the trachea are 15 to 20 C-shaped pieces of hyaline cartilage, called the **tracheal cartilage**, that keep the airway open. The trachea is lined with pseudostratified ciliated columnar epithelium. Palpate your trachea for the tracheal cartilage rings.

The trachea divides into the left and right **primary bronchi**. Each bronchus branches into increasingly smaller passageways to conduct air into the lungs (see Figure 16.5). Objects that are accidentally aspirated (inhaled) often enter the right primary bronchus because it is wider and more vertical than the left bronchus. The primary bronchi branch into as many secondary bronchi as there are lobes of each lung. The right lung has three lobes, and each lobe receives a secondary bronchus to supply it with air. The left lung has two lobes, and two secondary bronchi branch off the left primary bronchus. This branching pattern formed by the divisions of the bronchial structures is called the **bronchial tree**.

The secondary bronchi divide into **tertiary bronchi** that enter a bronchopulmonary segment of a lobe. Smaller divisions called **bronchioles** enter lobules and branch into **respiratory bronchioles**. These narrow tubules divide into **alveolar ducts** that connect to clusters of **alveoli** called **alveolar sacs**. Exchange of gases occur across the membranes of the alveoli and pulmonary capillaries.

As the bronchial tree branches from the primary bronchi to the terminal bronchioles, cartilage is gradually replaced with smooth muscle tissue. The epithelial lining of the bronchial tree also changes from pseudostratified ciliated columnar in the primary bronchus to simple squamous epithelium in the alveoli.

Laboratory Activity   THE TRACHEA AND BRONCHIAL TREE _____

MATERIALS

head model	torso model
lung model	respiratory system chart

PROCEDURES

1. Review the gross anatomy of the trachea in Figures 16.1 and 16.5. Locate these structures on the models and charts available in the laboratory.
2. Study the bronchial tree on the laboratory models, and identify the primary bronchi, secondary bronchi, tertiary bronchi, bronchioles, and respiratory bronchioles.

D. The Lungs

The lungs are a pair of cone-shaped organs lying in the pleural cavities (see Figure 16.6). The **apex** is the conical top of each lung, and the broad inferior portion is the

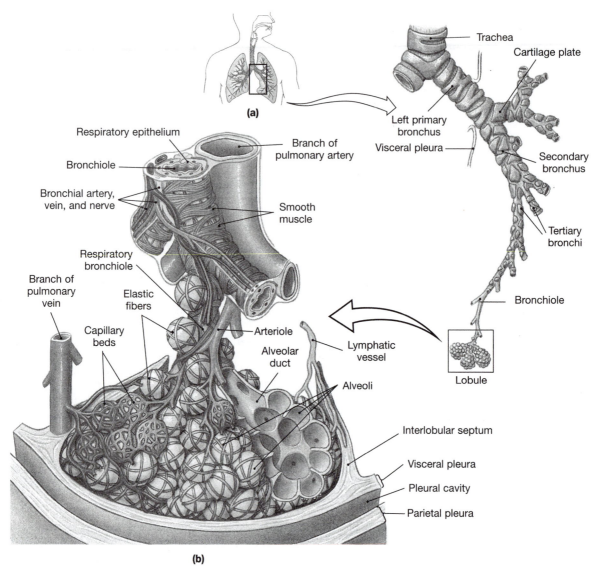

●Figure 16.5
The Bronchi and Lobules of the Lung
(a) Branching pattern of bronchi in the left lung, simplified. **(b)** The structure of a single lobule, part of a bronchopulmonary segment.

base. The heart lies on a medial concavity of the left lung called the **cardiac notch**. Each lung has a **hilus**, a medial slit where the bronchial tubes, vascularization, lymphatics, and nerves reach the lung. Each lung is partitioned into lobes by deep fissures. The left lung is divided by an **oblique fissure** into the **superior** and **inferior lobes**. The right lung has three lobes, the **superior**, **middle**, and **inferior lobes**. The superior and middle lobes are separated by a **horizontal fissure**, and the **oblique fissure** separates the inferior lobe. The mediastinum divides the thoracic cavity into two pleural cavities, each containing a lung. The pleural cavity is lined with a serous membrane, the **pleura**. The **parietal pleura** lines the thoracic wall, and the **visceral pleura** covers the superficial surface of the lung. The pleurae produce a slippery **serous fluid** that reduces friction and adhesion between the lungs and the thoracic wall during breathing.

1. _____
2. _____
3. _____
4. _____
5. _____
6. _____
7. _____
8. _____
9. _____
10. _____

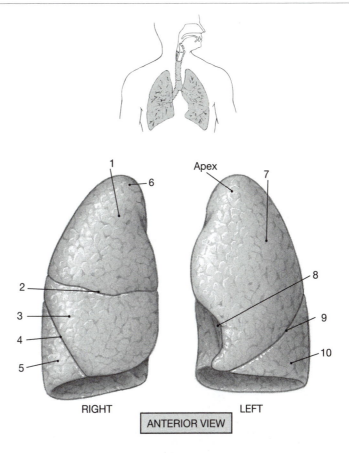

●Figure 16.6
Gross Anatomy of the Lungs

Laboratory Activity   THE LUNGS_____

MATERIALS

lung model
torso model
respiratory system chart

compound microscope
prepared slide:
   lung tissue

PROCEDURES

1. Label and review the gross anatomy of the lungs in Figure 16.6. Locate these structures on the models and charts available in the laboratory.

2. Observe a prepared slide of lung.

a. Notice the thin, fragile appearance of the walls of the alveoli.

b. Identify the alveoli and an alveolar duct using Figure 16.7 as reference.

PART TWO   PHYSIOLOGY OF THE RESPIRATORY SYSTEM

WORD POWER	MATERIALS
spirometer (spiro—to breath)	dry spirometers
emphysema (em—in, physema-a blowing)	disposable mouthpieces
apnea (apnoia—want of breath)	

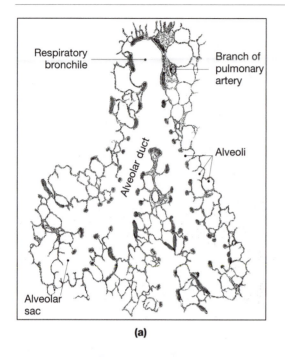

(a)

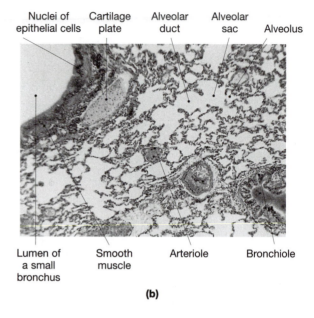

(b)

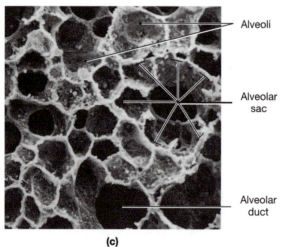

(c)

●Figure 16.7
The Bronchioles and Alveoli of the Lungs

(a) Distribution of a respiratory bronchiole supplying
a portion of a lobule. (b) The alveolar sacs and alveoli.
(LM × 42) (c) An SEM of the lung. Note the open,
spongy appearance of the lung tissue.

OBJECTIVES

On completion of this part of the exercise, you should be able to:

- Discuss the processes of pulmonary ventilation, internal respiration, and external respiration.
- Describe the activities of the respiratory muscles during inspiration and expiration.
- Define the various lung capacities, and explain how they are measured.
- Demonstrate how to use a dry spirometer.

The respiratory process involves three distinct phases: pulmonary ventilation, external respiration, and internal respiration. Breathing or **pulmonary ventilation** is the movement of air into and out of the lungs. This requires coordinated contractions of the diaphragm, the intercostal muscles, and the abdominal muscles. **External respiration** is the exchange of gases between the lungs and the blood. Inhaled air is rich in oxygen, and this gas constantly diffuses through the alveolar wall of the lungs and into the blood of the pulmonary capillaries. Simultaneously, carbon dioxide diffuses out of the blood and into the lungs, where it is exhaled. The freshly oxygenated blood is pumped to the tissues to deliver the oxygen and take up carbon dioxide from the tissues. **Internal respiration** is the exchange of gases between the blood and the tissues. Cellular respiration is the process by which cells use oxygen to produce ATP, carbon dioxide, and water.

Pulmonary **ventilation** or breathing consists of two separate processes, inspiration and expiration. **Inspiration** is inhalation, the replenishing of oxygen-rich air into the lungs. **Expiration**, or exhalation, involves emptying the carbon-dioxide-laden air from the lungs into the atmosphere. The average respiratory rate is approximately 12 breaths per minute. This rate is modified by many factors, such as exercise and stress, which increase respiration rates, or sleep and depression, which slow respiration rates.

To ventilate the lungs, a pressure difference must exist between the surrounding air pressure, the atmospheric pressure, and the pressure in the thoracic cavity. Atmospheric pressure is normally 760 mm of mercury (mm Hg) or approximately 15 pounds per square inch. To inhale, the pressure in the thoracic cavity must be lower than the surrounding air pressure. Boyle's law explains how changing the size of the thoracic cavity and the lungs modifies pulmonary pressures to create a pressure gradient for breathing. **Boyle's law** states the pressure of a gas, in a closed container, is inversely proportional to the volume of the container. Simply put, if a sealed container is made smaller, the gas in the container is pushed together and therefore has a higher pressure.

Figure 16.8 illustrates the mechanisms of pulmonary ventilation. When the diaphragm is relaxed, it is dome shaped. On contraction, it lowers and flattens the floor of the thoracic cavity. This results in an increase in thoracic volume and a subsequent decrease in the thoracic pressure. Simultaneously, the external intercostal muscles contract and elevate the rib cage and further increase the thoracic volume. An expansion of the thoracic walls and a decrease in thoracic pressure cause a concurrent expansion of the lungs and a decrease in lung pressure, the **intrapulmonic pressure**. Once intrapulmonic pressure falls below atmospheric pressure, air will begin to flow into the lungs.

Inspiration is an active process because it requires the contraction of several muscles to change pulmonary volumes and pressures. Expiration is essentially a passive process caused when the muscles for inspiration relax and the thoracic wall and elastic lung tissue recoil. During exercise, however, air may be actively exhaled by the combined contractions of the internal intercostal muscles and the abdominal muscles. The internal intercostal muscles depress the rib cage, and the abdominal muscles push the diaphragm higher into the thoracic cavity. Both actions decrease the thoracic volume and increase the thoracic pressure that forces more air out during exhalation.

A. Lung Volumes and Capacities

The respiratory system must accommodate the oxygen demands of the body. During exercise, the respiratory system must supply the muscular system with more oxygen. The respiratory rate increases, as does the volume of inhaled and exhaled gases. In this section, you will measure a variety of your respiratory volumes.

An instrument called a **spirometer** is used to measure respiratory volumes. The dry spirometer is a hand-held unit with disposable mouthpieces. This unit employs a small turbine that spins to measure the amount of air you **exhale**. Keep in mind during your volume measurements that the volumes vary according to the individual's sex, height, age, and general physical condition. Refer to Figure 16.9, and compare the relationships of the following lung volumes.

Tidal volume (TV) is the amount of air one inspires and exhales during normal resting breathing. Tidal volume averages 500 ml. Additional air can be inhaled or exhaled beyond the tidal volume. The **inspiratory reserve volume** (IRV) is the amount of air that can be forcibly inspired above a normal inhalation. This volume averages 3,300 ml. The amount of air that can be forcibly exhaled after a normal exhalation is the **expiratory reserve volume** (ERV). The ERV averages 1,000 ml.

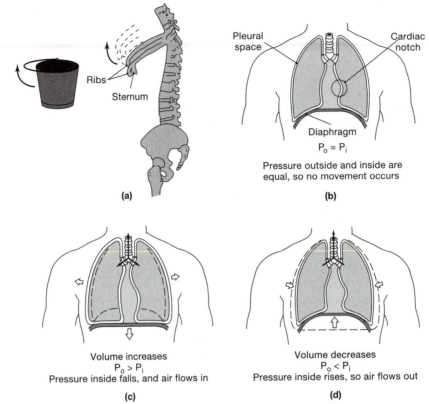

● **Figure 16.8**
Mechanisms of Pulmonary Ventilation
(a) As the ribs are elevated or the diaphragm depressed, the thoracic cavity increases in volume. **(b)** Anterior view at rest, with no air movement. **(c)** *Inhalation:* Elevation of the rib cage and contraction of the diaphragm increase the size of the thoracic cavity. Pressure decreases, and air flows into the lungs. **(d)** *Exhalation:* When the rib cage returns to its original position, the volume of the thoracic cavity decreases. Pressure rises, and air moves out of the lungs.

Vital capacity (VC) is the maximum amount of air that may be exhaled from the lungs after a maximum inhalation. This volume averages 4,500 ml to 5,500 ml and includes the combined volumes of the IRV, TV, and ERV: VC = IRV + TV + ERV.

The respiratory system always contains some air. The **residual volume** is the amount of air that cannot be forcibly exhaled from the lungs. Surfactant produced by the septal cells of the alveoli prevents the alveoli from collapsing completely during exhalation. Since the alveoli are not allowed to completely empty, they will always maintain a resident volume of air. This volume averages 1,200 ml.

To calculate the **total lung capacity** (TLC), the vital capacity is added to the residual volume: TLC = VC + RV. Total lung capacity averages 5,900 ml.

Respiratory rate (RR) is the number of breaths taken per minute. If the RR is multiplied by the tidal volume, the **minute volume** (MV) is obtained, the amount of air exchanged between the lungs and the environment in one minute: MV = TV x RR.

Laboratory Activity   SPIROMETRY

MATERIALS

dry spirometer
disposable mouthpieces

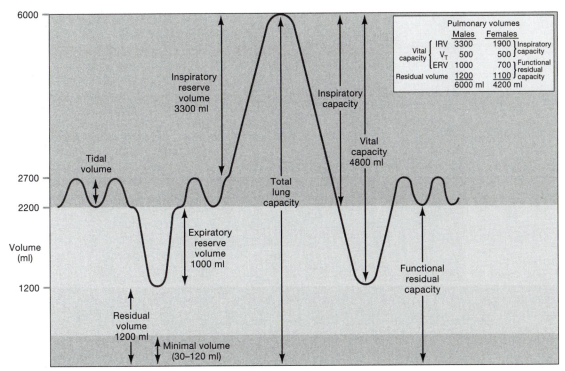

●Figure 16.9
Respiratory Volumes and Capacities
The graph diagrams the relationships between the respiratory volumes and capacities
of an average male.

Safety Tips

- Do not use the spirometer if you have a cold or a communicable disease.
- Always use a clean mouthpiece on the spirometer. Do not reuse a mouthpiece.
- Only exhale into the dry spirometer.

PROCEDURES

1. Insert a new, clean plastic mouthpiece onto the breathing tube of the spirometer.

2. Only *exhale* into the dry spirometer; the instrument cannot measure inspiratory volumes. To obtain as accurate a reading as possible, use a nose clip or your fingers to pinch your nose closed while exhaling into the spirometer.

3. Before using a spirometer, you will need to reset the dial face to "zero." This is accomplished by turning the silver ring around the face of the dial (see Figure 16.10). As you turn this ring, you will notice the scale moving also. Set the zero even with the needle of the dial.

4. You will conduct the test for each volume three times and then calculate the average volume of that respiratory volume.

I. Tidal Volume (TV)

PROCEDURES

1. Set the dial face to 1,000. Tidal volume is a small volume, and the scale on most dry spirometers is not graded before the 1,000 ml setting.

**●Figure 16.10
Dry Spirometer**

2. Take a normal breath and exhale normally into the spirometer. The exhalation should not be forcibly exhaled but should be more of a sigh. Repeat this test twice, and then calculate your average tidal volume. Record your data in Table 16.1.

II. Expiratory Reserve Volume (ERV)

PROCEDURES

1. Set the dial face to zero.
2. After a normal exhalation:
   a. Stop breathing.
   b. Pinch nose with fingers or nose clip.
   c Forcibly exhale all remaining air.
3. Repeat twice, and then record your readings in Table 16.1.

TABLE 16.1	Spirometry Data			
Volume	*Reading 1*	*Reading 2*	*Reading 3*	*Average*
Tidal volume				
Expiratory reserve volume				
Vital capacity				
Respiratory rate				
Minute volume (calculated)				
Inspiratory reserve volume (calculated)				

III. Vital Capacity (VC)

PROCEDURES

1. Set the dial face to "zero."
2. Inhale maximally once, and then exhale maximally.

3. Inhale maximally and then:
   a. Insert the mouthpiece into your mouth.
   b. Pinch nose with fingers or nose clip.
   c. Forcibly exhale all air from your lungs.
4. Repeat the measurement twice, and then record your data in Table 16.1.
5. Compare your vital capacity with the predicted vital capacities for females and males in Table 16.2. To use these tables, find your height across the top of the chart and your age along the side of the chart. (Multiply your height in inches times 2.54 to get your height in centimeters.) Follow the age row across and the height column down. The box where the two intersect gives the value of the average vital capacity of the population of individuals of your height and sex.

Your vital capacity might differ from the average listed in the chart. Genetics has an influence on your potential lung capacities. Damage from smoking or environmental factors such as air pollution decrease vital capacity. Cardiovascular exercises such as swimming and jogging increase lung volumes.

IV. Respiratory Rate (RR) Determination

PROCEDURES

1. Have your partner observe your breathing rate while you read a textbook. Record the number of respirations that occur in 60 seconds.
2. Record your respiratory rate in Table 16.1.
3. Calculate the minute volume by multiplying your tidal volume by your respiratory rate. Enter the calculation in Table 16.1.

V. Inspiratory Reserve Volume (IRV) Calculation

Although the dry spirometer cannot measure your IRV, it can be calculated from other volumes: IRV = VC–(TV + ERV).

PROCEDURES

1. Calculate your IRV and enter it in Table 16.1.

TABLE 16.2A   **Predicted Vital Components—Females**

Height in centimeters and inches

Age	cm 152 in. 59.8	154 60.6	156 61.4	158 62.2	160 63.0	162 63.7	164 64.6	166 65.4	168 66.1	170 66.9	172 67.7	174 68.5	176 69.3	178 70.1	180 70.9	182 71.7	184 72.4	186 73.2	188 74.0
16	3,070	3,110	3,150	3,190	3,230	3,270	3,310	3,350	3,390	3,430	3,470	3,510	3,550	3,590	3,630	3,670	3,715	3,755	3,800
17	3,055	3,095	3,135	3,175	3,215	3,255	3,295	3,335	3,375	3,415	3,455	3,495	3,535	3,575	3,615	3,655	3,695	3,740	3,780
18	3,040	3,080	3,120	3,160	3,200	3,240	3,280	3,320	3,360	3,400	3,440	3,480	3,520	3,560	3,600	3,640	3,680	3,720	3,760
20	3,010	3,050	3,090	3,130	3,170	3,210	3,250	3,290	3,330	3,370	3,410	3,450	3,490	3,525	3,565	3,605	3,645	3,695	3,720
22	2,980	3,020	3,060	3,095	3,135	3,175	3,215	3,255	3,290	3,330	3,370	3,410	3,450	3,490	3,530	3,570	3,610	3,650	3,685
24	2,950	2,985	3,025	3,065	3,100	3,140	3,180	3,220	3,260	3,300	3,335	3,375	3,415	3,455	3,490	3,530	3,570	3,610	3,650
26	2,920	2,960	3,000	3,035	3,070	3,110	3,150	3,190	3,230	3,265	3,300	3,340	3,380	3,420	3,455	3,495	3,530	3,570	3,610
28	2,890	2,930	2,965	3,000	3,040	3,070	3,115	3,155	3,190	3,230	3,270	3,305	3,345	3,380	3,420	3,460	3,495	3,535	3,570
30	2,860	2,895	2,935	2,970	3,010	3,045	3,085	3,120	3,160	3,195	3,235	3,270	3,310	3,345	3,385	3,420	3,460	3,495	3,535
32	2,825	2,865	2,900	2,940	2,975	3,015	3,050	3,090	3,125	3,160	3,200	3,235	3,275	3,310	3,350	3,385	3,425	3,460	3,495
34	2,795	2,835	2,870	2,910	2,945	2,980	3,020	3,055	3,090	3,130	3,165	3,200	3,240	3,275	3,310	3,350	3,385	3,425	3,460
36	2,765	2,805	2,840	2,875	2,910	2,950	2,985	3,020	3,060	3,095	3,130	3,165	3,205	3,240	3,275	3,310	3,350	3,385	3,420
38	2,735	2,770	2,810	2,845	2,880	2,915	2,950	2,990	3,025	3,060	3,095	3,130	3,170	3,205	3,240	3,275	3,310	3,350	3,385
40	2,705	2,740	2,775	2,810	2,850	2,885	2,920	2,955	2,990	3,025	3,060	3,095	3,135	3,170	3,205	3,240	3,275	3,310	3,345
42	2,675	2,710	2,745	2,780	2,815	2,850	2,885	2,920	2,955	2,990	3,025	3,060	3,100	3,135	3,170	3,205	3,240	3,275	3,310
44	2,645	2,680	2,715	2,750	2,785	2,820	2,855	2,890	2,925	2,960	2,995	3,030	3,060	3,095	3,130	3,165	3,200	3,235	3,270
46	2,615	2,650	2,685	2,715	2,750	2,785	2,820	2,855	2,890	2,925	2,960	2,995	3,030	3,060	3,095	3,130	3,165	3,200	3,235
48	2,585	2,620	2,650	2,685	2,715	2,750	2,785	2,820	2,855	2,890	2,925	2,960	2,995	3,030	3,060	3,095	3,130	3,160	3,195
50	2,555	2,590	2,625	2,655	2,690	2,720	2,755	2,785	2,820	2,855	2,890	2,925	2,955	2,990	3,025	3,060	3,090	3,125	3,155
52	2,525	2,555	2,590	2,625	2,655	2,690	2,720	2,755	2,790	2,820	2,855	2,890	2,925	2,955	2,990	3,020	3,055	3,090	3,125
54	2,495	2,530	2,560	2,590	2,625	2,655	2,690	2,720	2,755	2,790	2,820	2,855	2,885	2,920	2,950	2,985	3,020	3,050	3,085
56	2,460	2,495	2,525	2,560	2,590	2,625	2,655	2,690	2,720	2,755	2,790	2,820	2,855	2,885	2,920	2,950	2,980	3,015	3,045
58	2,430	2,460	2,495	2,525	2,560	2,590	2,625	2,655	2,690	2,720	2,750	2,785	2,815	2,850	2,880	2,920	2,945	2,975	3,010
60	2,400	2,430	2,460	2,495	2,525	2,560	2,590	2,625	2,655	2,685	2,720	2,750	2,780	2,810	2,845	2,875	2,915	2,940	2,970
62	2,370	2,405	2,435	2,465	2,495	2,525	2,560	2,590	2,620	2,655	2,685	2,715	2,745	2,775	2,810	2,840	2,870	2,900	2,935
64	2,340	2,370	2,400	2,430	2,465	2,495	2,525	2,555	2,585	2,620	2,650	2,680	2,710	2,740	2,770	2,805	2,835	2,865	2,895
66	2,310	2,340	2,370	2,400	2,430	2,460	2,495	2,525	2,555	2,585	2,615	2,645	2,675	2,705	2,735	2,765	2,800	2,825	2,860
68	2,280	2,310	2,340	2,370	2,400	2,430	2,460	2,490	2,520	2,550	2,580	2,610	2,640	2,670	2,700	2,730	2,760	2,795	2,820
70	2,250	2,280	2,310	2,340	2,370	2,400	2,425	2,455	2,485	2,515	2,545	2,575	2,605	2,635	2,665	2,695	2,725	2,755	2,780
72	2,220	2,250	2,280	2,310	2,335	2,365	2,395	2,425	2,455	2,480	2,510	2,540	2,570	2,600	2,630	2,660	2,685	2,715	2,745
74	2,190	2,220	2,245	2,275	2,305	2,335	2,360	2,390	2,420	2,450	2,475	2,505	2,535	2,565	2,590	2,620	2,650	2,680	2,710

TABLE 16.2B   Predicted Vital Capacities—Males

Height in centimeters and inches

cm	152	154	156	158	160	162	164	166	168	170	172	174	176	178	180	182	184	186	188
Age in.	59.8	60.6	61.4	62.2	63.0	63.7	64.6	65.4	66.1	66.9	67.7	68.5	69.3	70.1	70.9	71.7	72.4	73.2	74.0
16	3,920	3,975	4,025	4,075	4,130	4,180	4,230	4,285	4,335	4,385	4,440	4,490	4,540	4,590	4,645	4,695	4,745	4,800	4,850
18	3,890	3,940	3,995	4,045	4,095	4,145	4,200	4,250	4,300	4,350	4,405	4,455	4,505	4,555	4,610	4,660	4,710	4,760	4,815
20	3,860	3,910	3,960	4,015	4,065	4,115	4,165	4,215	4,265	4,320	4,370	4,420	4,470	4,520	4,570	4,625	4,675	4,725	4,775
22	3,830	3,880	3,930	3,980	4,030	4,080	4,135	4,185	4,235	4,285	4,335	4,385	4,435	4,485	4,535	4,585	4,635	4,685	4,735
24	3,785	3,835	3,885	3,935	3,985	4,035	4,085	4,135	4,185	4,235	4,285	4,330	4,380	4,430	4,480	4,530	4,580	4,630	4,680
26	3,755	3,805	3,855	3,905	3,955	4,000	4,050	4,100	4,150	4,200	4,250	4,300	4,350	4,395	4,445	4,495	4,545	4,595	4,645
28	3,725	3,775	3,820	3,870	3,920	3,970	4,020	4,070	4,115	4,165	4,215	4,265	4,310	4,360	4,410	4,460	4,510	4,555	4,605
30	3,695	3,740	3,790	3,840	3,890	3,935	3,985	4,035	4,080	4,130	4,180	4,230	4,275	4,325	4,375	4,425	4,470	4,520	4,570
32	3,665	3,710	3,760	3,810	3,855	3,905	3,950	4,000	4,050	4,095	4,145	4,195	4,240	4,290	4,340	4,385	4,435	4,485	4,530
34	3,620	3,665	3,715	3,760	3,810	3,855	3,905	3,950	4,000	4,045	4,095	4,140	4,190	4,225	4,285	4,330	4,380	4,425	4,475
36	3,585	3,635	3,680	3,730	3,775	3,825	3,870	3,920	3,965	4,010	4,060	4,105	4,155	4,200	4,250	4,295	4,340	4,390	4,435
38	3,555	3,605	3,650	3,695	3,745	3,790	3,840	3,885	3,930	3,980	4,025	4,070	4,120	4,165	4,210	4,260	4,305	4,350	4,400
40	3,525	3,575	3,620	3,665	3,710	3,760	3,805	3,850	3,900	3,945	3,990	4,035	4,085	4,130	4,175	4,220	4,270	4,315	4,360
42	3,495	3,540	3,590	3,635	3,680	3,725	3,770	3,820	3,865	3,910	3,955	4,000	4,050	4,095	4,140	4,185	4,230	4,280	4,325
44	3,450	3,495	3,540	3,585	3,630	3,675	3,725	3,770	3,815	3,860	3,905	3,950	3,995	4,040	4,085	4,130	4,175	4,220	4,270
46	3,420	3,465	3,510	3,555	3,600	3,645	3,690	3,735	3,780	3,825	3,870	3,915	3,960	4,005	4,050	4,095	4,140	4,185	4,230
48	3,390	3,435	3,480	3,525	3,570	3,615	3,655	3,700	3,745	3,790	3,835	3,880	3,925	3,970	4,015	4,060	4,105	4,150	4,190
50	3,345	3,390	3,430	3,475	3,520	3,565	3,610	3,650	3,695	3,740	3,785	3,830	3,870	3,915	3,960	4,005	4,050	4,090	4,135
52	3,315	3,353	3,400	3,445	3,490	3,530	3,575	3,620	3,660	3,705	3,750	3,795	3,835	3,880	3,925	3,970	4,010	4,055	4,100
54	3,285	3,325	3,370	3,415	3,455	3,500	3,540	3,585	3,630	3,670	3,715	3,760	3,800	3,845	3,890	3,930	3,975	4,020	4,060
56	3,255	3,295	3,340	3,380	3,425	3,465	3,510	3,550	3,595	3,640	3,680	3,725	3,765	3,810	3,850	3,895	3,940	3,980	4,025
58	3,210	3,250	3,290	3,335	3,375	3,420	3,460	3,500	3,545	3,585	3,630	3,670	3,715	3,755	3,800	3,840	3,880	3,925	3,965
60	3,175	3,220	3,260	3,300	3,345	3,385	3,430	3,470	3,500	3,555	3,595	3,635	3,680	3,720	3,760	3,805	3,845	3,885	3,930
62	3,150	3,190	3,230	3,270	3,310	3,350	3,390	3,440	3,480	3,520	3,560	3,600	3,640	3,680	3,730	3,770	3,810	3,850	3,890
64	3,120	3,160	3,200	3,240	3,280	3,320	3,360	3,400	3,440	3,490	3,530	3,570	3,610	3,650	3,690	3,730	3,770	3,810	3,850
66	3,070	3,110	3,150	3,190	3,230	3,270	3,310	3,350	3,390	3,430	3,470	3,510	3,550	3,600	3,640	3,680	3,720	3,760	3,800
68	3,040	3,080	3,120	3,160	3,200	3,240	3,280	3,320	3,360	3,400	3,440	3,480	3,520	3,560	3,600	3,640	3,680	3,720	3,760
70	3,010	3,050	3,090	3,130	3,170	3,210	3,250	3,290	3,330	3,370	3,410	3,450	3,480	3,520	3,560	3,600	3,640	3,680	3,720
72	2,980	3,020	3,060	3,100	3,140	3,180	3,210	3,250	3,290	3,330	3,370	3,410	3,450	3,490	3,530	3,570	3,610	3,650	3,680
74	2,930	2,970	3,010	3,050	3,090	3,130	3,170	3,200	3,240	3,280	3,320	3,360	3,400	3,440	3,470	3,510	3,550	3,590	3,630

Predicted Vital Capacities (a) Female (b) Male. From Archives of Environmental Health, Volume 12, pp 146–189, February 1966. Reprinted with permission of the Helen Dwight Reid Educational Foundation. Published by Heldref Publications, 4000 Albemarle St., N.W., Washington, D.C. 20016. Copyright © 1966.

ANATOMY AND PHYSIOLOGY OF THE RESPIRATORY SYSTEM
Exercise 16 Laboratory Report

A. Short Answer Questions

1. List the parts of the upper and lower respiratory system.

2. What are the functions of the superior, middle, and inferior meatuses?

3. Where are the pharyngeal tonsils located?

4. What is the function of goblet cells?

5. Why are the oropharynx and laryngopharynx lined with stratified squamous epithelium?

6. Which type of epithelium lines the trachea and bronchial tubes? What specialized function does this tissue perform?

B. Matching

Match the answer on the right with the short definition on the left.

1. _____ C rings	A.	voice box
2. _____ internal intercostals	B.	tissue of epiglottis
3. _____ cricoid cartilage	C.	serous membrane of lung
4. _____ pleura	D.	left lung
5. _____ elastic cartilage	E.	elevate rib cage
6. _____ larynx	F.	tracheal cartilage
7. _____ diaphragm	G.	right lung
8. _____ cardiac notch	H.	muscular thoracic floor
9. _____ external intercostals	I.	depress ribs cage
10. _____ has three lobes	J.	base of larynx

C. Short Answer Questions

1. Define the following:
   a. pulmonary ventilation
   b. intrapulmonary pressure
   c. intrapleural pressure

2. Define the following, and give the normal value for each:
    a. tidal volume
    b. expiratory reserve volume
    c. vital capacity
    d. residual volume
    e. minute volume
    f. respiratory rate
3. Describe how to calculate the inspiratory reserve volume.

4. Use Boyle's law to explain the process of pulmonary ventilation.

D. Completion

1. Complete Figure 16.11 by labeling the structures of the thorax and lungs.

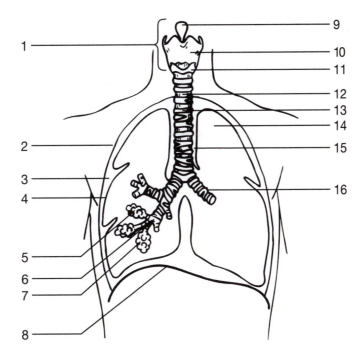

1. _____	7. _____	13. _____
2. _____	8. _____	14. _____
3. _____	9. _____	15. _____
4. _____	10. _____	16. _____
5. _____	11. _____	
6. _____	12. _____	

●**Figure 16.11**
Thorax and Lungs

17 ANATOMY AND PHYSIOLOGY OF THE DIGESTIVE SYSTEM

The digestive system processes food into usable forms for cellular metabolism. The digestive process includes (1) ingestion of food into the mouth, (2) movement of food through the digestive tract, (3) digestion of food by mechanical and enzymatic activities, (4) absorption of nutrients into the blood, and (5) formation and elimination of indigestible material and waste. To accomplish these functions, the digestive system is highly specialized. Food passes through a tubular **digestive tract** that extends from the mouth to the anus. Each organ of the tract has a specific function in the digestive process. **Accessory organs**, which include the salivary glands, liver, gallbladder, and pancreas, occur outside the digestive tract. These organs manufacture enzymes and other compounds that help in the digestive process. In this exercise, you will study the structure of the digestive organs. The digestive physiology section includes enzyme experiments to simulate the chemical breakdown of food.

PART ONE   ANATOMY OF THE DIGESTIVE SYSTEM

WORD POWER

mucosa (muco—mucus)
deciduous (decid—falling off)
omentum (oment—fat skin)
plicae (plico—a fold)
cecum (cec—blind)

MATERIALS

anatomical charts
torso and head model
dentition model set
stomach and intestinal model
liver and gallbladder model
pancreas model
compound microscope

prepared slides
 stomach
 ileum
 liver
 pancreas
 dissected cat for demonstrations

OBJECTIVES

On completion of this part of the exercise, you should be able to:

- Identify the major layers and tissues of the digestive tract.
- Identify all digestive anatomy on lab models, charts, and exhibits.
- Identify the histological structure of the various digestive organs.
- Trace the secretion of bile from the liver to the duodenum.
- List the organs of the digestive tract and the accessory organs.

A. Organization of the Digestive Tract

Examine the inside of your cheek with your tongue. What do you feel? The lining of your mouth and the rest of your digestive tract is a mucous membrane that is kept wet. Glands drench the lining epithelium with enzymes, mucus, pH buffers, and other compounds to orchestrate the sequential breakdown of food as it passes through your mouth, pharynx, esophagus, stomach, small intestine, large intestine, and finally, rectum. Hormones secreted by the digestive tract regulate the activity of digestive organs and glands.

Four major tissue layers compose the digestive tract, the **mucosa**, **submucosa**, **muscularis externa**, and **serosa**. Although the basic histological structure of the tract is similar along its entire length, each region has anatomical specializations that reflect that region's role in the digestive process. Figure 17.1 illustrates the anatomy of the digestive tract.

The **mucosa** lines the lumen of the digestive tract and is in contact with food passing through the tract. The mucosa consists of epithelial and connective tissues and a thin sheet of smooth muscle tissue. The mucosa is the only layer exposed to the lumen of the tract. From the mouth to the esophagus, the epithelium is **stratified squamous epithelium**. Food passing from the mouth to the stomach is still in a solid condition, and the stratified epithelium protects the mucosa from abrasion. The stomach, small intestine, and large intestine are lined with **simple columnar epithelium**. Food in these parts of the tract is liquid and less abrasive to the digestive epithelium. Beneath the epithelium is a layer of connective tissue that attaches the epithelium and contains blood vessels, lymphatic vessels, and nerves.

Superficial to the mucosa is the **submucosa**, a loose connective tissue layer containing large blood and lymphatic vessels and many nerves. Surrounding the submucosa are several layers of smooth muscle tissue in the **muscularis externa** that move and process materials in the digestive tract. The inner layer is **circular** muscle that wraps around the digestive tract, causing the tract to narrow or constrict on contraction. The outer layer is **longitudinal** muscle with fibers oriented parallel to the length of the tract. Contraction of this muscle layer shortens and widens the tract. Between the muscle layers are nerves that control the activity of the muscularis externa.

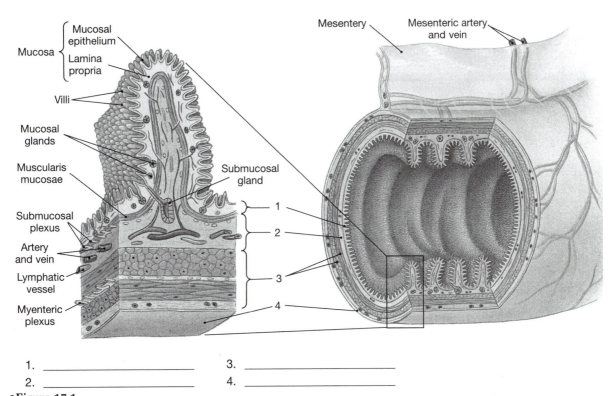

1. _____   3. _____

2. _____   4. _____

●**Figure 17.1**
Structure of the Digestive Tract

This is a diagrammatic view of a representative portion of the digestive tract. The features illustrated are those of the small intestine.

The superficial layer of the digestive tract is the **serosa**, a loose connective tissue covering that attaches and holds the tract in position. In the abdominal cavity, the serosa is called the **peritoneum**, a serous membrane covering of the stomach and intestines.

Laboratory Activity   ORGANIZATION OF THE DIGESTIVE TRACT_____

MATERIALS

intestine model   prepared microscope slide
compound microscope   ileum

PROCEDURES

1. Label and review the anatomy presented in Figure 17.1.

2. Identify the four layers of the digestive tract from a cross section of intestine or on an intestinal model.

3. Focus on the slide of the ileum using low magnification of a compound microscope.

   a. Using Figure 17.1 as reference, locate the lumen and the lining epithelium that would be in contact with digestive contents. This epithelium and the underlying connective tissue constitute the mucosa. Also locate the submucosa, identified by its loose connective tissue and many blood and lymphatic vessels.

   b. Identify the circular and longitudinal muscles of the muscularis externa and the serosa.

   c. In the space provided, draw a section of the microscope field, including all four main layers of the tract.

B. The Mouth

The mouth, or **oral cavity**, is the site of ingestion and the initial processing of food. Figure 17.2 presents the anatomy of the oral cavity in sagittal and anterior views. The oral cavity extends from the **labia** (lips) to the **fauces**, the opening between the mouth and the throat. The roof of the oral cavity consists of the **hard** and **soft palates**. A conical **uvula** is suspended from the posterior soft palate just anterior to the fauces. The floor of the mouth is muscular, mostly from muscles of the **tongue**, and the **cheeks** form the sides. A fold of tissue, the **lingual frenulum**, anchors the tongue yet allows free movement for food processing and speech.

Accessory structures of the mouth include the salivary glands and the teeth. Three pairs of salivary glands, illustrated in Figure 17.3, produce a watery fluid called saliva that contains mucus and enzymes. The **parotid gland** is the large salivary gland in front of the ear between the skin and the masseter muscle. The **parotid duct** pierces through the buccinator muscle and enters the oral cavity near the upper second molar. The **submandibular gland** is located under the mandible. The **submandibular duct**

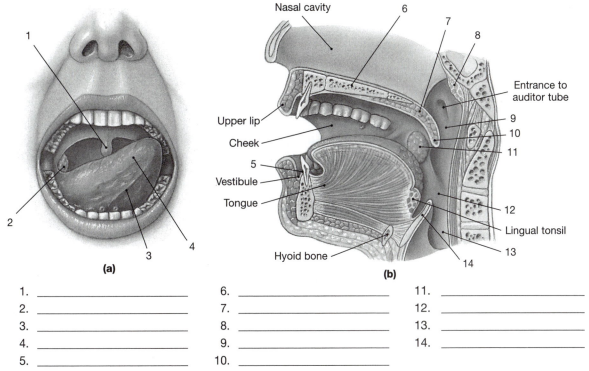

(a)

(b)

1. _____	6. _____	11. _____
2. _____	7. _____	12. _____
3. _____	8. _____	13. _____
4. _____	9. _____	14. _____
5. _____	10. _____	

●**Figure 17.2**
The Oral Cavity

(a) An anterior view of the oral cavity, as seen through the open mouth. **(b)** The oral
cavity as seen in sagittal section.

passes through the **lingual frenulum** and opens at the swelling on the central margin
of this tissue. The **sublingual glands** are located under the tongue at the floor of the
mouth. Many **lesser sublingual ducts** open along the base of the tongue.

The **teeth** are used to mechanically process food into smaller pieces. Figure 17.4a
details the anatomy of a typical adult tooth. The tooth is anchored into the alveolar
bone of the jaw by a strong **periodontal ligament** that lines the embedded part of the

●**Figure 17.3**
The Salivary Glands

Lateral view, showing the relative
positions of the salivary glands and
ducts on the left side of the head.

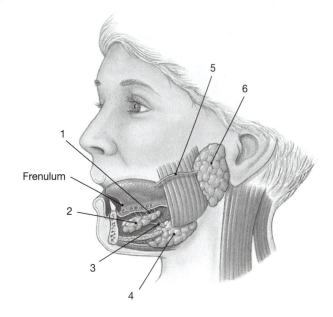

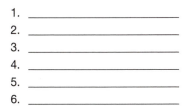

1. _____
2. _____
3. _____
4. _____
5. _____
6. _____

tooth, the **root**. The **crown** is the portion of the tooth above the **gingiva,** or gums. Although a tooth has many distinct layers, only the inner **pulp cavity** is filled with a living tissue, the **pulp**. Supplying the pulp tissue are blood vessels, lymphatic vessels, and nerves, all of which enter the pulp cavity through the **root canal**. Surrounding the pulp cavity is **dentin**, a hard, nonliving solid that composes most of the structural mass of a tooth. At the root, the dentin is covered by **cementum**, which attaches to the periodontal ligament. The exposed crown is covered with **enamel**, the hardest substance produced by living organisms.

Humans have two sets of teeth during life. The first set are the **deciduous**, which are replaced starting around the age of six by the **permanent** dentition. The permanent dentition consists of 32 teeth (see Figure 17.4b). Each quadrant or jaw quarter has a **central incisor**, a **lateral incisor**, a **bicuspid** or **canine**, two **premolars**, and three **molars**. The last molar is also called the wisdom tooth. The incisors are chisellike for snipping food such as a carrot. The bicuspids are pointed for tearing actions. The premolars and molars have large flat surfaces for grinding food into small pieces.

Laboratory Activity   THE MOUTH

MATERIALS

head and torso models
teeth models

digestive system chart
mirror

PROCEDURES

1. Label and review the anatomy presented in Figures 17.2, 17.3, and 17.4.
2. Identify the anatomy of the mouth on all laboratory models and charts. Identify each salivary gland and duct on the laboratory models.

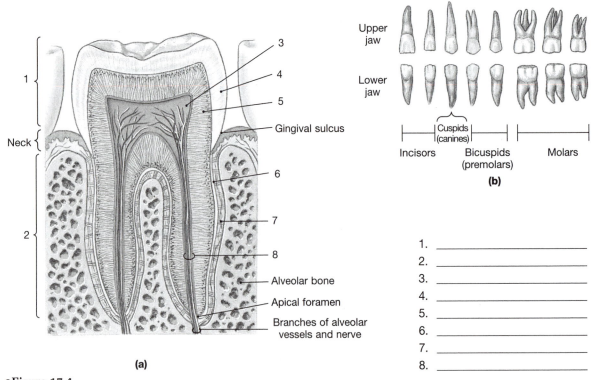

1. _____
2. _____
3. _____
4. _____
5. _____
6. _____
7. _____
8. _____

●**Figure 17.4**
Teeth

(a) Diagrammatic section through a typical adult tooth. **(b)** The adult teeth.

3. Examine your mouth in a mirror. Locate your uvula, fauces, and lingual frenulum.

_____

_____

4. Study a laboratory model of a sectioned tooth. Next, examine your teeth in a mirror. Locate your incisors, cuspids, bicuspids, and molars. How many teeth do you have? Are you missing any owing to extractions? Do you have any wisdom teeth?

_____

_____

C. The Pharynx and Esophagus

The throat or **pharynx** is a common passageway for nutrients and air. The pharynx is divided into three anatomical regions, the nasopharynx, oropharynx, and laryngopharynx, which we discussed in Exercise 16. During swallowing, or **deglutition**, muscles of the soft palate contract and close the passageway to the nasopharynx to prevent food from entering the nasal cavity. The inferior portion of the oropharynx area branches into the larynx of the respiratory system and the **esophagus** that leads to the stomach. To prevent food from entering the respiratory passageways, the larynx has a flap called the **epiglottis**, which closes the larynx during swallowing.

The food tube or **esophagus** connects the pharynx to the stomach (see Figure 17.5). It is inferior to the pharynx and posterior to the trachea, or windpipe. The esophagus is approximately 25 cm (12 inches) long. The esophagus pierces through the diaphragm and connects with the stomach in the abdominal cavity. At the stomach, the esophagus ends in a lower esophageal sphincter, a muscular valve that prevents stomach contents from backwashing into the esophagus.

Laboratory Activity   THE PHARYNX AND ESOPHAGUS _____

MATERIALS

torso model
digestive system chart

PROCEDURES

1. Identify the pharynx and esophagus on all laboratory models and charts.

D. The Stomach

The stomach is a J-shaped organ located inferior to the diaphragm in the epigastric and lower left hypochondriac regions. Locate in Figure 17.5 the four major regions of the stomach: the **cardia** is where the stomach connects with the esophagus, the **fundus** is the upper rounded area, the **body** is the middle region, and the **pylorus** is the narrowed distal end that connects to the small intestine. The **pyloric sphincter** or **valve** controls movement of material from the stomach into the small intestine. Extending from the stomach is the **greater omentum**, a part of the serosa commonly called the fatty apron. This fatty layer protects the abdominal organs and attaches the transverse colon of the large intestine to the abdominal wall. The **lesser omentum** suspends the stomach from the liver.

The inner lining of the stomach has **rugae** or folds that enable the stomach to expand as it fills with food. Unlike other regions of the digestive tract, the **muscularis externa** of the stomach contains three layers of smooth muscle instead of two. The inner layer is an oblique layer, followed by a circular layer, then an external longitudinal layer. The three muscle layers contract and churn the stomach contents into a soupy mixture called **chyme**.

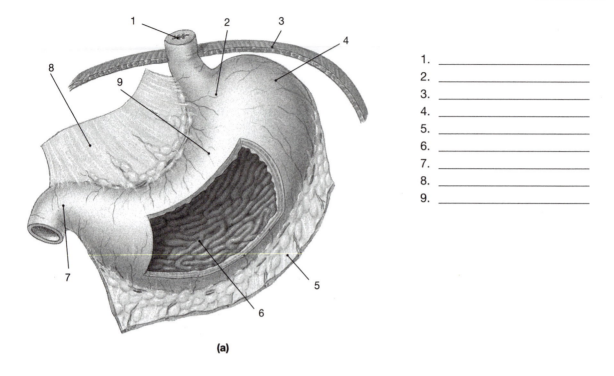

1. _____
2. _____
3. _____
4. _____
5. _____
6. _____
7. _____
8. _____
9. _____

(a)

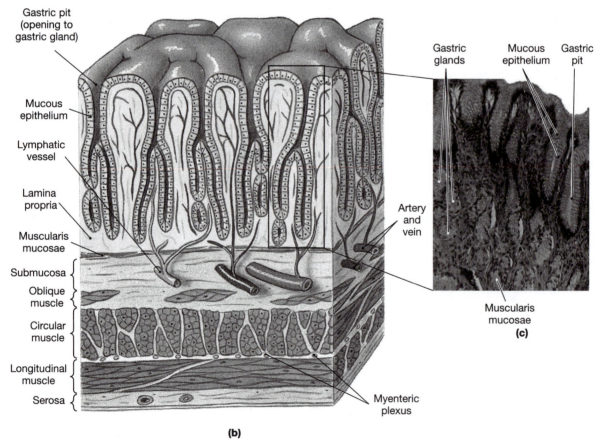

Gastric pit (opening to gastric gland)

Mucous epithelium

Lymphatic vessel

Lamina propria

Muscularis mucosae

Submucosa

Oblique muscle

Circular muscle

Longitudinal muscle

Serosa

Artery and vein

Myenteric plexus

Gastric glands   Mucous epithelium   Gastric pit

Muscularis mucosae

(c)

(b)

●Figure 17.5
Gross Anatomy of the Stomach

(a) Anterior view of the stomach showing superficial landmarks. **(b)** Diagrammatic view of the organization of the stomach wall. **(c)** The lining of the stomach. (LM × 90)

Laboratory Activity   THE STOMACH _____

MATERIALS

torso model
stomach model
digestive system chart

PROCEDURES

1. Label the anatomy of the stomach in Figure 17.5. Identify the gross anatomy of the stomach on all laboratory models and charts.

2. If available, examine the stomach of a cat or other animal. Locate the rugae, main regions of the stomach, omenta, and sphincters.

E.   The Small Intestine and Large Intestine

The small intestine is approximately 21 feet long and consists of three segments: the duodenum, the jejunum, and the ileum (see Figure 17.6). The first 10 inches, the **duodenum**, is attached to the distal region of the pylorus. The duodenum receives chyme from the stomach and digestive secretions from the liver, gallbladder, and pancreas. The **jejunum** is approximately 12 feet long and is the site of most nutrient absorption. The last region, the **ileum**, ends at the **ileocecal valve**, where it empties into the cecum of the large intestine. The serosa or peritoneum of the small intestine is modified into sheets of tissue called **mesentery**, which support and attach the small intestine to the abdominal wall.

The small intestine is the site of most digestive and absorptive activities and has specialized folds to increase surface area for these functions (see Figure 17.6b). The submucosa and mucosa are creased together into large folds called **plicae**. Along the plicae, the mucosa is convoluted into fingerlike projections called **villi**.

The large intestine is the site of water absorption and waste compaction. It is approximately 5 feet long and is divided into two groups, the **colon** and the **rectum**. Figure 17.7 details the gross anatomy of the large intestine. The first part of the colon, a pouchlike **cecum**, is located in the right iliac region. The **ileocecal valve** is at the junction of the ileum and cecum. This sphincter muscle controls the movement of materials from the small intestine into the cecum. Attached to the cecum is the wormlike **appendix**. Past the cecum, the colon is divided into several regions. The **ascending colon** travels up the right side of the abdomen. It bends left to cross the abdomen below the stomach as the **transverse colon**, which then turns downward as the **descending colon**. The S-shaped **sigmoid colon** passes through the pelvic cavity to join the **rectum**. The rectum, detailed in Figure 17.7b, is the last 6 inches of the large intestine and the end of the digestive tract. The opening of the rectum is the **anus**, which is controlled by internal and external **anal sphincters**.

The wall of the colon has many specializations. The longitudinal layer of the muscularis externa is reduced into three separate bands of muscle called the **taenia coli**. Muscle tone of the taenia coli constrict the colon wall into pouches called **haustrae**. The haustrae allow expansion and stretching of the colon wall.

Laboratory Activity   THE SMALL INTESTINE AND LARGE INTESTINE _____

MATERIALS

torso and intestinal models
digestive system chart
dissected cat

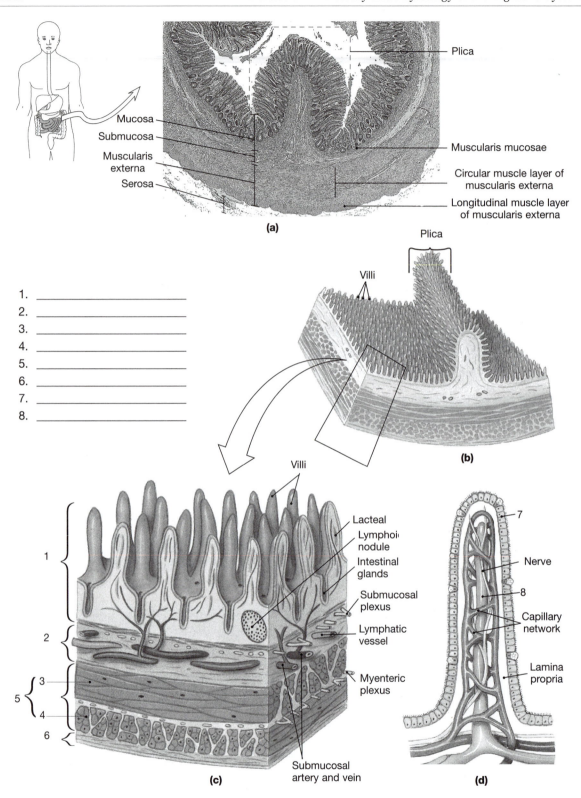

(a)

Plica

Mucosa
Submucosa
Muscularis externa
Serosa

Muscularis mucosae

Circular muscle layer of muscularis externa

Longitudinal muscle layer of muscularis externa

Plica

Villi

(b)

1. _____
2. _____
3. _____
4. _____
5. _____
6. _____
7. _____
8. _____

Villi

Lacteal
Lymphoid nodule
Intestinal glands
Submucosal plexus
Lymphatic vessel

Myenteric plexus

Submucosal artery and vein

(c)

Nerve

Capillary network

Lamina propria

(d)

●**Figure 17.6**
The Intestinal Wall

(a) Diagrammatic view of the intestine in section. (LM × 120,382) (b) A single plica and multiple villi. (c) Diagrammatic view of the organization of the intestinal wall. (d) Internal structures in a single villus, showing the capillary and lymphatic supply.

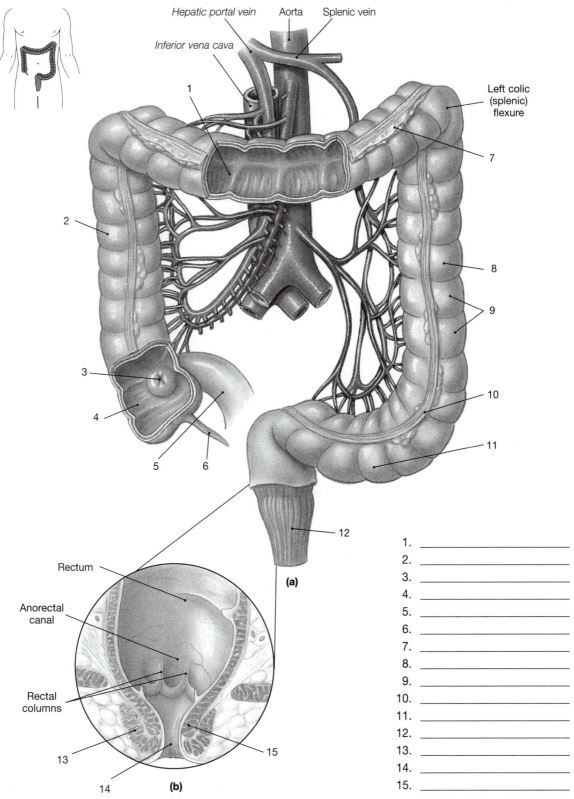

Hepatic portal vein   Aorta   Splenic vein

Inferior vena cava

Left colic
(splenic)
flexure

Rectum

Anorectal
canal

Rectal
columns

(a)

(b)

1. _____
2. _____
3. _____
4. _____
5. _____
6. _____
7. _____
8. _____
9. _____
10. _____
11. _____
12. _____
13. _____
14. _____
15. _____

●**Figure 17.7**
The Large Intestine

(a) Gross anatomy and regions of the large intestine. **(b)** Detailed anatomy of the rectum and anus.

PROCEDURES

1. Label and review the regions of the small intestine in Figure 17.6. Identify the anatomy of the small intestine on all laboratory models and charts.
2. Label and review the anatomy of the large intestine in Figure 17.7. Identify the gross anatomy of the large intestine on all laboratory models and charts.
3. If available, examine a segment of small and large intestine of a cat or other animal. Locate each intestinal region, the plicae, taenia coli, and haustrae.

F.   The Liver and Gallbladder

The liver is located mostly in the right hypochondriac region, inferior to the diaphragm. As shown in Figure 17.8, the liver has four major lobes. Each lobe is divided into thousands of smaller **lobules**, which we discuss in an upcoming section. The liver has a wide range of functions. It manufactures bile, a salt involved in fat digestion. Bile flows into **right** and **left hepatic ducts**, which merge to form a **common hepatic duct**.

The **gallbladder** is located inferior to the right lobe of the liver. It is a small muscular sac that stores and concentrates bile salts used in digestion of lipids. Figure 17.9 details the ducts of the liver and gallbladder. The common hepatic duct from the liver meets the **cystic duct** of the gallbladder to form the **common bile duct**. This duct passes through the lesser omentum and joins the pancreatic duct of the pancreas. These ducts enter the duodenum of the small intestine. A sphincter muscle regulates the flow of pancreatic juice and bile into the duodenum.

●**Figure 17.8**
Anatomy of the Liver
(a) The anterior surface of the liver. **(b)** The posterior surface of the liver.

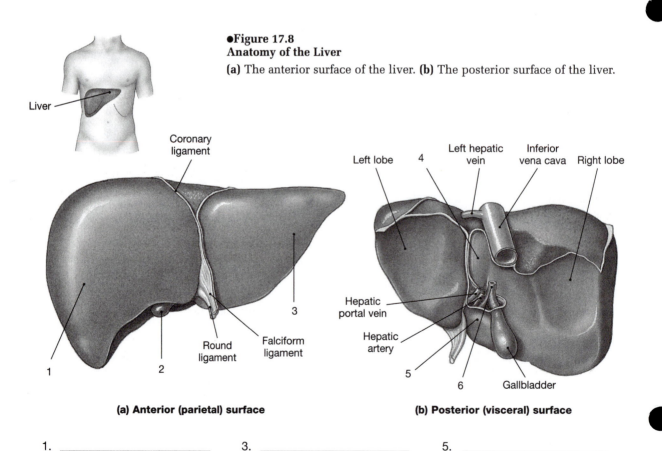

Liver

Coronary ligament

Left lobe   4   Left hepatic vein   Inferior vena cava   Right lobe

Hepatic portal vein

Hepatic artery

Round ligament   Falciform ligament

3

1   2

5   6   Gallbladder

(a) Anterior (parietal) surface   **(b) Posterior (visceral) surface**

1. _____   3. _____   5. _____
2. _____   4. _____   6. _____

Laboratory Activity   THE LIVER AND GALLBLADDER_____

MATERIALS

torso model
liver and gallbladder models
digestive system chart

PROCEDURES

1. Label and review the anatomy of the liver and gallbladder in Figures 17.8 and 17.9.

2. Identify the gross anatomy of the liver and gallbladder on all laboratory models and charts.

3. If available, examine the liver and gallbladder of a cat or other animal. Locate the liver lobes and the hepatic, cystic, and common bile ducts.

G.   The Pancreas

The pancreas lies between the stomach and the duodenum (see Figure 17.10). The pancreas is a "double" gland with endocrine and exocrine functions. The endocrine cells occur in **pancreatic islets** and secrete hormones for sugar metabolism. Most of the pancreatic cells consist of exocrine glands. **Acinar cells** secrete pancreatic juice rich in enzymes and buffers into the pancreatic duct that meets the common bile duct in the duodenal ampulla.

Laboratory Activity   THE PANCREAS_____

MATERIALS

torso model
pancreas model
dissected cat

●Figure 17.9
The Gallbladder

A view of the inferior surface of the liver, showing the position of the gallbladder and ducts that transport bile from the liver to the gallbladder and duodenum.

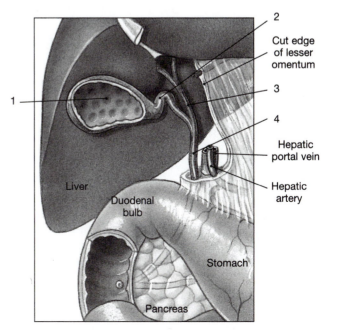

1. _____
2. _____
3. _____
4. _____

PROCEDURES

1. Review the anatomy of the pancreas in Figure 17.10. Identify the anatomy of the pancreas on all laboratory models and charts.

2. If available, examine the pancreas of a cat or other animal. Locate the pancreatic duct.

H. Histology of the Digestive System

Complete the anatomical studies before proceeding to the histological organization of the digestive organs. Obtain a prepared microscope slide of the digestive organs designated by your lab instructor. Observe each slide at various magnifications and locate the anatomy discussed.

Laboratory Activity   HISTOLOGY OF THE DIGESTIVE SYSTEM _____

MATERIALS

compound microscope
prepared slides
stomach, small intestine, liver, pancreas

●**Figure 17.10**
The Pancreas

(a) Gross anatomy of the pancreas. The head of the pancreas is tucked into a curve of the duodenum that begins at the pylorus of the stomach. (b) Light micrograph of pancreatic tissue. (LM × 168)

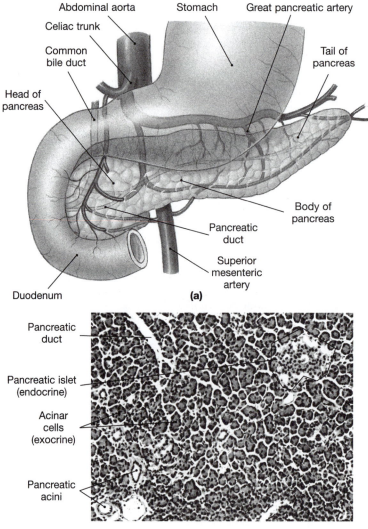

Study Tips

- Start with low magnification to scan each slide.
- Center the specimen in your field of view, and then increase magnification.
- Examine individual cell populations. Keep in mind that the purpose of your microscopic observations is to examine the regional similarities and differences within the digestive system.

HISTOLOGY OF THE STOMACH

PROCEDURES

1. Figure 17.5d highlights the histology of the stomach wall in cross section. At low power, observe the large folds of the mucosa called **rugae** (see Figure 17.5a). Examine the mucosa lined with **simple columnar epithelium**. Interspersed in the mucosal lining are **goblet cells** that secrete a thick mucus to protect the gastric wall.

2. Increase the magnification, and locate the many **gastric pits**, the shallow pockets along the surface of the rugae. Glands in the gastric pits secrete enzymes and acid for the chemical digestion of protein molecules.

HISTOLOGY OF THE SMALL INTESTINE

PROCEDURES

1. On the slide, locate the lumen of the section. Examine the folding of the mucosa at low magnification, and scan the entire section. Note the large folds called **plicae**. Along the plicae are smaller fingerlike extensions, the **villi**.

2. Increase magnification, and examine the histology of the major layers. Refer to Figures 17.1 and 17.6 for details of the intestinal wall.

3. Examine a single villus, and locate the **columnar epithelium** and **goblet cells** lining the mucosa. In the middle of a villus, locate the **lacteal**, a lymphatic vessel that absorbs fatty acids and monoglycerides from lipid digestion. These structures are detailed in Figure 17.6(c).

4. In the **submucosa** are scattered mucus-producing **submucosal** (Brunner's) **glands**.

5. Identify the **muscularis externa** and the **serosa**. If the intestinal slide is of the ileum, the **submucosa** layer possesses **Peyer's patches**, large lymphatic nodules that prevent bacteria from entering the blood.

HISTOLOGY OF A LIVER LOBULE

PROCEDURES

1. The liver comprises approximately 100,000 small lobules. Figure 17.11 is a micrograph of a single lobule similar in appearance to what you should locate on the liver slide. At low magnification, notice the many oval holes in the tissue. Most of these are **central veins** that drain blood from the lobules.

2. Pick a central vein, and increase magnification to observe a **lobule**. Depending on the quality of your slide, you may be able to define the outline of an individual lobule, as in Figure 17.11. Note the open **sinusoids**, open spaces where blood circulates, and the **hepatocyte** cells lining the sinusoids. Hepatocytes produce bile for fat digestion.

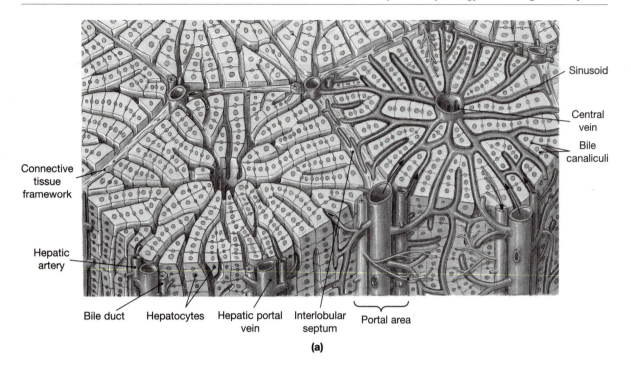

Sinusoid

Central vein

Bile canaliculi

Connective tissue framework

Hepatic artery

Bile duct   Hepatocytes   Hepatic portal vein   Interlobular septum   Portal area

(a)

Hepatic artery   Hepatic vein   Central vein   Sinusoid

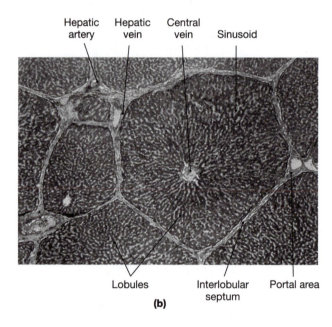

Lobules   Interlobular septum   Portal area

(b)

●**Figure 17.11**
Liver Histology

(a) Diagrammatic view of lobular organization.
(b) Light micrograph showing a section through a liver lobule. (LM × 38)

HISTOLOGY OF THE PANCREAS

PROCEDURES

1. Using Figure 17.9 as reference, observe the pancreas slide at low magnification, and locate the many oval-shaped ducts. The dark-stained **acini cells** are the exocrine cells that secrete enzymes and buffers into the ducts.

2. Surrounded by the acini cells are clusters of light-stained **pancreatic islets**, the endocrine cells of the pancreas that produce hormones for blood sugar regulation. The open structures in the tissue are blood vessels and pancreatic ducts.

PART TWO   THE PHYSIOLOGY OF DIGESTION

WORD POWER	MATERIALS	
catabolism (cata—to break down)	water bath	amylase enzyme solution
monomer (mono—one)	thermometer	Lugol's solution (IKI)
saccharide (sacchar—sugar)	test tube racks and holders	Benedict's solution
	ten test tubes	litmus
	wax marker	whipping cream
	squeeze bottles for reagents	pancreatic lipase enzyme solution
	hot plate	protein solution
	beakers	Biuret's solution
	starch solution	

OBJECTIVES

On completion of this part of the exercise, you should be able to:

- Describe why enzymes are used to digest your food.
- Describe the chemical composition of carbohydrates, lipids, and proteins.
- List the enzymes and substrates for each digestion experiment.
- Describe each chemical test performed on the substrates, and analyze the results of each experiment.

Before nutrients can be converted into usable energy by cells, the large organic **macromolecules** in food must be **catabolized** or broken down into **monomers**, the building blocks of macromolecules. These chemical reactions are controlled by enzymes produced by the digestive system. **Enzymes** are protein (organic) catalysts that lower **activation energy**, the energy required for a chemical reaction. Without enzymes, the body would have to heat up to dangerous temperatures to provide the activation energy necessary to decompose ingested food. Enzymes have a narrow range of physical conditions in which they operate at maximum efficiency. Temperature and pH are two important factors of enzymatic reactions. For example, the enzymes involved in protein digestion require a different pH. The enzyme pepsin is most active in acidic conditions of the stomach. The next protein-digesting enzyme in the sequence, trypsin, requires an alkaline environment.

An enzymatic reaction involves reactants, called the **substrate**, and results in a **product**. The enzyme has an **active site** where the substrate binds. Only a substrate that is compatible with an enzyme's active site will be metabolized, and the enzyme is said to have **specificity** for compatible substrates. On completion of the chemical reaction, the product is released and the enzyme, unaltered in the reaction, may bind to another substrate and repeat the reaction.

In this laboratory exercise, you will study the conditions necessary for the enzymatic reactions for carbohydrate, lipid, and protein digestion. All enzyme experiments in this exercise will be incubated in a warm water bath set at body temperature, 37° C. If the temperature becomes too high, the enzyme will **denature**, destroying its chemical structure and function.

Table 17.1 summarizes the time and materials required for each enzyme experiment. Use this table to help manage your laboratory time.

A. Carbohydrate Digestion

Starch and sugar molecules are classified as carbohydrates. These molecules are made up of **saccharide** molecules. A carbohydrate composed of a single saccharide is a

TABLE 17.1	Summary of Enzyme Experiments		
	Carbohydrate	*Lipid*	*Protein*
Incubation time	One-half hour	One hour	One hour
Number of test tubes required	Six	Two	Two
Solutions required	• Starch solution • Amylase • Lugol's solution • Benedict's reagent	• Litmus cream • Lipase	• Protein solution • Pepsinogen • Hydrochloric acid • Biuret's solution

monosaccharide, a simple sugar. Glucose and fructose are examples of monosaccharides. Monosaccharides are the monomers used to build larger sugar and starch molecules. A **disaccharide** is formed when two monosaccharides bond together by a dehydration synthesis reaction. Table sugar, sucrose, is a common disaccharide. The complex carbohydrates are **polysaccharides**. These molecules consist of long chains of monosaccharides. Cells store polysaccharides for future energy sources. Plants store polysaccharides as **starch**, the molecules found in potatoes and grains. Animal cells, such as muscle and liver cells, store polysaccharides as **glycogen**.

In this experiment, you will start with a polysaccharide substrate, a starch solution prepared by your instructor, and a carbohydrate-digesting enzyme called amylase. Although amylase is easily obtainable from saliva, your lab instructor may provide amylase produced by a nonhuman source.

Laboratory Activity   Carbohydrate Digestion _____

Safety Tips

- Read all procedures before starting an experiment.
- Wear gloves and safety glasses while pouring reagents and working near water baths.
- Report all spills and broken glass to the laboratory instructor.

MATERIALS

Refer to Table 17.1 for solutions and chemicals

     wax marker   test tube rack
     37° C water bath   boiling water bath

I. Preparation

PROCEDURES

1. With a wax marker, label six test tubes 1 through 6. Fill test tubes 1 and 2 approximately 1/4 full with the starch solution. Be sure to shake the starch solution before pouring.
2. Add 20 ml of amylase enzyme to test tube 1 only.
3. Place tubes 1 and 2 into a water bath set at body temperature, 37° C. Incubate the solutions for a minimum of 30 minutes, then remove both from the water bath.

●Figure 17.12
**Overview of Carbohydrate
Digestion Procedures**

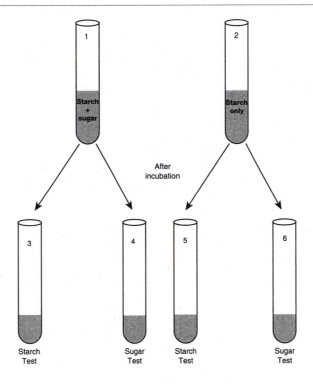

II. Analysis

PROCEDURES

1. Equally divide the solution in test tube 1 into tubes 3 and 4 (see Figure 17.12).
2. Equally divide the solution in test tube 2 into tubes 5 and 6.
3. Use the iodine test for the presence of starch in test tubes 3 and 5 as follows:
   a. Place one or two drops of Lugol's solution (IKI) in each test tube.
   b. A dark blue color shows the presence of starch.
   c. Record your observations in Table 17.2.
4. Use the Benedict's test for monosaccharides in test tubes 4 and 6 as follows:
   a. Place 10 ml of Benedict's solution in each test tube and mix by gently swirling the test tubes.
   b. Set each tube in a boiling water; bath for four to five minutes. Do not point the test tubes toward anyone, the solution could splatter and burn.
   c. A color change shows presence of monosaccharides. Benedict's will turn from light olive to dark orange, depending on the quantity of reducing sugars.
   d. Record your observations in Table 17.2.

TABLE 17.2	**Digestion of Starch by Amylase**			
	Tube 1 *Starch and Amylase*		*Tube 2* *Starch Only*	
	Tube 3	Tube 4	Tube 5	Tube 6
Lugol's test for starch				
Benedict's test				
Conclusions				

B. Lipid Digestion

There are many classes of fats or lipids. Most dietary lipids are **monoglycerides** and **triglycerides**. These lipids are constructed of one or more **fatty acid** molecules bonded to a **glycerol** molecule. Pancreatic juice contains **pancreatic lipase**, a lipid-digesting enzyme that hydrolyses triglycerides into monoglycerides and free fatty acids that turn a solution acidic. The lipid substrate you will use in this experiment is whipping cream to which a pH indicator, litmus, has been added. Litmus is blue in basic (alkaline) conditions and pink in acidic conditions.

Laboratory Activity   LIPID DIGESTION_____

MATERIALS

Refer to Table 17.1 for solutions and chemicals.

> wax marker
> test tube rack
> 37° C water bath

I. Preparation

PROCEDURES

1. Label two test tubes 7 and 8. Fill each test tube 1/4 full with litmus cream.
2. Add the same amount of pancreatic lipase as there is litmus cream to test tube 7. Do not add the enzyme to test tube 8.
3. Record the smell and color of the solution in each test tube in Table 17.3.
4. Incubate the test tubes for one hour in a water bath set at body temperature.

II. Analysis

PROCEDURES

1. After one hour, remove the test tubes from the water bath.
2. Record the color of the litmus cream in each tube in Table 17.3.
3. Carefully smell the solutions by wafting fumes from the tube up to your nose. Record the smell in Table 17.3.

C. Protein Digestion

Proteins comprise long chains of **amino acids** bonded together by **peptide bonds**. There are 20 different amino acids, and their abundance and position in a protein molecule determines the function of the protein. In this experiment, you will use an inactive enzyme, classified as a proenzyme, called pepsinogen. The proenzyme alone will not

TABLE 17.3	**Digestion of Lipid by Pancreatic Lipase**	
	Tube 7 *Litmus Cream and Enzyme*	*Tube 8* *Litmus Cream Only*
Smell		
Color		
Conclusions		

cause digestion; it must be activated by certain chemical conditions. By doing this experiment, you will not only learn about protein digestion but will also discover the importance of the environment in which the enzyme operates. The protein substrate you will use was obtained from gelatin.

Laboratory Activity   PROTEIN DIGESTION

MATERIALS

Refer to Table 17.1 for solutions and chemicals.

wax marker
test tube rack
37° C water bath

I. Preparation

PROCEDURES

1. Label two test tubes 9 and 10. Fill each tube 1/4 full with the protein solution.
2. To test tube 9 add 10 ml of the proenzyme pepsinogen.
3. To test tube 10, add 10 ml of the proenzyme pepsinogen and 10 ml of 0.5M HCl.
4. Incubate both tubes for one hour in a body temperature water bath, then remove.

II. Analysis

PROCEDURES

1. After one hour, remove test tubes 9 and 10 from the water bath. To test for protein digestion, use Biuret's solution, which turns purple in the presence of free amino acids.
2. Add 10 ml of Biuret's solution to each test tube.
3. Observe the color of the protein solution. Record your observations in Table 17.4.

TABLE 17.4	**Digestion of Protein by Pepsin**	
	Tube 9 *Protein and Enzyme*	*Tube 10* *Protein and Enzyme and HCl*
Color after Biuret's test		
Occurrence of digestion		
Conclusions		

ANATOMY AND PHYSIOLOGY OF THE DIGESTIVE SYSTEM
Exercise 17 Laboratory Report

A. Matching

Match each digestive structure on the left with the correct description on the right.

1. _____ pyloric sphincter	A. fatty apron hanging off stomach
2. _____ greater omentum	B. outer layer of digestive tract
3. _____ incisor	C. muscle folds of stomach wall
4. _____ esophagus	D. opening between mouth and pharynx
5. _____ taenia coli	E. hard outer layer of tooth
6. _____ haustra	F. pouch in colon wall
7. _____ mesentery	G. folds of intestinal wall
8. _____ muscularis externa	H. tooth used for snipping food
9. _____ common bile duct	I. longitudinal muscle of colon
10. _____ serosa	J. gum surrounding tooth
11. _____ fauces	K. tooth used for grinding
12. _____ lingual frenulum	L. valve between stomach and duodenum
13. _____ gingiva	M. connective membrane of intestines
14. _____ enamel	N. the food tube
15. _____ rugae	O. major muscle layer of digestive tract
16. _____ plicae	P. middle segment of small intestine
17. _____ molar	Q. common hepatic and cystic duct join here
18. _____ jejunum	R. anchors tongue to floor of mouth

B. Short Answer Questions

1. Discuss the specializations of the large intestine's wall.

2. List the features of the small intestine that increase surface area for absorption of nutrients.

3. List the accessory organs of the digestive system.

4. Trace a drop of bile through the ducts from the liver to the duodenum.

5. Describe the histology of the pancreas.

6. Describe the location of the appendix.

C. Drawing

1. Draw the anatomy of a typical tooth in a longitudinal section.

2. Draw a cross section of the duodenal wall detailing the structures that increase surface area for digestion and absorption.

D. Fill in the Blanks

1. An organic catalyst that lowers the activation energy of a chemical reaction is called an _____.

2. The molecules an enzyme reacts with are called the _____.

3. Macromolecules may be catabolized into smaller units called _____.

4. The three primary nutrients are _____, _____, and _____.

5. The reagent used to test for the presence of starch was _____(IKI).

6. The reagent used to test for the presence of sugar was _____.

E. Discussion

1. In each digestion experiment, you used a control tube that had no enzyme or other reagent added to the substrate. What was the function of the control?

2. Why were both test tubes (1 and 2) used in the starch experiment tested for starch and sugar at the conclusion of the incubation period?

   3. Briefly outline the chemical composition of carbohydrates, lipids, and proteins.

   4. Why was a warm water bath used rather than a boiling bath to incubate test tubes
      containing substrates and enzymes?

F. **Completion**

   1. Complete Figure 17.13 by labeling the organs of the digestive system.

1. _____
2. _____
3. _____
4. _____
5. _____
6. _____
7. _____
8. _____
9. _____
10. _____
11. _____
12. _____
13. _____
14. _____
15. _____
16. _____
17. _____
18. _____
19. _____
20. _____
21. _____
22. _____
23. _____
24. _____
25. _____
26. _____
27. _____
28. _____
29. _____

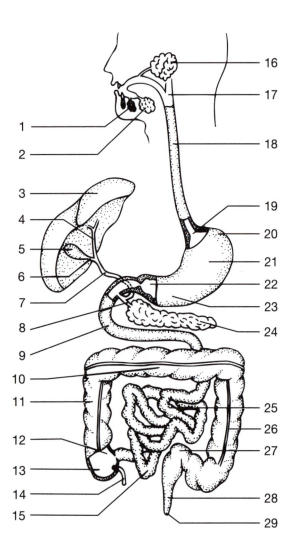

●**Figure 17.13**
The Digestive System

18 ANATOMY AND PHYSIOLOGY OF THE URINARY SYSTEM

All cells produce waste products through their metabolic activities. Waste products such as urea, creatinine, carbon dioxide, nitrogen wastes, and excess electrolytes must be eliminated from the body to maintain homeostasis. Several organ systems eliminate wastes from the body. The lungs remove carbon dioxide during exhalation; the digestive tract eliminates undigested solid wastes; and the skin removes salts and urea in sweat. The primary function of the urinary system is to control the composition, volume, and pressure of blood by removing and restoring selected amounts of water and solutes. These eliminated products are collectively called urine. The urinary system, highlighted in Figure 18.1, comprises a pair of kidneys and ureters, a urinary bladder, and a urethra.

PART ONE   ANATOMY OF THE URINARY SYSTEM

WORD POWER

hilus (hilus—a depression)
renal (ren—kidney)
rugae (rugae—a wrinkle)
trigone (trigone—triangle)
arcuate (arcuate—bowed, arched)
afferent (afferent—to bring to)
glomerulus (glomus—a ball of yarn)
efferent (efferent—to bring out)
vasa (vas—a vessel)
recta (recta—straight)

MATERIALS

human torso models
kidney models
model of the nephron
preserved sheep kidneys
dissecting instruments
latex gloves
compound microscope
prepared slide
   kidney c.s.

OBJECTIVES

On completion of this part of the exercise, you should be able to:

- Identify and describe the basic anatomy of the urinary system.
- Trace the blood flow through the kidney.
- Explain the function of the kidney.
- Identify the basic components of the nephron.
- Define glomerular filtration, tubular reabsorption, and tubular secretion.
- Describe the differences between the male and female urinary tracts.

A. The Kidneys

The kidneys lie on the posterior surface of the abdomen between the waist and the 11th and 12th pairs of ribs. The right kidney is positioned lower than the left kidney owing to the position of the liver. The kidneys, adrenal glands, and most of the ureters are **retroperitoneal** because they are located behind the parietal peritoneum.

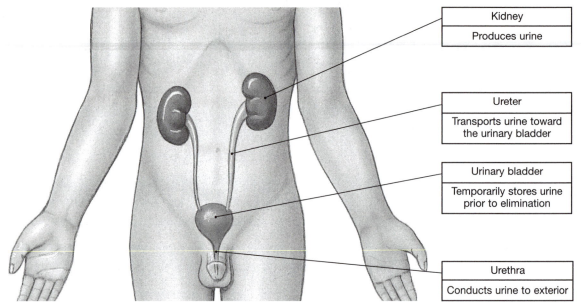

Kidney
Produces urine

Ureter
Transports urine toward the urinary bladder

Urinary bladder
Temporarily stores urine prior to elimination

Urethra
Conducts urine to exterior

●**Figure 18.1**
Components of the Urinary System

Figure 18.2 illustrates the sectional anatomy of the kidney. Each kidney is covered in a sheet of connective tissue called the **renal capsule**. Each kidney is about 13 cm (5.12 inches) long and 2.5 cm (1 inch) thick. The medial aspect of the kidney contains a notch called the **hilus**. Blood vessels, nerves, lymphatics, and the ureters enter and exit the kidney through the hilus. The hilus also leads to a cavity in the kidney called the **renal sinus**. The **cortex** is the outer light red layer or area of the kidney. Deep to the cortex is a region called the **medulla**, which consists of triangular **renal pyramids** that project toward the center of the kidney. Areas of the cortex extending between the renal pyramids are **renal columns**. At the apex or tip of each renal pyramid is a **renal papilla** that empties urine into a small cuplike space called the **minor calyx**. Several minor calyces empty into a common space, the **major calyx**. These larger calyces merge to form the **renal pelvis**. The cavity surrounding the renal pelvis is the **renal sinus**. Each kidney has a single ureter, a muscular tube that transports urine from the renal pelvis to the urinary bladder.

Laboratory Activity   OBSERVATIONS OF THE KIDNEY _____

MATERIALS
kidney models

PROCEDURES
1. Complete Figure 18.2 by labeling each structure of the kidney.
2. Examine the kidney models and charts, and locate each structure shown in Figure 18.2.

B. The Nephron
The basic functional unit of the kidney is called the **nephron**. Approximately 1.25 million nephrons occur in each kidney. In each nephron, materials are removed from the blood by a process called **filtration** to produce a fluid called **filtrate**. The filtrate circulates through a series of convoluting or twisting tubules and a U-shaped section to reclaim essential materials and remove wastes and excess ions. The remaining filtrate is

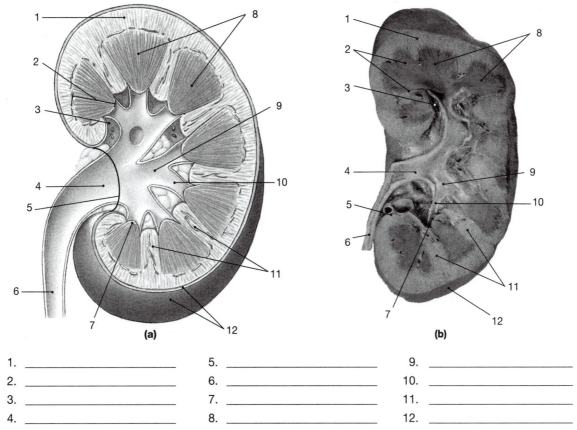

1. _____   5. _____   9. _____
2. _____   6. _____   10. _____
3. _____   7. _____   11. _____
4. _____   8. _____   12. _____

●**Figure 18.2**
Structure of the Kidney
(a) Diagrammatic and **(b)** sectional views of a frontal section through the left kidney.

excreted as urine. We cover the process of urine formation in more detail in Part Two of this exercise.

Each nephron consists of two portions: a **renal corpuscle** and a **renal tubule**, detailed in Figure 18.3. The renal corpuscle consists of a double-walled capsule called the **Bowman's** (glomerular) **capsule**. Inside this capsule is a capillary called the **glomerulus**. The glomerular capsule has an outer **parietal layer** and an inner **visceral layer** that reflects over the surface of the glomerulus. Between these two layers is the **capsular space**. An afferent arteriole passes into the Bowman's capsule and supplies blood to the glomerulus.

The renal tubule consists of twisted and straight tubules composed mainly of cuboidal epithelium. The first segment after the Bowman's capsule is a coiled tubule called the **proximal convoluted tubule**. The next portion of the tubule, the **loop of Henle**, consists of a descending limb, a loop of the limb and an ascending limb. The ascending limb leads to a second convoluted section, the **distal convoluted tubule**. Surrounding the convoluted tubules are the peritubular capillaries, whereas the loop of Henle is covered with the vasa recta capillary. The nephron ends where the distal convoluted tubule empties into a **collecting duct**. Several nephrons drain into the same collecting duct. These ducts merge with other collecting ducts into a common papillary duct that drains into a minor calyx. There are between 25 and 35 papillary ducts per renal pyramid.

Two types of nephrons occur in the kidney, cortical and juxtamedullary nephrons. Approximately 85% of the nephrons are **cortical nephrons** and have their glomeruli and most of their tubules in the cortex. **Juxtamedullary nephrons** have their glomeruli at the junction of the cortex and the medulla. These nephrons have long loops that extend deep into the medulla before turning back toward the cortex.

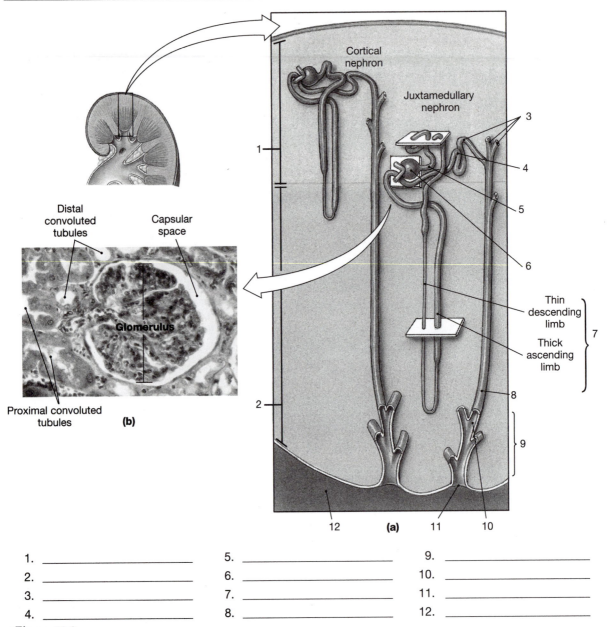

1. _____   5. _____   9. _____
2. _____   6. _____   10. _____
3. _____   7. _____   11. _____
4. _____   8. _____   12. _____

●**Figure 18.3**
Sectional Views of the Nephron

(a) Diagrammatic view of nephrons. **(b)** The association of proximal and distal
convoluted tubules with a glomerus. (LM × 370)

Laboratory Activity   OBSERVATIONS OF THE NEPHRON_____

MATERIALS

kidney and nephron models
compound microscope
prepared slide of kidney

PROCEDURES

1. Examine the kidney models, and locate all structures of the nephron. Complete
the labeling of the nephron in Figure 18.3.

2. Obtain a microscope and a slide of the kidney. Use the photomicrograph in Figure 18.3 for reference.

   a. Use 10x and 40x magnifications, and locate the renal cortex and renal medulla. Examine several renal tubules, visible as ovals on the kidney slide. Is the renal capsule visible on the outer margin of the kidney tissue?

   b. Locate a renal corpuscle that appears as a small knot in the cortex. Distinguish between the parietal layer of Bowman's capsule, the capsular space, and the glomerulus.

   c. Draw a section of the kidney slide in the space provided. Label the cortex, a renal corpuscle, and a renal tubule. The medulla is usually not present on most kidney slides.

C. Blood Supply

Every minute, approximately 25% of the total blood volume travels through the kidneys. This blood is delivered to a kidney by the **renal artery**, which branches off the abdominal aorta (see Figure 18.4). Once it enters the hilus of the kidney, the renal artery divides into five **segmental arteries**. The segmental arteries branch into **interlobar arteries**, which pass through the renal columns. The interlobar arteries divide into **arcuate arteries**, which cross the bases of the pyramids and enter the cortex as **interlobular arteries**. These arteries branch into **afferent arterioles**, which enter the nephron, the site of urine formation.

When an afferent arteriole enters a nephron, it forms a capillary tuft called the **glomerulus**. A smaller **efferent arteriole** exits the glomerulus and forms two additional capillary beds around the tubular portion of the nephron. The capillaries surrounding cortical nephrons are called **peritubular capillaries**. In juxtamedullary nephrons, the peritubular capillaries surround only the proximal and distal convoluted tubules. The loop of Henle in juxtamedullary nephrons is surrounded by **vasa recta** capillaries. Both capillary networks are involved in the reabsorption of materials from the filtrate of the renal tubules back into the blood. The vasa recta and peritubular capillaries drain into **interlobular veins**, which then drain into **arcuate veins** along the base of the renal pyramids. **Interlobar veins** pass through the renal columns and empty into **segmental veins**. These veins join the **renal vein**, which drains into the **inferior vena cava**.

Laboratory Activity   BLOOD SUPPLY TO THE KIDNEY _____

MATERIALS

kidney and nephron models

PROCEDURES

1. Label each blood vessel in Figure 18.4.

2. On the kidney models and charts, trace the blood vessels that supply and drain the kidneys.

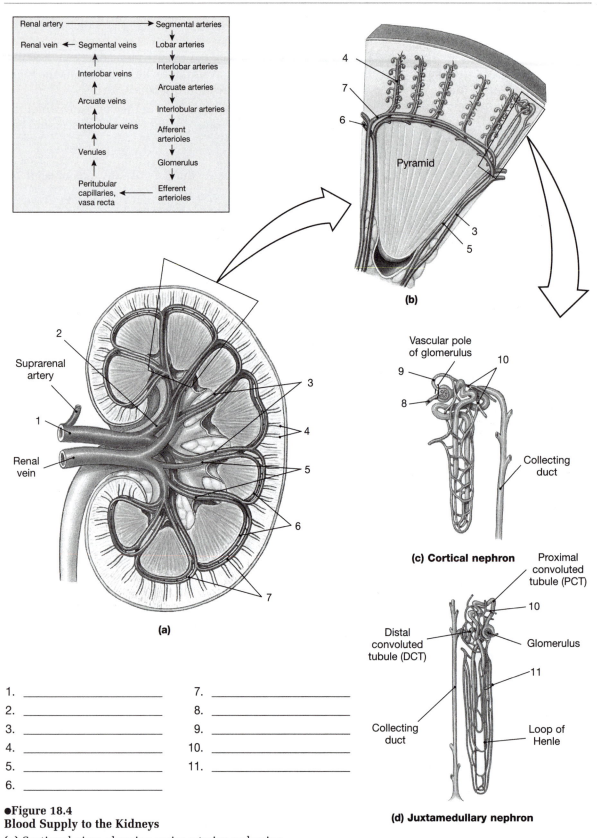

Renal artery ──────────→ Segmental arteries

Renal vein ◄── Segmental veins   Lobar arteries

Interlobar veins   Interlobar arteries

Arcuate veins   Arcuate arteries

Interlobular veins   Interlobular arteries

Venules   Afferent arterioles

Glomerulus

Peritubular capillaries, vasa recta ◄── Efferent arterioles

(b)

Pyramid

(a)

Suprarenal artery

Renal vein

Vascular pole of glomerulus

Collecting duct

(c) Cortical nephron

Proximal convoluted tubule (PCT)

Distal convoluted tubule (DCT)

Glomerulus

Collecting duct

Loop of Henle

(d) Juxtamedullary nephron

1. _____
2. _____
3. _____
4. _____
5. _____
6. _____
7. _____
8. _____
9. _____
10. _____
11. _____

●**Figure 18.4**
Blood Supply to the Kidneys

(a) Sectional view, showing major arteries and veins.
(b) Circulation in the cortex. **(c)** Circulation to a cortical nephron.
(d) Circulation to a juxtamedullary nephron.

D. The Ureters, Urinary Bladder, and Urethra

The two **ureters** conduct urine, by way of gravity and peristalsis, to the urinary bladder. The ureters, detailed in Figure 18.5, are lined with a mucus-producing epithelium to protect the ureteral walls from the acidic urine. The ureters enter the urinary bladder on the posterior surface at the top two corners of the **trigone**, a smooth, triangular-shaped area of the urinary bladder floor.

The **urinary bladder** is a hollow, muscular organ that functions in the temporary storage of urine. In males, the urinary bladder lies between the pubic symphysis and the rectum. In females, it is posterior to the pubic symphysis, inferior to the uterus, and superior to the vagina. The inner wall contains folds called **rugae** that allow expansion and shrinkage of the bladder as it fills and empties with urine.

A single duct, the **urethra**, drains urine from the bladder out of the body. Around the opening to the urethra are two sphincter muscles, the internal and external urethral sphincters. The **internal urethral sphincter** consists of smooth muscle and is therefore under involuntary nerve control. The **external urethral sphincter** is skeletal muscle tissue and under voluntary control. This sphincter enables you to control the release of urine from the urinary bladder. As the bladder fills and expands with urine, stretch receptors signal motor neurons in the spinal cord to relax the internal urethral sphincter and contract the **detrusor muscle** of the bladder wall. This reflex automatically causes closing of the external urethral sphincter. When convenient, the external urethral sphincter is relaxed and urination, or **micturition**, occurs.

Laboratory Activity THE URETERS, URINARY BLADDER, AND URETHRA _____

MATERIALS

urinary system models and charts

PROCEDURES

1. Locate the ureters on a urinary tract model. Trace the transport of urine into the ureter from the renal papilla. Label all structures in Figure 18.5.
2. Examine the wall of the urinary bladder. Identify the trigone and the rugae. Which structures control emptying of the urinary bladder?

   _____

   _____

3. Examine the urethra on the models. How does this structure differ between males and females? _____

   _____

E. Dissection of the Sheep Kidney

The sheep kidney is very similar to the human kidney in both size and structure. Dissection of a sheep kidney will reinforce your observations of kidney models in the laboratory. Use the cadaver photo in Figure 18.2 for reference during your sheep dissection.

Laboratory Activity DISSECTION OF THE SHEEP KIDNEY _____

MATERIALS

preserved sheep kidneys
dissection kit and dissection pan
disposable gloves, safety glasses

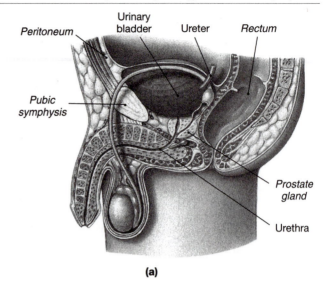

(a)

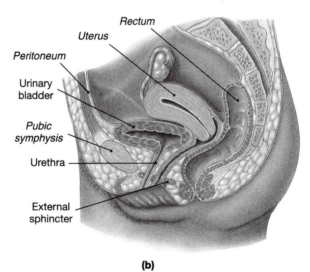

(b)

1. _____
2. _____
3. _____
4. _____
5. _____

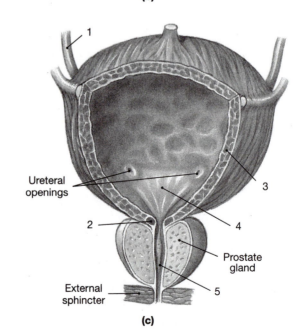

(c)

●Figure 18.5
Organs Responsible for the
Conduction and Storage of Urine

(a) The ureter, urinary bladder, and urethra in the male. **(b)** The same organs in the female. **(c)** The urinary bladder in a male.

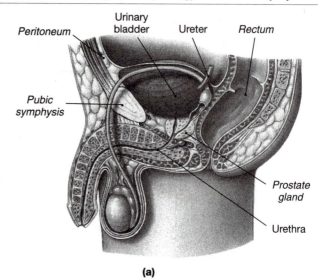

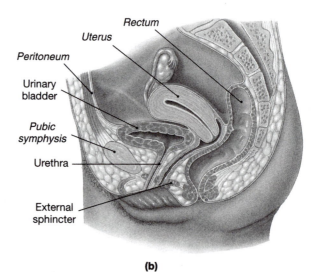

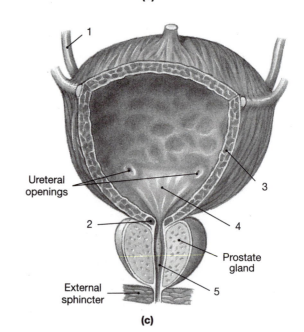

(c)

PROCEDURES

1. Wear surgical gloves while dissecting to protect your skin from the fixatives used to preserve the kidneys. Also wear your safety glasses to prevent preservatives from entering your eyes. Rinse the kidney with water to remove excess preservatives.

2. Examine the external features of the kidney. Locate the **hilus**, a concave slit on the medial border where the renal vessels and the ureters pass through the kidney. Locate the outer renal capsule, and gently lift it with a teasing needle. Below this capsule is a light pink area called the **cortex**.

3. With a scalpel, make a longitudinal cut to divide the kidney into anterior and posterior portions. A single long, smooth cut is less damaging to the internal anatomy than a sawing motion with the scalpel.

4. Distinguish between the cortex and the darker **medulla** that is organized into many triangular **renal pyramids**. The bases of the pyramids face the cortex, whereas the tips or apices narrow into **renal papillae**.

5. The **renal pelvis** is the large expanded end of the ureter. Extending from this area are structures called **major calyces** that have smaller channels called **minor calyces** into which the renal papillae project. The minor calyces collect urine from the renal papillae and channel it into the major calyx, which then delivers it to the renal pelvis. From here, the urine leaves the kidney by way of the **ureter**.

PART TWO   PHYSIOLOGY OF THE URINARY SYSTEM

WORD POWER

urine (urin, uria, urino—urine)
bilirubinuria (bili—bile, rubin—red, uria—urine)
hematuria (hemato—blood)
diabetes (diabetes—a siphon)
mellitus (mellitus—sweetened with honey)
pyuria (py—pus)

MATERIALS

500 ml urine cups
urinometer
dip sticks (Chemstix or Multistix)
BioHazard disposal container

OBJECTIVES

On completion of this part of the exercise, you should be able to:

- Describe the normal physical characteristics of urine.
- Recognize normal and abnormal constituents.
- Conduct a urinalysis test with special dip sticks.
- Define several urinary conditions.

The kidneys remove metabolic waste products such as toxins, excess water, and electrolytes from the blood plasma. Three physiological processes occur in the nephrons to maintain the composition of the blood and produce urine: filtration, reabsorption, and secretion. **Filtration** occurs in the renal corpuscle as blood pressure in the glomerulus forces water, small solutes, and ions out of the blood plasma and into the capsular space. The resulting fluid in the capsular space is called **filtrate**. Solute small enough to pass through the membranes of the renal corpuscle will appear in the filtrate. This results in a removal of both wastes and essential solutes from the blood. The essential solutes and most of the water are returned to the blood by a process called **reabsorption**. Additionally, **secretion** by cells of the distal convoluted tubule occurs to add excess materials such as hydrogen and ammonium ions to the filtrate for excretion in the urine. As the filtrate passes through the tubules of the nephron, reabsorption and secretion occur. The modified filtrate, now called **urine**, drips out of the renal papillae into the minor calyxes.

As the filtrate moves through the proximal convoluted tubule, 60% to 70% of the water, nutrients, and ions in the filtrate are reabsorbed into the blood. The cells of this tubule are **simple cuboidal epithelium** with **microvilli** to increase the lumen's surface area for reabsorption.

The kidneys filter 25% of the body's blood per minute, 24 hours a day. Approximately 180 liters of filtrate are formed by the glomerulus per day, which eventually results in the production of an average daily urinary output of between 1.0 and 2.0 liters of urine. The composition of urine can change daily depending on one's metabolic rate and urinary output. Water accounts for about 95% of the volume of urine. The other 5% contains excess water soluble vitamins, drugs, electrolytes, and nitrogenous wastes.

A. Composition of Urine

Normal constituents of urine include water; urea; creatinine; uric acid; many electrolytes; and possibly small amounts of hormones, pigments, carbohydrates, fatty acids, mucin, and enzymes.

Urea is produced during **deamination** reactions that remove ammonia (NH_3) from amino acids. The ammonia then combines with CO_2 to form urea. About 1,800 milligrams of urea are produced daily. Urea makes up about 70% to 90% of all nitrogenous material in urine. Creatinine is formed from the breakdown of creatine phosphate, a molecule found associated with muscle tissue as an energy storage molecule. This material is entirely secreted and not reabsorbed. Uric acid is produced from the breakdown of the nucleic acids DNA and RNA from foods or cellular destruction.

Several inorganic ions and molecules are also found in the urine. Their presence is a reflection of diet and general health. Calcium, potassium, and magnesium ions are cations that form salts with chloride, sulfate, and phosphate ions. Na^+ and Cl^- are ions from sodium chloride, the principal salt of the body. Excretion of salt varies with their dietary intake. Ammonium (NH_4^+) is a product of protein catabolism and must be removed from the blood before it reaches toxic concentrations. Many types of ions bind with sodium and form a buffer in the blood and urine to stabilize fluid pH. Other substances such as hormones, enzymes, carbohydrates, fatty acids, pigments, and mucin occur in small quantities in the urine.

Abnormal materials in the urine suggest a disease process or an injury to the kidneys. Excessive consumption of certain substances causes a high concentration of that substance in the filtrate that saturates the transport mechanisms of reabsorption. Because the tubular cells cannot reclaim all the substance, it will appear in the urine.

Ketones in the urine (ketosis) may be the result of starvation, diabetes mellitus, or a very low carbohydrate diet. When carbohydrate concentration in the blood is low, cells begin to catabolize fats. The products of fat catabolism are glycerol and fatty acids. Liver cells convert these fatty acids into ketone bodies that diffuse out of the liver cells and into the blood where they are filtered by the kidneys. In diabetes mellitus, commonly called sugar diabetes, sufficient amounts of glucose cannot enter cells. As a result, cells use fatty acids to produce ATP. This increase in fatty acid catabolism results in ketone bodies appearing in the urine.

Glucose in the urine (glucosuria) is usually an indication of diabetes mellitus (usually Type II diabetes). In this form of the disease, the individual produces an adequate amount of insulin. The body's cells, however, are less sensitive to the insulin because of a reduction in the number of insulin membrane receptors. As a result, excess levels of glucose are removed from the blood by the kidneys and voided in the urine.

Albumin is a large protein that normally cannot pass through filtering membranes of the glomerulus. A trace amount of albumin in the urine is considered normal. Excessive albumin in the urine, a condition called albuminuria, suggests an increase in the permeability of the glomerular membrane. Reasons for the increased permeability could be the result of an injury, high blood pressure, disease, or the effect of bacterial toxins on the cells of the kidneys.

Erythrocytes in the urine (hematuria) are usually an indication of an inflammation or infection of the urinary tract with the leakage of blood into the urinary tract. Blood in the urine could also indicate a pathological condition such as kidney disease, trauma, or kidney stones (renal calculi). **Leukocytes** in the urine (pyuria) indicates a urinary tract infection.

Bilirubin in large amounts in the urine, or bilirubinuria, is the result of the breakdown of hemoglobin from old red blood cells that are being removed from the circulatory system by phagocytic cells in the liver.

Urobilinogen in the urine is called urobilinogenuria. Small amounts of urobilinogen in the urine are normal. It is a product of the breakdown of bilirubin by the intestines, and it is responsible for the normal brown color of feces. Greater than trace levels in the urine may be due to infectious hepatitis, cirrhosis, congestive heart failure, or a variety of other diseases.

Microbes in large numbers in the urine usually indicate an infection. Generally, urine is sterile, but microbe content is possible in a urine sample for several reasons. Microbes may contaminate a urine sample owing to their presence at the urethral opening and in the urethra.

The **pH** of normal urine ranges from 4.5 to 8.0. Diet greatly affects the pH of urine. A high vegetable fiber diet results in an alkaline pH, whereas a high protein diet contributes to an acidic pH.

Specific gravity of a fluid is a comparison of the density of that fluid to the density of water. The density of a fluid is the ratio of the weight of the solutes compared to the volume of the solvent. The specific gravity of water is 1.000. The average specific gravity for a normal urine sample is between 1.003 and 1.030, a density greater than water. Normal constituents of urine include water, urea, creatinine, sodium, chloride, potassium, sulfates, phosphates, and ammonium salts. These solutes determine the specific gravity of urine. High fluid intake results in frequent urination of a dilute urine with a lower specific gravity. Low fluid intake results in a more concentrated urine with a higher specific gravity. Excessively concentrated urine results in the crystallization of solutes into renal calculi or kidney stones.

B. **Urinalysis**

Abnormal substances in urine can usually be detected by urinalysis, an analysis of the chemical and physical properties of urine. **Test strips** are a fast and inexpensive method of detecting the chemical composition of urine. Multistix and Chemstix are two popular brands of test strips. These sticks may contain from one to nine testing pads. Single test strips usually are used to detect if glucose or ketones are in the urine. The multiple test strips are used for a detailed analysis of the urine's chemical content.

During this urinalysis, you will examine only your urine sample. Physical characteristics such as volume, color, cloudiness, and odor will be observed. The test strips will determine the chemical characteristics of your sample. Both physical and chemical properties of urine vary according to fluid intake and diet.

Alternatively, your laboratory instructor may elect to provide your class with a mock urine sample for analysis. This artificial sample will probably include several abnormal constituents of urine for instructional purposes.

Laboratory Activity   URINALYSIS _____

MATERIALS

500 ml sample cup
Chemstix or Multistix test strips
BioHazard disposal container

Safety Tip

Handle and test only *your* urine. Dispose of all urine-contaminated materials in the BioHazard disposal container as outlined by your laboratory instructor. If a urine spill occurs, wear gloves to clean your workspace with a mild bleach solution.

I. Physical Characteristics

PROCEDURES

1. Obtain a sterile 500 ml cup for the collection of your urine sample. A midstream sample should be collected. By avoiding the collection of the first few milliliters of the urine, materials such as bacteria and pus from the urethra will not contaminate the sample.

2. Observe the physical characteristics of the sample. Record your observations in Table 18.1.

   a. **Volumes** in excess of 2 liters or less than 500 ml per day are considered abnormal. The volume of a single urine sample varies depending on an individual's hydration and fluid intake.

   b. The **color** of urine varies from clear to amber. Urochrome is a byproduct of the breakdown of hemoglobin, which gives urine its yellowish color. Color also varies because of ingested food. Vitamin supplements, certain drugs, and the amount of solutes also contribute to the color of urine. A dark red or brown color indicates blood in the urine.

   c. **Turbidity** or cloudiness is related to the amount of solids in the urine. Contributing factors include bacteria, mucus, cell casts, crystals, and epithelial cells. Observe and describe the turbidity of the urine sample. Use descriptive words such as "clear," "clouded," and "hazy."

   d. Freshly voided urine has no **odor**. As chemical breakdown of substances occurs, the sample may acquire an ammonia odor. Odor serves as a diagnostic tool in freshly voided urine. Starvation causes the body to break down fats and produce ketones that give urine a fruity or acetonelike smell. Individuals with diabetes mellitus often produce a sweet-smelling urine. To smell the

---

TABLE 18.1	**Physical Observations of Urine Sample**
Characteristic	*Observation*
Volume	
Color	
Turbidity	
Odor	

sample, place it approximately 12 inches from your face and wave your hand over it toward your nose.

II. Chemical Characteristics

PROCEDURES

1. Swirl your sample of urine before doing this test.
2. Holding a test strip by the end that does not have any test pads, immerse all the pads of the dip stick into the urine, and then withdraw the dip stick. Lay the strip on a clean, dry paper towel to absorb any excess urine that may be forming droplets on the dip stick.
3. Reading of the tests varies from immediately to two minutes. Use the color chart on the side of the bottle to interpret the tests.
4. Record your data from the test strip in Table 18.2.

TABLE 18.2	Chemical Evaluation of Urine Sample
Substance	*Remark*
pH	
Specific gravity	
Glucose	
Ketone	
Protein	
Erythrocytes	
Bilirubin	
Other	

ANATOMY AND PHYSIOLOGY OF THE URINARY SYSTEM
Exercise 18 Laboratory Report

A. Matching
Match the anatomy on the left with the correct description on the right.

1. _____ renal papilla	A.	drains into collecting duct
2. _____ cortex	B.	located at base of pyramid
3. _____ Bowman's capsule	C.	covers surface of kidney
4. _____ loop of Henle	D.	surrounds glomerulus
5. _____ renal pelvis	E.	surrounds renal pelvis
6. _____ distal convoluted tubule	F.	entrance for blood vessels
7. _____ efferent arteriole	G.	extends into minor calyx
8. _____ renal sinus	H.	tissue between pyramids
9. _____ renal pyramid	I.	U-shaped tubule
10. _____ arcuate artery	J.	transports urine to bladder
11. _____ hilus	K.	functional unit of kidney
12. _____ renal columns	L.	drains glomerulus
13. _____ renal capsule	M.	outer layer of kidney
14. _____ nephron	N.	comprises the medulla
15. _____ ureter	O.	drains major calyces

B. Short Answer Questions

1. What is the normal pH range of urine?

2. What is the specific gravity reading of a normal sample of urine?

3. List five abnormal components of urine.

4. What substances in urine might indicate a person has diabetes?

5. Explain why ketones may appear in the urine.

6. What might affect the odor, color, and pH of a sample of urine?

7. Describe the three physiological processes of urine production.

B. Competition

1. Complete Concept Map I detailing the urinary system.

Concept Map I

Using the following terms, fill in the circled, numbered, blank spaces to complete the concept map. Follow the numbers that comply with the organization of the map.

Renal sinus	Urinary bladder	Medulla
Ureters	Minor calyces	Nephrons
Glomerulus	Proximal convoluted tubule	Collecting tubules

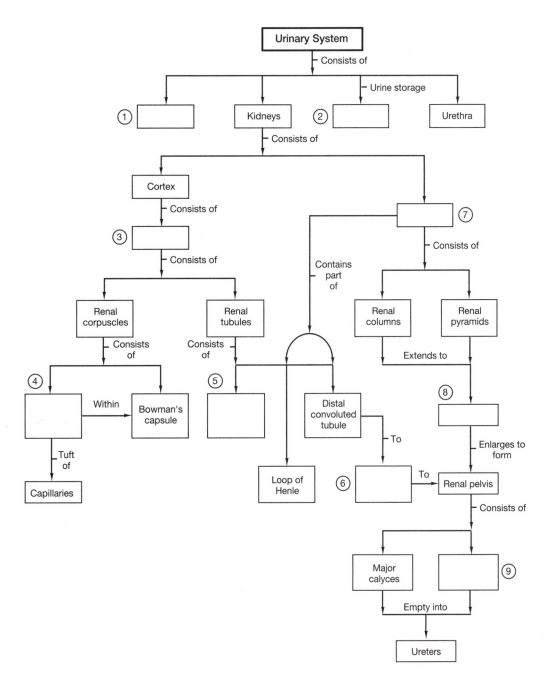

19 HUMAN REPRODUCTIVE SYSTEMS

Whereas all the other systems of the body function to support the continued life of the organism, the reproductive system functions to assure the continued perpetuation of the species. The primary sex organs or **gonads** of the male and female are the testes and ovaries, respectively. They function in the production of **gametes**, the sperm and eggs (ova). Gonads secrete hormones that support the maintenance of the male and female sex characteristics. Since these glands secrete their gametes into ducts and their hormones into the blood, they have both exocrine and endocrine functions. The ducts function in receiving and storing gametes. Several accessory glands secrete products to protect and support the gametes.

PART ONE   THE MALE REPRODUCTIVE SYSTEM

WORD POWER

gametes (gamete—husband, wife)
sperm (sperm—seed)
epididymis (epi—upon, didymos—twin)
corpus (corpus—body)
glans penis (glans—acorn)
corona (corona—crown)
cremaster (a suspender)

MATERIALS

male urogenital organs model and chart
male pelvis model
compound microscope
prepared slide
  testis

OBJECTIVES

On completion of this part of the exercise, you should be able to:

- Describe the location of the male gonad.
- Identify the epididymis and the ductus deferens.
- Identify the three accessory glands in males.
- Describe the composition of semen.
- Identify the three regions of the male urethra.
- Identify the structures of the penis.
- Trace the formation and transport of semen in the male reproductive tract.

The male reproductive system, shown in Figure 19.1, consists of a pair of testes, ducts to convey sperm and fluids secreted by accessory glands, and the penis. The **scrotum** is a cutaneous pouch that houses the testes. It is divided into two compartments each containing a testicle. The **dartos** muscle forms part of the septum separating the testes and is responsible for the wrinkling of the skin of the scrotum. The **cremaster muscles** encase the testes. They raise or lower the testes to maintain a cooler temperature for optimum sperm production. Locate the dartos and cremaster muscles in Figure 19.2.

A. The Testes

The **testes** or testicles are a pair of oval organs about 5 cm (2 inches) in length by 2.5 cm (1 inch) in diameter. They are the primary sex organ of the male and function in the production of sperm and the hormone testosterone. Compartments in the testicle called **lobules** contain highly coiled **seminiferous tubules**, shown in Figure 19.2. Millions of sperm are produced each day by the seminiferous tubules in a process called **spermatogenesis**. This process is presented in detail in Exercise 20. Follicle stimulating hormone (FSH) from the anterior pituitary gland regulates spermatogenesis. Between the seminiferous tubules are small clusters of cells called **interstitial cells** that secrete **testosterone**, the male sex hormone. Testosterone is responsible for the development and maintenance of the male secondary sex characteristics and the male sex drive.

Laboratory Activity   THE TESTES _____

MATERIALS

male urogenital model and chart
compound microscope
prepared slide of testis

PROCEDURES

1. Label and review the male reproductive anatomy in Figures 19.1 and 19.2.
2. Locate the scrotum, testes, and associated anatomy on the laboratory models.

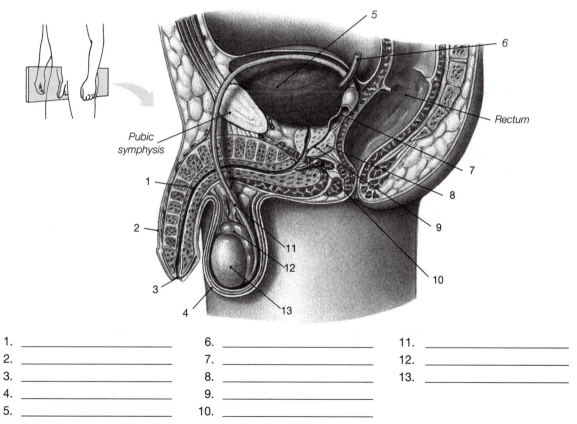

1. _____	6. _____	11. _____
2. _____	7. _____	12. _____
3. _____	8. _____	13. _____
4. _____	9. _____	
5. _____	10. _____	

●**Figure 19.1**
Components of the Male Reproductive System
The male reproductive organs as seen in sagittal section.

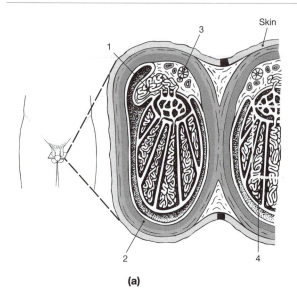

(a)

●**Figure 19.2**
Structure of the Testes

(a) Diagrammatic sketch and anatomical relationships of the testes. **(b)** A section through a coiled
seminiferous tubule. (LM × 786) **(c)** Cellular organization of a seminiferous tubule.

1. _____

2. _____

3. _____

4. _____

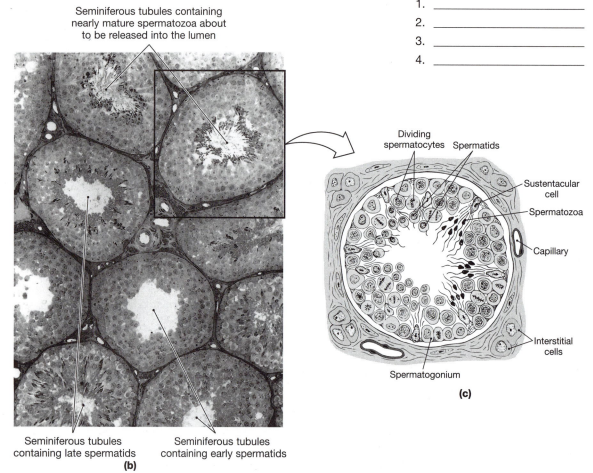

Seminiferous tubules containing
nearly mature spermatozoa about
to be released into the lumen

Seminiferous tubules
containing late spermatids

Seminiferous tubules
containing early spermatids

(b)

Dividing
spermatocytes   Spermatids

Sustentacular
cell

Spermatozoa

Capillary

Interstitial
cells

Spermatogonium

(c)

3. Examine a microscope slide of the testis in transverse section. Use the photomicrograph in Figure 19.2 for reference. Scan the slide at low magnification, and
   observe the many seminiferous tubules. Increase magnification, and locate the
   interstitial cells between the tubules. In the space provided, draw and label a section of the testis.

B. The Epididymis and the Ductus Deferens

After spermatozoa are produced in seminiferous tubules, they move into the **epididymis**, a highly coiled tubule located on the posterior of the testis, visible in Figure 19.1. Spermatozoa mature in the epididymis and are stored until ejaculation out of the male reproductive system. Peristalsis by the smooth muscle of the epididymis propels the sperm into the ductus deferens, the duct that empties into the urethra.

The **ductus deferens**, Figure 19.1, or vas deferens ascends as part of the **spermatic cord** into the pelvic cavity. Other structures in the spermatic cord include blood and lymphatic vessels, nerves, and the cremaster muscle. The duct is 46 to 50 cm (approximately 18 to 20 inches) long and is lined with pseudostratified columnar epithelium. Peristaltic waves propel spermatozoa toward the urethra. The ductus deferens passes through the **inguinal canal** in the lower abdominal wall to enter the body cavity. This canal is a weak area and is frequently injured. An **inguinal hernia** occurs when portions of intestine protrude through the canal and slide into the scrotum. The ductus deferens continues around the posterior of the urinary bladder and widens into the **ampulla** before joining the seminal vesicle at the ejaculatory duct.

Laboratory Activity   THE EPIDIDYMIS AND DUCTUS DEFERENS _____

MATERIALS

male urogenital model and chart

PROCEDURES

1. Locate the epididymis and ductus deferens on the laboratory models. Label these structures in Figures 19.1 and 19.2.
2. Locate the spermatic cord, cremaster muscles, and inguinal canal.

C. The Accessory Glands

Three accessory glands, the seminal vesicles, prostate, and bulbourethral glands, produce fluids that nourish, protect, and support the spermatozoa. Locate these glands in Figures 19.1 and 19.3. The sperm and fluids from the seminal vesicles, prostate gland, and bulbourethral glands form a mixture called **semen**. The semen is propelled by peristalsis and contractions of skeletal muscles through the urethra of the penis and expelled into the vagina as the **ejaculate**. The average volume of an ejaculate is between 2 ml and 5 ml. The average number of sperm per milliliter of semen is between 50 million and 150 million.

The **seminal vesicles** are a pair of glands on the posterior of the urinary bladder. Each gland is approximately 15 cm (6 inches) long and merges with the ductus deferens into an **ejaculatory duct**. The seminal vesicles contribute about 60% of the total volume of semen. They secrete a viscous, alkaline fluid containing the sugar fructose. The alkaline nature of this fluid neutralizes the acidity of the male urethra and the female vagina. The fructose provides the sperm an energy source necessary to beat the flagellum.

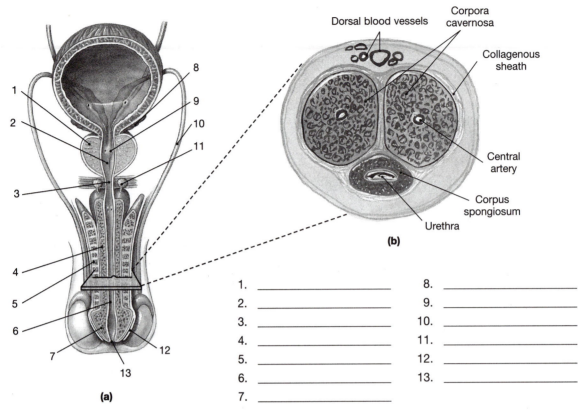

Corpora
cavernosa

Dorsal blood vessels

Collagenous
sheath

Central
artery

Corpus
spongiosum

Urethra

(b)

1. _____   8. _____
2. _____   9. _____
3. _____   10. _____
4. _____   11. _____
5. _____   12. _____
6. _____   13. _____
7. _____

(a)

●**Figure 19.3**
The Penis

(a) Frontal section through the penis and associated organs.
(b) Sectional view through the penis.

The **prostate gland** is a single gland about 1.5 inches in diameter at the base of the urinary bladder. The ejaculatory duct passes into the prostate gland and empties into the first segment of the urethra, the **prostatic urethra**. The prostate gland secretes a milky white, acidic fluid that contains clotting enzymes to coagulate the semen. These secretions contribute about 30% of the semen volume.

The prostatic urethra exits the prostate gland and passes through the floor of the pelvis as the **membranous urethra**. A pair of **bulbourethral** (Cowper's) **glands** on each lateral side of the membranous urethra adds an alkaline mucus to the semen. Before ejaculation, the bulbourethral secretions neutralize the acidity of the urethra and lubricate the end of the penis for sexual intercourse. These glands contribute about 5% of the volume of semen.

Laboratory Activity   THE ACCESSORY GLANDS _____

MATERIALS

male urogenital model and chart

PROCEDURES

1. Locate each accessory gland on the laboratory models. Label these glands in Figures 19.1 and 19.3.
2. Locate the urinary bladder and the prostatic and membranous urethra.

D. The Penis

The penis, detailed in Figure 19.3, is the male copulatory organ and serves to deliver semen into the vagina. It is cylindrical and has an enlarged, acorn-shaped head called the **glans**. Around its base is a margin, the **corona** (crown). On an uncircumcised penis, the glans penis is covered with a loose-fitting skin called the **prepuce** or **foreskin**. A circumcision is the surgical removal of the prepuce. The **penile urethra** transports semen or urine through the penis and ends at the **external urethral meatus** in the tip of the glans. The **root** of the penis anchors the penis to the pelvis. The **body** consists of three cylinders of erectile tissue, the paired dorsal **corpora cavernosa** and the ventral **corpus spongiosum**. During sexual arousal, the three erectile tissues become engorged with blood and cause the penis to stiffen into an **erection**.

Laboratory Activity   THE PENIS _____

MATERIALS

male urogenital model and chart

PROCEDURES

1. Identify the glans, body, and root of the penis on the laboratory models. Complete the labeling of Figure 19.3.
2. Examine a model of the penis in transverse section, and identify the corpora cavernosa and the corpus spongiosum.

PART TWO   The Female Reproductive System

WORD POWER	MATERIALS

gametes (gamete—husband, wife)
gonads (gone—seed)
corpus luteum (luteum—yellow)
fimbriae (fimbrae—fringe)
vulva (vulva—a wrapper or covering)
mons pubis (mons—a mountain)
	female reproductive organs model and chart
	female pelvis model
	breast model
	compound microscope
	prepared slides
	   ovary
	   uterus

OBJECTIVES

On completion of this part of the exercise, you should be able to:

- Identify the structures of the female reproductive tract.
- Identify the ovaries and its ligaments.
- Identify the structures of the uterine tubes.
- Identify the three main regions of the uterus.
- Describe and recognize the three main layers of the uterine wall.
- Identify the vagina and the features of the vulva.
- Identify the structures of the mammary glands.

The female reproductive system, highlighted in Figure 19.4, includes the ovaries, uterine tubes, uterus, vagina, external genitalia, and the mammary glands. **Gynecology** is the branch of medicine that deals with the care and treatment of the female reproductive system.

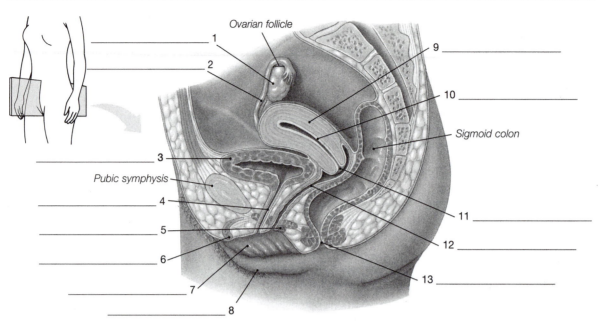

Ovarian follicle

1

2

9

10

Sigmoid colon

Pubic symphysis

3

4

5

6

7

8

11

12

13

●Figure 19.4
The Female Reproductive System

A. The Ovaries

The ovaries are the primary sex organs of the female. They are suspended in the pelvic cavity on each side of the uterus by several ligaments. A double-layered fold of peritoneum called the **mesovarium** holds the ovaries to the **broad ligament** of the uterus (see Figure 19.5). The **suspensory ligaments** hold the ovaries to the wall of the pelvis, and the **ovarian ligament** holds the ovaries to the uterus.

Each almond-sized ovary contains from 100,000 to 200,000 undeveloped eggs clustered into **egg nests**. Within the nests are **primordial follicles**, which consist of ova surrounded by follicular cells. Figure 19.6 details the monthly **ovarian cycle**, during which hormones stimulate follicular cells to proliferate and produce several **primary follicles**. These follicles increase in size, and a few become **secondary follicles**. Eventually, one secondary follicle will develop into a **tertiary follicle**, also called a **Graafian follicle**. This follicle fills with fluid and eventually ruptures, casting out the ovum in a process called **ovulation**. The Graafian follicle secretes **estrogen**, which stimulates rebuilding of the spongy lining of the uterus. After ovulation, the tertiary follicle becomes the **corpus luteum** and primarily secretes the hormone **progesterone**, which prepares the uterus for pregnancy. If fertilization of an ovum does not occur, the corpus luteum degenerates into the **corpus albicans**, and most of the rebuilt lining of the uterus is shed as the menstrual flow. Observe the prepared slide of the ovary. Locate primary and secondary follicles, and a tertiary follicle.

Laboratory Activity   THE OVARIES _____

MATERIALS

female reproductive system model and chart   prepared slide:
compound microscope   ovary

PROCEDURES

1. Locate the ovaries and ligaments on the laboratory models. Label these structures in Figures 19.4 and 19.5.

2. Examine a microscope slide of an ovary section. Using Figure 19.6 as reference, locate the following structures:

   a. Scan the slide at low magnification, and locate an egg nest along the periphery of the ovary.

   b. Identify the primary follicles that are larger than the primordial follicles in the egg nests. The ovum has increased in size and is surrounded by follicular cells.

   c. Secondary follicles are larger than primary follicles and have a separation between the outer and inner follicular cells.

   d. Tertiary follicles are easy to distinguish by their large, fluid-filled space called the **antrum**.

B. The Uterine Tubes and Uterus

The ends of the **uterine** or fallopian **tubes** have fingerlike projections called **fimbriae** (see Figure 19.5). These fimbriae sweep over the surface of the ovary to capture an ovum released during ovulation and draw it into the expanded **infundibulum**. Once inside the uterine tube, ciliated epithelium transports the ovum toward the uterus. The tube widens midway along its length in the **ampulla**, then narrows at the **isthmus** to enter the uterus. Fertilization of the egg usually occurs in the upper third of the uterine tube.

The **uterus** is a pear-shaped muscular organ located between the urinary bladder and the rectum. It serves as a site for implantation of a fertilized ovum and the subsequent development of the fetus during pregnancy. The uterus consists of three major regions, the fundus, body, and cervix. The superior dome-shaped portion of the uterus is the **fundus**. Most of the uterus is called the **body**, and the inferior narrow portion is called the **cervix**, which opens into the vagina. Within the uterus is a space called the

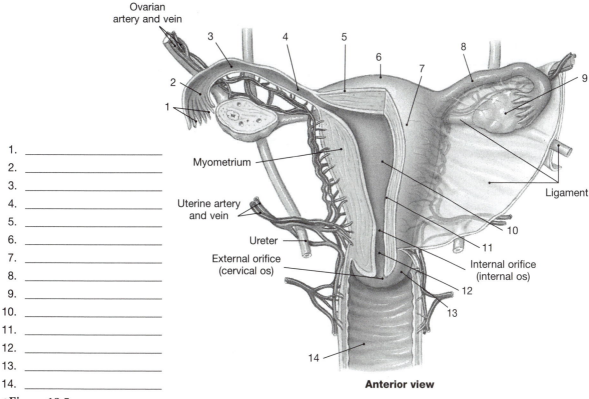

1. _____
2. _____
3. _____
4. _____
5. _____
6. _____
7. _____
8. _____
9. _____
10. _____
11. _____
12. _____
13. _____
14. _____

Anterior view

●**Figure 19.5**
The Ovaries and Uterus

Anterior view with right portion of uterus, uterine tube, and ovary shown in section.

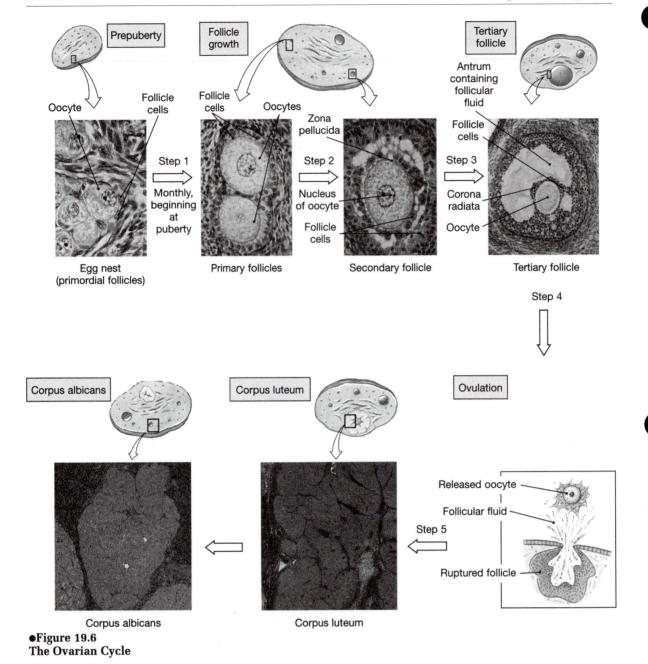

●**Figure 19.6**
The Ovarian Cycle

uterine cavity, which narrows to the **cervical canal** in the cervix. Locate these regions of the uterus in Figure 19.5.

The uterine wall consists of three main layers: the perimetrium, the myometrium, and the endometrium (see Figure 19.7). The **perimetrium** is the outer covering of the uterus. It is an extension of the visceral peritoneum and is therefore also called the serosa. The thick middle layer, the **myometrium**, comprises three layers of smooth muscle and is responsible for the powerful contractions during labor. The **endometrium** consists of two layers, the **stratum basalis** (basilar zone) and the **stratum functionalis** (functional zone). The stratum basalis faces the myometrium and produces a new stratum functionalis each month. The stratum functionalis is very glandular and is highly vascularized to support an implanted embryo. This is also the layer that is shed each cycle during menstruation. Observe the prepared slide of the uterus, and identify the three major uterine layers.

Laboratory Activity   THE UTERINE TUBE AND THE UTERUS _____

MATERIALS

female reproductive system model and chart   prepared slide
compound microscope   uterus

PROCEDURES

1. Label the uterus and associated structures in Figures 19.4 and 19.5.
2. Identify the uterine tubes, ampulla, and isthmus on the laboratory models.
3. Distinguish between the fundus, body, and cervix of the uterus on available laboratory models.
4. Using Figure 19.7 for reference, examine a prepared microscope slide of the uterus in transverse section. Scan the slide at low and medium magnifications, and locate the thick myometrium composed of smooth muscle tissue. The endometrium lines the uterine cavity and is rich with blood vessels and glands.
5. Draw a section of the uterine wall from your microscopic studies in the space provided.

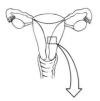

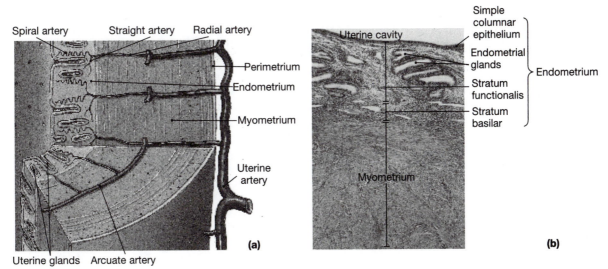

(a)   (b)

●**Figure 19.7**
The Uterine Wall

(a) A sectional view of the uterine wall, showing the endometrial regions and the circula-

C. The Vagina and the Vulva

The **vagina** is a muscular tube approximately 10 cm (4 inches) long. Locate the vagina in the various views of Figures 19.4, 19.5, and 19.8. It is lined with stratified squamous epithelium and serves as the female's copulatory organ, the pathway for menstrual flow, and the lower birth canal. The **vaginal orifice** is the external opening of the vagina. This opening may be partially or totally occluded by a thin fold of vascularized mucus membrane called the **hymen**. On either side of the vaginal orifice are openings of the **greater vestibular glands**, which produce a mucous secretion to lubricate the vaginal entrance for sexual intercourse. These glands are similar to the bulbourethral glands of the male. A vaginal pocket called the **fornix** is formed where the cervix protrudes into the vagina.

The **vulva** consists of the external genitalia of the female (see Figure 19.8). It includes the following structures:

- The **mons pubis** is a fatty pad over the pubic symphysis. It is covered with skin and pubic hair and serves as a cushion for the pubic symphysis during sexual intercourse.
- The **labia majora** are two fatty folds of skin that extend from the mons pubis and continue posteriorly. They are homologous to the scrotum of the male. They usually have pubic hair and contain many sudoriferous (sweat) and sebaceous (oil) glands.
- The **labia minora** are two smaller parallel folds of skin that contain many sebaceous glands. This pair of labia lacks hair.
- The **clitoris** is a small cylindrical mass of erectile tissue similar to the penis. It also contains a small fold of skin that covers it called the **prepuce**. The exposed portion of the clitoris is called the glans.
- The **vestibule** is the area between the labia minora that contains the vaginal orifice, hymen, and external urethral orifice.

1. _____
2. _____
3. _____
4. _____
5. _____
6. _____
7. _____
8. _____
9. _____

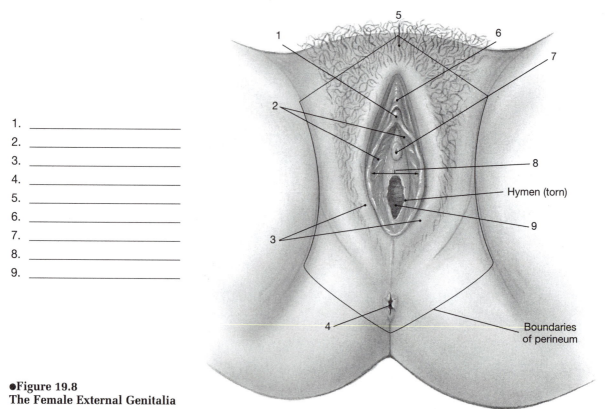

●Figure 19.8
The Female External Genitalia

Laboratory Activity   THE VAGINA AND THE VULVA _____

MATERIALS

female reproductive system model and chart

PROCEDURES

1. Locate the vagina and vaginal orifice on the laboratory models. Examine the fornix where the cervix and vagina connect.
2. Locate each component of the vulva. How is the urethra positioned compared with the vagina and clitoris?
3. Label all structures of the female external genitalia in Figure 19.8.

D. The Mammary Glands

The **mammary glands**, Figure 19.9, are modified sweat glands that produce milk to nourish the newborn infant. At puberty, release of estrogens stimulates an increase in the size of the mammary glands. Fat deposition is the major contributor to the size of the breast. Development of the ducts of the mammary glands also occurs. The mammary glands consist of 15 to 20 **lobes** separated by fat and connective tissue. Each lobe contains smaller compartments called **lobules** that contain milk-secreting cells called **alveoli**. **Lactiferous ducts** drain milk from the lobules toward the **lactiferous sinuses**. These sinuses empty milk at the raised portion of the breast called the **nipple**. A circular, pigmented area called the **areola** surrounds the nipple.

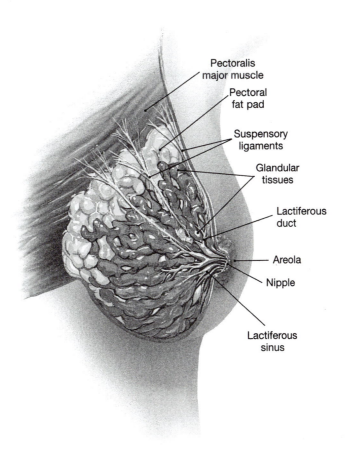

Pectoralis
major muscle

Pectoral
fat pad

Suspensory
ligaments

Glandular
tissues

Lactiferous
duct

Areola

Nipple

Lactiferous
sinus

●Figure 19.9
The Mammary Glands

Structure of the breast.

Laboratory Activity   THE MAMMARY GLANDS _____

MATERIALS

breast model

PROCEDURES

1. Review the structure of the mammary glands in Figure 19.9.
2. Examine the models and charts of the mammary glands. Trace the pathway of milk from the lobules to the surface of the nipple.

HUMAN REPRODUCTIVE SYSTEMS
Exercise 19 Laboratory Report

A. Short Answer Questions

1. The primary sex organs of the male are the _testes_____.

2. The _____ respond to changes in temperature by raising or lowering the testes.

3. The _____ produce the male sex hormone _____.

4. The three regions of the male urethra are:

   a.

   b.

   c.

5. The _____ is formed by the union of the ampulla of the ductus deferens and the duct from the seminal vesicle.

6. List the three accessory glands of the male reproductive system.

   a.

   b.

   c.

B. Matching

1.	_____	epididymis	A.	site of sperm storage
2.	_____	ductus deferens	B.	site of sperm production
3.	_____	bulbourethral glands	C.	enlarged tip of penis
4.	_____	glans	D.	paired erectile cylinder
5.	_____	corpora cavernosa	E.	small glands in pelvic floor
6.	_____	prepuce	F.	first segment of urethra
7.	_____	prostatic urethra	G.	also called the foreskin
8.	_____	scrotum	H.	transports sperm to urethra
9.	_____	membranous urethra	I.	pouch that contains testes
10.	_____	seminiferous tubules	J.	urethra in pelvic floor

C. Short Answer Questions

1. After ovulation the tertiary follicle becomes the _____.

2. List the three layers of the uterus from superficial to deep.

   a.

   b.

   c.

3. The pigmented portion of the breast is called the _____.

4. List the components of the vulva.

_____

_____

5. Describe how the clitoris is similar to the penis.

_____

_____

D. Matching

1. _____	labia minora	A.	space between labia minora
2. _____	myometrium	B.	flared end of uterine tube
3. _____	mons pubis	C.	domed portion of uterus
4. _____	fundus	D.	female external genitalia
5. _____	infundibulum	E.	uterine protrusion into vagina
6. _____	isthmus	F.	narrow portion of uterine tube
7. _____	vestibule	G.	small fold lacking pubic hair
8. _____	labia majora	H.	fatty cushion
9. _____	vulva	I.	muscular layer of uterine wall
10. _____	cervix	J.	large fold with pubic hair

E. Completion

1. Complete Concept Map I by outlining the male reproductive organs. Use your lecture textbook as an additional resource if necessary.

2. Complete Concept Map II by outlining the female reproductive organs. Use your lecture textbook as an additional resource if necessary.

Concept Map I

Using the following terms, fill in the circled, numbered, blank spaces to complete the concept map. Follow the numbers that comply with the organization of the map.

Urethra   Seminiferous tubules   Penis
Product testosterone   FSH   Seminal vesicles
Ductus deferens   Bulbourethral glands

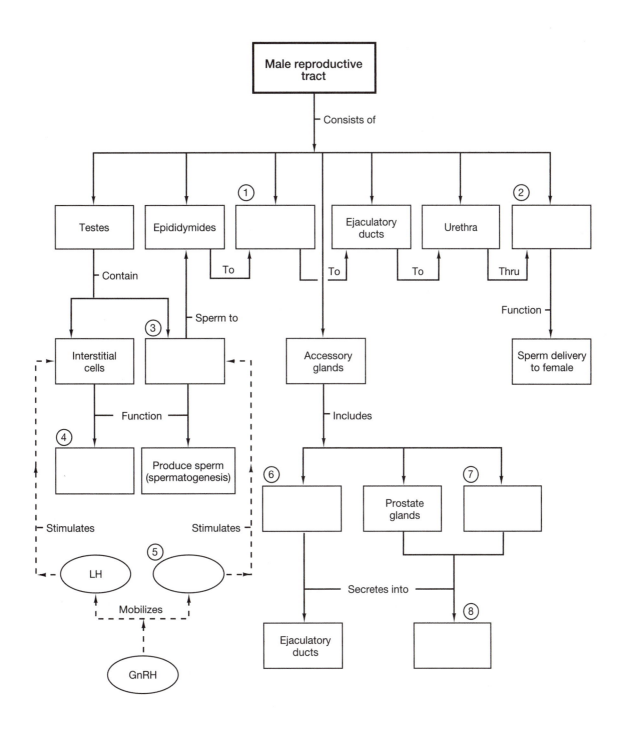

Concept Map II

Using the following terms, fill in the circled, numbered, blank spaces to complete the concept map. Follow the numbers that comply with the organization of the map.

nutrients   supports fetal development   endometrium
vulva   clitoris   vagina
uterine tubes   labia majora and minora   follicles

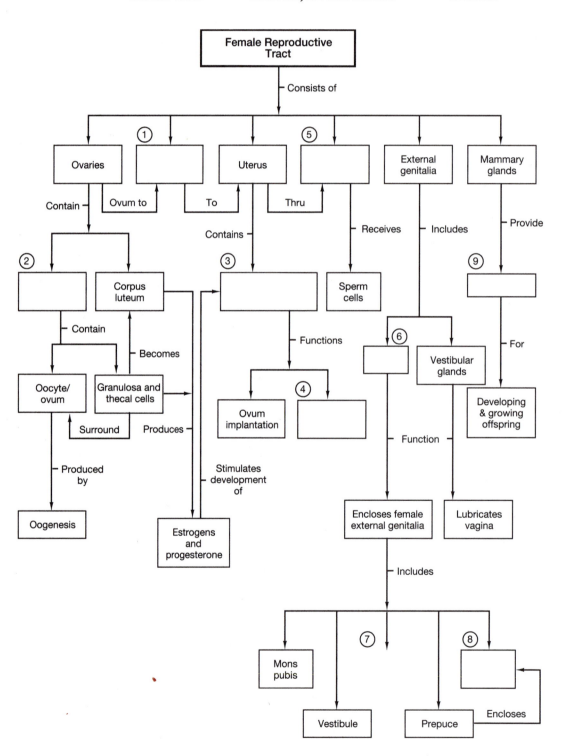

20 HUMAN DEVELOPMENT

An intriguing topic in biology is human development. Two cells called **gametes** from the parents fuse and create a new life, a single cell called the **zygote**, which has its own genetic identity. A phenomenal number of cell divisions and migration of cells occur during the early development of the embryo. Tissue layers and organ systems form in an elaborate sequence designed to progress in complexity until a functional human is born. This exercise will survey the major stages of early development. Some individuals debate when life begins; however, biologically, there is no question: life begins at conception.

The figures in this exercise provide a visual guide to the developmental process. If available, supplement your lab studies with observations of development models to gain a three-dimensional representation of each developmental process.

WORD POWER

gamete (gamet—spouse)
zygote (zygo—a yoke)
meiosis (meio—less)
diploid (diplo—double)
haploid (haplo—single)
spermatogenesis (...genesis—origin)
blastomere (blast—a sprout)
morula (morul—small mulberry)
trophoblast (tropho—food)

syncytial (syn—without, cytial—cell)
ectoderm (ecto—outer, derm—skin)
endoderm (endo—inner)
mesoderm (meso—middle)
chorion (chorio—a membrane)
placenta (placent—round, flat cake)
decidua (decid—falling off)
parturition (partur—to give birth)

MATERIALS

embryology models
embryology charts
placenta model
parturition model
compound microscope
prepared slides
   seminiferous tubules

OBJECTIVES

On completion of this exercise, you should be able to:

- Discuss the mechanics of meiosis, and compare gamete formation in males and females.
- Describe the process of fertilization and early cleavage to the blastocyst stage.
- Describe the process of implantation and placenta formation.
- List the three germ layers and the embryonic fate of each.
- List the four extraembryonic membranes and the function of each.
- Describe the general developmental events of the first, second, and third trimesters.
- List the three stages of labor.

A. Meiosis: The Formation of Gametes

Sexual reproduction involves the union of gametes from different individuals. Human somatic cells (nonreproductive cells) contain 23 pairs of chromosomes for a total of 46 individual chromosomes. If the number of chromosomes in gametes is not reduced by one-half, each new generation of humans would have twice the number of chromosomes as their parents. Cell division, called **meiosis**, produces gametes that are **haploid**; that is, they have a chromosome from each pair of chromosomes. Cells that have both chromosomes from all pairs are **diploid** cells. A human gamete has 23 individual chromosomes, 1 chromosome from each pair. When fertilization occurs, the gametes fuse their genetic material, and the diploid condition is restored in the zygote. The zygote does not represent half the genes from each parent; the zygote has a unique genetic constitution. In many ways, the process of meiosis is remarkably similar to mitosis, cell division that produces identical diploid daughter cells. If necessary, review mitosis in Exercise 5 before continuing in this exercise

Males produce spermatozoa in a process called **spermatogenesis**. Four haploid sperm cells are produced from each cell that started in meiosis. Formation of the female gamete, the ovum, is called **oogenesis** and results in only one viable haploid cell. Figure 20.1 is a comparative overview of spermatogenesis and oogenesis. For simplicity, the figure uses a cell with 3 diploid chromosomes instead of 23 diploid as in humans. As in mitosis, cells prepare for meiosis by duplicating their genetic material. After replication, each chromosome consists of two **chromatids**. Meiosis begins as the nuclear membrane dissolves and the chromosomes become visible. The replicated chromosomes match into pairs in a process called **synapsis**, and the four chromatids of the pair are collectively called a **tetrad**. Notice in Figure 20.1 that in the first cell, a primary spermatocyte, the chromosome pairs have joined to form tetrads. Since each chromosome has replicated, a total of four potential chromosomes, each called a **chromatid**, align in a **tetrad**. Because each chromatid in a tetrad belongs to the same chromosome pair, genetic information may be exchanged between chromatids. This **crossing over** or mixing of the genes increases genetic diversity of the population.

When a male reaches puberty, or sexual maturity, hormones stimulate the testes to begin the process of spermatogenesis (see Figures 20.1 and 20.2). **Spermatogonia** cells in the walls of the seminiferous tubules divide by mitosis and produce diploid **primary spermatocytes**. During **meiosis I**, chromosome pairs synapse and form tetrads along the equatorial plane of the primary spermatocytes. Chromosome pairs separate, and the cell divides into two haploid **secondary spermatocytes**. This step is called the **reduction division** of meiosis because haploid cells are produced. Each secondary spermatocyte still possesses chromosomes consisting of 2 chromatids and must go through meiosis II, which results in 4 **spermatids**, each with 23 single-stranded chromosomes. The spermatids undergo a maturation process and develop into **spermatozoa**. The entire process from spermatogonia to mature spermatozoa takes approximately ten weeks.

The process of oogenesis begins in the female's uterus before she is born. **Oogonia** divide and produce **primary oocytes**, which remain suspended in this stage until puberty, when, each month, a primary oocyte divides into two **secondary oocytes**. One of the secondary oocytes is much smaller than its sister cell and is a nonfunctional cell called the **first polar body** (see Figure 20.1). The remaining secondary oocyte will continue to divide if fertilization occurs. Then, the secondary oocyte goes through meiosis II for the separation of double-stranded chromosomes. When this cell divides, another polar body is formed, the **second polar body**. The remaining cell is the fertilized ovum, the zygote. Females produce only a single functional cell by oogenesis, whereas in males, spermatogenesis results in four spermatozoa.

Laboratory Activity   Meiosis _____

MATERIALS

meiosis models
compound microscope
prepared slides of seminiferous tubules

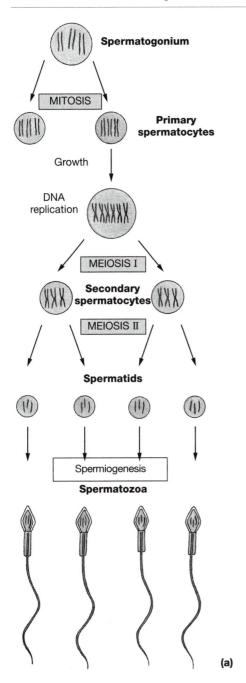

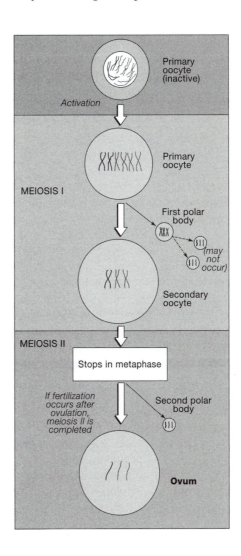

●Figure 20.1
Spermatogenesis and Oogenesis

(a) This figure contrasts spermatogenesis and oogenesis. In spermatogenesis a single primary spermatocyte produces four spermatids that mature into spermatozoa. **(b)** In oogenesis a single primary oocyte produces an ovum and two or three polar bodies. (The first polar body may or may not undergo an equational division.)

PROCEDURES

1. Label and review the process of meiosis in Figures 20.1 and 20.2.
2. Examine the models of meiosis. When does the reduction division occur to produce haploid cells? Why must meiosis II occur if the cells from meiosis I are already haploid?

3. Obtain a slide of mammalian seminiferous tubules prepared from the testis. Start at low magnification, and locate an oval seminiferous tubule that has distinct cells on the interior. Increase magnification, and using Figure 20.2 as a guide,

Seminiferous tubules containing
nearly mature spermatozoa about
to be released into the lumen

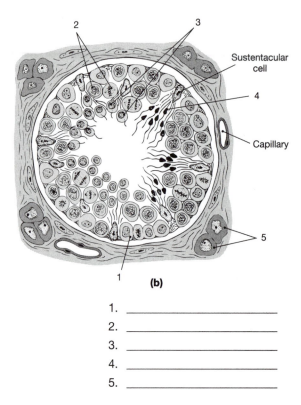

Seminiferous tubules   Seminiferous tubules
containing late spermatids   **(a)**   containing early spermatids

Sustentacular
cell

Capillary

(b)

1. _____

2. _____

3. _____

4. _____

5. _____

●**Figure 20.2**
Spermatogenesis in Seminiferous Tubules

(a) A section through a coiled seminiferous tuble.
(b) A cross-section through a single seminiferous
tubule. (LM × 786)

identify the spermatogonia, spermatocytes in various stages of division, and
spermatids. Do you see any spermatozoa in the middle of the tubule?

_____

_____

4. In the space provided, draw a section of a seminiferous tubule, and label the
spermatogonia, primary spermatocytes, secondary spermatocytes, spermatids,
and spermatozoa.

B. Survey of Human Development: The First Trimester

In humans, the prenatal period of development occurs over a nine-month **gestation**. Human gestation is divided into three trimesters, each three months in length. During the first trimester, the **embryo** develops cell layers that are precursors to organ systems. Special membranes, such as the amnion, support the growing embryo and fetus until birth. By the end of the second month, most organ systems have started to form, and the embryo is then called a **fetus**. The second trimester is characterized by growth in length, weight gain, and the appearance of functional organ systems. In the third trimester, increases in length and weight occur, and all organ systems become functional, or are prepared to become operational at birth. After 38 weeks of gestation, the uterus begins to rhythmically contract to deliver the fetus into the world. Although maternal changes occur during the gestation period, this exercise will focus on the development of the fetus.

Development is the process of morphogenesis, growth, and development of offspring to achieve maturity. Morphogenesis involves the division of cells to achieve specialized cells and the migration of cells to produce anatomical form and function. Ultimately, organ systems appear and become functional as the offspring gets closer to birth.

Fertilization is the act of the sperm and egg joining their haploid nuclei to produce a diploid **zygote**, a genetically unique cell that develops into an individual. The male ejaculates approximately 300 million spermatozoa into the female's reproductive tract during intercourse. Once exposed to the female's reproductive tract, the sperm complete a process called **capacitation**, during which the spermatozoa increase their motility and become capable of fertilizing an egg. Most sperm do not survive the journey through the vagina and uterus, and only an estimated 100 sperm reach the upper uterine tube where fertilization occurs. Normally, only a single egg is released from a single ovary during one ovulation cycle.

Figure 20.3 illustrates the fertilization process. Ovulation releases a secondary oocyte from the ovary. The oocyte is surrounded by a layer of cells called the **corona radiata** that the sperm must pass through to reach the cell membrane of the oocyte. Spermatozoa swarming around the oocyte release an enzyme called hyaluronidase from their acrosomal caps. The combined action of enzymes contributed by all the spermatozoa eventually creates a gap between some coronal cells, and a single spermatozoan slips into the oocyte. The membrane of the oocyte instantly undergoes chemical and electrical changes that prevent additional sperm from entering the cell. The oocyte, suspended in meiosis II since ovulation, now completes meiosis while the sperm prepares the paternal chromosomes for the union with the maternal chromosomes. Each set of nuclear material is called a **pronucleus**. Within 30 hours of fertilization, the male and female pronuclei come together in **amphimixis** and undergo the first **cleavage**, a mitotic division resulting in two cells, each called a **blastomere**. During cleavage, the existing cell mass of the egg is subdivided by each mitosis.

As the zygote slowly descends in the uterine tube toward the uterus, cleavages occur approximately every 12 hours. By the third day, the blastomeres are organized into a solid ball of nearly equal cells called a **morula**, shown in Figure 20.4. Around day six, the morula has entered the uterus and changed into a **blastocyst**, a hollow ball of cells with an internal cavity called the **blastocoele**. The process of **differentiation** or specialization begins. The blastomeres are now different sizes and have migrated into two regions. Cells on the outside compose the **trophoblast**, which will burrow into the uterine lining and eventually form part of the placenta. Cells clustered inside the blastocoele form the **inner cell mass**, which will develop into the embryo.

Implantation begins on day seven or eight when the blastocyst touches the spongy uterine lining. The trophoblast near the inner cell mass faces the uterus and burrows into the functional zone of the endometrium. The plasma membranes of these trophoblast cells dissolve, and the cells mass together as a cytoplasmic layer of multiple nuclei called the **syncytial trophoblast**. The cells secrete hyaluronidase to erode a path

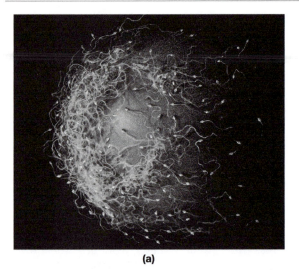

(a)

●Figure 20.3
Fertilization

(a) An oocyte at the time of fertilization: note the difference in size between the gametes.
(b) Fertilization and the preparations for cleavage.

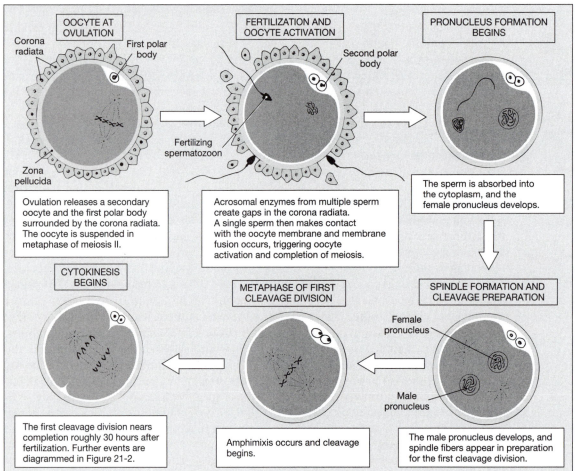

(b)

for implantation of the blastocyst (see Figure 20.5). Implantation continues until the embryo is completely covered by the stratum functionalis of the endometrium, which occurs by the 14th day. To establish a diffusional link with maternal circulation, the syncytial trophoblast grows extensions called **villi** into spaces in the endometrium called **lacunae**. Maternal blood from the endometrium seeps into the lacunae and bathes the villi with nutrients and oxygen. These materials diffuse into the blastocyst

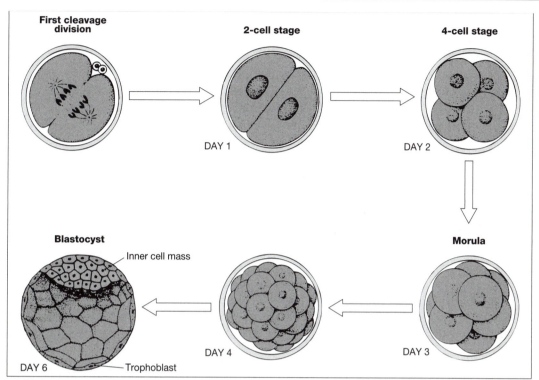

First cleavage division

2-cell stage

DAY 1

4-cell stage

DAY 2

Morula

DAY 3

Blastocyst

Inner cell mass

DAY 4

DAY 6

Trophoblast

●**Figure 20.4**
Cleavage and Blastocyst Formation

to support the inner cell mass. Beneath the syncytial layer is the **cellular trophoblast**, which will soon help form the placenta.

On the ninth or tenth day, the middle layers of the inner cell mass drop away from the upper layer next to the cellular trophoblast. This movement of cells forms the **amniotic cavity**. The inner cell mass organizes into a **blastodisc**, which contains two cell layers, the **epiblast** on top and the **hypoblast** facing the blastocoele. The blastodisc is the early embryo and is illustrated in Figure 20.6.

Within the next few days a process of cell migration called **gastrulation** begins (see Figure 20.6). Cells of the epiblast move toward the medial plane of the blastodisc to a region known as the **primitive streak**. As cells arrive at the primitive streak, infolding, or **invagination**, occurs and cells are liberated between the epiblast and the hypoblast, producing three cell layers in the embryo. The epiblast becomes the **ectoderm**, the hypoblast is now the **endoderm**, and the cells proliferating between the two layers form the **mesoderm**. These three layers, called **germ layers**, each produce specialized tissues that contribute to the formation of the organ systems. For example, the ectoderm forms the nervous system, the skin, hair, and nails. The mesoderm contributes to the development of the skeletal and muscular systems, and the endoderm forms part of the lining of the respiratory and digestive systems.

By the end of the fourth week of development, the embryo is distinct and has a **tail fold** and a **head fold**. The dorsal and ventral surfaces and the right and left sides are well defined. The process of **organogenesis** begins as organ systems develop from the germ layers. Observe Figure 20.7a, a fiber-optic view of the embryo at four weeks. The heart is clearly visible and has beat since the 21st day in utero. **Somites**, embryonic precursors of skeletal muscles, appear. Elements of the nervous system are also developing. Buds for the arms and legs and small discs for the eyes and ears are also present. By week eight, individual fingers and toes are present (see Figure 20.7b). At the end of the second month, the embryo is called a **fetus**. At the end of the third month, the first trimester is completed, and each organ system has appeared in the fetus.

1. _____
2. _____
3. _____
4. _____
5. _____
6. _____
7. _____

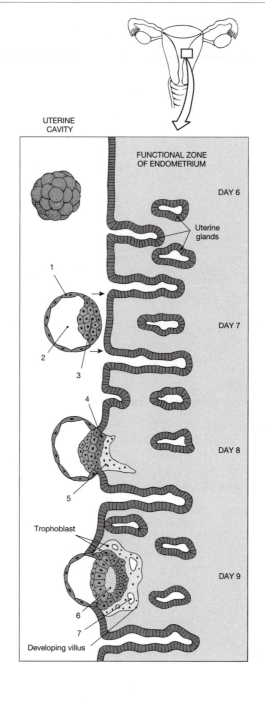

●Figure 20.5
Stages in the Implantation Process

Laboratory Activity   THE FIRST TRIMESTER _____

MATERIALS

embryology models
embryology charts

PROCEDURES

1. Label and review the developmental processes illustrated in Figures 20.3, 20.4, 20.5, 20.6, and 20.7.

2. Examine the models of fertilization and early embryologic development.

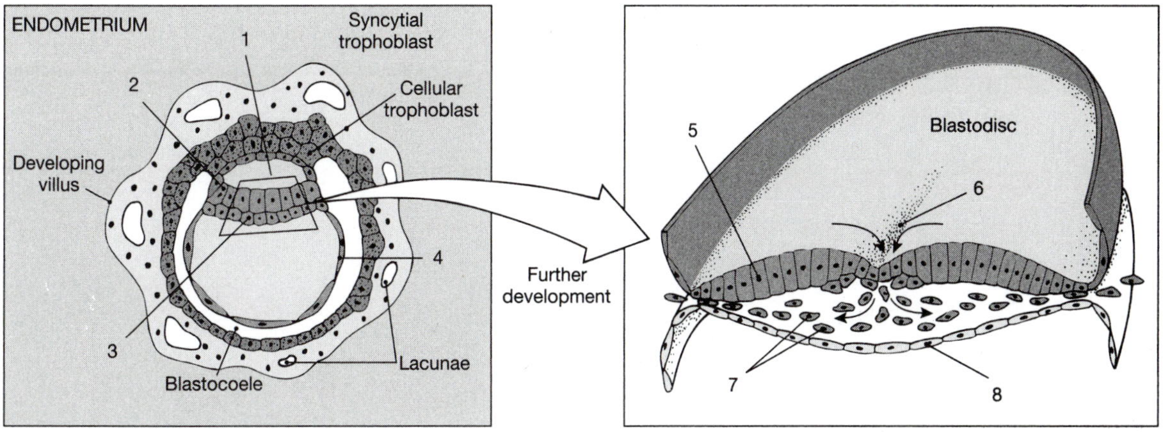

ENDOMETRIUM — Syncytial trophoblast — Cellular trophoblast — Developing villus — Blastocoele — Lacunae — Further development — Blastodisc

Day 10: The blastodisc begins as two layers, an epiblast facing the amniotic cavity and the hypoblast exposed to the blastocoele. Migration of epiblast cells around the amniotic cavity is the first step in the formation of the amnion. Migration of hypoblast cells creates a sac that hangs below the blastodisc. This is the first step in yolk sac formation.

Day 12: Migration of epiblast cells into the region between epiblast and hypoblast gives the blastodisc a third layer. From the time this process, called gastrulation, begins, the epiblast is called *ectoderm*, the hypoblast *endoderm*, and the migrating cells *mesoderm*.

1. _____   4. _____   7. _____

2. _____   5. _____   8. _____

3. _____   6. _____

●Figure 20.6
Blastodisc Organization and Gastrulation

Describe the fertilization process and the early cleavages that form the morula. Draw a sequence of cells from the zygote to the morula.

_____

_____

3. Identify the structures of the blastocyst at the time of implantation. How does the amniotic cavity form? What are the two cell layers called below the amniotic cavity? Do the models show the distinction between the cellular and syncytial trophoblast? In the space provided, draw a blastocyst with each trophoblast layer.

_____

_____

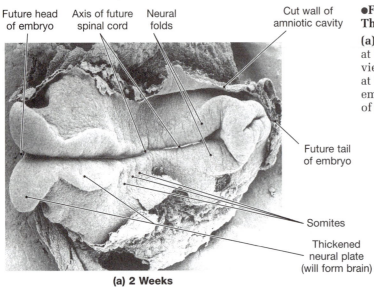

Future head of embryo
Axis of future spinal cord
Neural folds
Cut wall of amniotic cavity
Future tail of embryo
Somites
Thickened neural plate (will form brain)

(a) 2 Weeks

●**Figure 20.7**
The First Trimester

(a) Superior view of an SEM of an embryo at Week 2. **(b)** An SEM and a fiberoptic view of the lateral surface of embryos at Week 4-5. **(c)** Fiberoptic view of an embryo at 8 weeks. **(d)** Fiberoptic view of an embryo at 12 weeks.

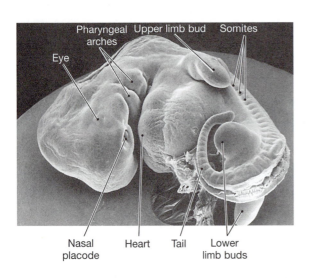

Pharyngeal arches
Upper limb bud
Somites
Eye
Nasal placode
Heart
Tail
Lower limb buds

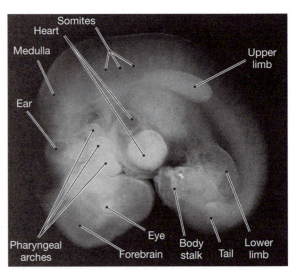

Somites
Heart
Medulla
Ear
Upper limb
Pharyngeal arches
Forebrain
Eye
Body stalk
Tail
Lower limb

(b) 4 Weeks

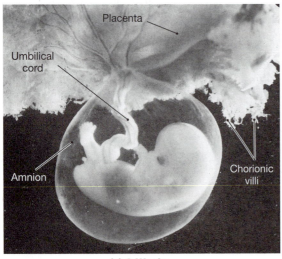

Placenta
Umbilical cord
Amnion
Chorionic villi

(c) 8 Weeks

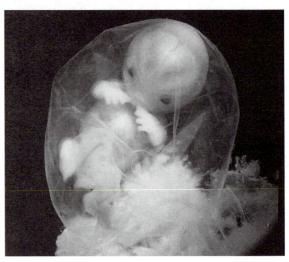

(d) 12 Weeks

4. If available, examine a model or chart of a 12-day-old embryo. Identify each germ layer if possible. Briefly list the fate of each germ layer.

_____

_____

_____

_____

C. Extraembryonic Membranes and the Placenta

Four extraembryonic membranes develop from the germ layers: the yolk sac, amnion, chorion, and allantois. These membranes lie outside the embryonic disc and provide protection and nourishment for the embryo and fetus. The **yolk sac** is the first membrane to appear around the tenth day (see Figures 20.6 and 20.8). Initially, cells from the hypoblast form a pouch under the blastodisc. Mesoderm reinforces the sac, and blood vessels appear. As the syncytial trophoblast develops more villi, the yolk sac's importance in providing nourishment for the embryo diminishes.

While the yolk sac is forming, cells in the epiblast migrate to line the inner surface of the amniotic cavity with a membrane called the **amnion**, the "water bag." As with the yolk sac, the amnion is soon reinforced with mesoderm tissue. Early embryonic growth continues, and the amnion mushrooms and envelops the entire embryo in a protective environment of **amniotic fluid**, as illustrated in Figure 20.8e.

The **allantois** develops from the endoderm and mesoderm near the base of the yolk sac (see Figure 20.8b). The allantois forms part of the urinary bladder and contributes to the **body stalk**, the tissue between the embryo and the developing chorion. Blood vessels pass through the body stalk and into the villi protruding into the lacunae of the endometrium.

The outer extraembryonic membrane is the **chorion**, formed by the cellular trophoblast and mesoderm. The chorion completely encases the embryo and the blastocoele. In the third week of growth, the chorion extends **chorionic villi** and blood vessels into the endometrial lacunae to establish the structural framework for the development of the **placenta** (see Figure 20.8).

The placenta, detailed in Figure 20.8e, is a temporary organ through which nutrients, blood gases, and wastes are exchanged between the mother and the embryo. The embryo is connected to the placenta by the body stalk constructed of the allantois and blood vessels. The **yolk stalk** where the yolk sac attaches and the body stalk together form the **umbilical cord**. Inside the umbilical cord are two **umbilical arteries**, which transport deoxygenated blood into the placenta, and a single **umbilical vein**, which returns oxygenated blood to the embryo.

The placenta does not completely surround the embryo. By the fourth week of development, the chorionic villi have enlarged only where they face the uterine wall, and villi that face the uterine cavity become insignificant. The placenta is in contact with the area of the endometrium called the **decidua basalis**, where the chorionic villi occur. The rest of the endometrium, where villi are absent, isolates the embryo from the uterine cavity and is called the **decidua capsularis**.

Laboratory Activity   EXTRAEMBRYONIC MEMBRANES AND THE PLACENTA _____

MATERIALS

embryology models
embryology chart
placenta model or biomount

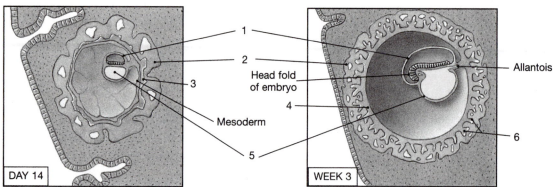

(a) Migration of mesoderm around the inner surface of the trophoblast creates the chorion. Mesodermal migration around the outside of the amniotic cavity, between the ectodermal cells and the trophoblast, creates the amnion. Mesodermal migration around the endodermal pouch below the blastodisc creates the definitive yolk sac.

(b) The embryonic disc bulges into the amniotic cavity at the head fold. The allantois, an endodermal extension surrounded by mesoderm, extends toward the trophoblast.

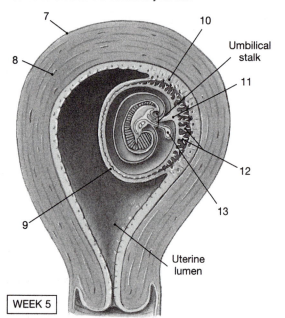

(c) The developing embryo and extraembryonic membranes bulge into the uterine cavity. The trophoblast pushing out into the uterine lumen remains covered by endometrium, but no longer participates in nutrient absorption and embryo support. The embryo moves away from the placenta, and the body stalk and yolk stalk fuse to form an umbilical stalk.

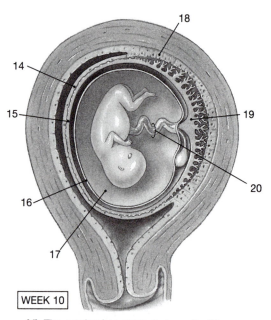

(d) The amnion has expanded greatly, filling the uterine cavity. The fetus is connected to the placenta by an elongate umbilical cord that contains a portion of the allantois, blood vessels, and the remnants of the yolk stalk.

1. _____	8. _____	15. _____
2. _____	9. _____	16. _____
3. _____	10. _____	17. _____
4. _____	11. _____	18. _____
5. _____	12. _____	19. _____
6. _____	13. _____	20. _____
7. _____	14. _____	

●**Figure 20.8**
Embryonic Membranes and Placenta Formation.

PROCEDURES

1. Label and review the extraembryonic membranes in Figure 20.8.

2. Examine a model or chart that details the extraembryonic membranes. Locate the yolk sac and the amnion. How do each of these membranes form, and what is the function of each?

   _____

   _____

3. Locate the allantois and the chorion. How do each of these membranes form? Describe the chorionic villi and their significance to the embryo.

   _____

   _____

4. Examine the model or biomount of a placenta. How do the surfaces of the placenta appear? Is the amniotic membrane attached to the placenta?

   _____

   _____

5. Examine the umbilical cord attached to the placenta. Describe the vascular anatomy in the cord.

   _____

   _____

D. The Second and Third Trimesters and Birth

Growth during the second trimester is fast, and the fetus doubles in size and substantially increases its weight. As the fetus grows, the uterus expands and displaces the maternal abdominal organs. Figure 20.9 details the uterine size compared with the gestational month. Quarters for the fetus become cramped toward the end of pregnancy. The fetus begins to move as the muscular system becomes functional, and articulations begin to form in the skeleton. The nervous system organizes the neural tissue that developed in the first trimester, and many sensory organs complete their formation. During the third trimester, weight increases, and all organ systems complete their development and become partially operational. The fetus responds to sensory stimuli such as a hand rubbing across the mother's swollen abdomen.

Birth, or **parturition**, involves muscular contractions of the uterine wall to expel the fetus into the outside world. Delivering the fetus is much like pulling on a turtle neck sweater. Muscle contractions must stretch the cervix of the uterus over the fetus's head, pulling the uterine wall thinner as the fetus passes into the lower birth canal, the vagina. Once true labor contractions begin, positive feedback mechanisms operate to increase the frequency and force of uterine contractions. Stretching of the uterine wall causes secretion of **oxytocin** from the posterior pituitary gland. Oxytocin stimulates contraction of the myometrium of the uterus. The hormone **relaxin** causes the symphysis pubis to become less rigid.

Labor is divided into three stages: the dilation, expulsion, and placental stages, each illustrated in Figure 20.10. The **dilation stage** begins at the onset of true labor contractions. The cervix of the uterus dilates, and the fetus begins to travel down the cervical canal. To be maximally effective at dilation, the contractions must be less than ten minutes apart. Each contraction lasts approximately one minute and spreads from the upper cervix downward to efface or thin the cervix for delivery. Contractions usually rupture the amnion, and amniotic fluid flows out of the uterus and the vagina.

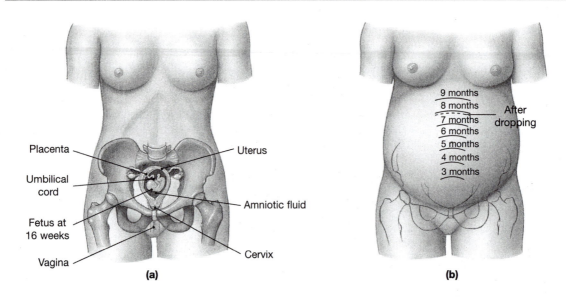

Placenta

Umbilical
cord

Fetus at
16 weeks

Vagina

Uterus

Amniotic fluid

Cervix

(a)

9 months
8 months
7 months
6 months
5 months
4 months
3 months

After
dropping

(b)

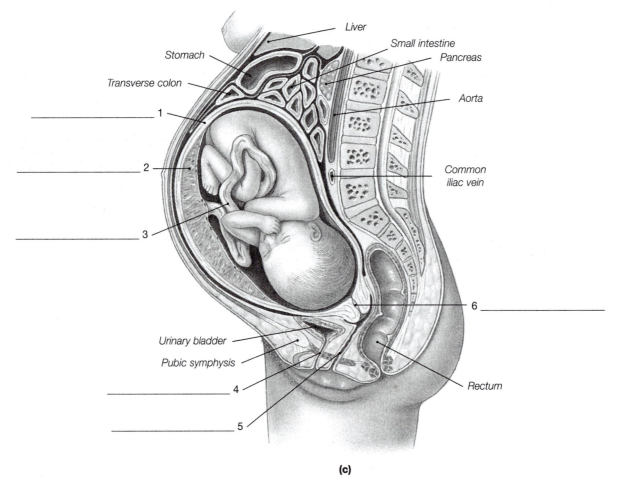

Liver

Small intestine

Pancreas

Stomach

Transverse colon

Aorta

1

2

Common
iliac vein

3

6

Urinary bladder

Pubic symphysis

4

Rectum

5

(c)

●**Figure 20.9**
Growth of the Fetus and Changes in Uterine Size

(a) Pregnancy at 16 weeks, showing position of uterus and placenta. **(b)** Pregnancy at
third month to full term, showing the position of the uterus within the abdomen.
(c) Pregnancy at full term. Note the position of the uterus and full-term fetus within
the abdomen and the displacement of abdominal organs.

The **expulsion stage** occurs when the cervix is dilated completely, usually to 10 cm, and the fetus passes through the cervix and the vagina. The expulsion stage usually lasts less than two hours and results in the **delivery** or birth of the newborn. Once the baby is breathing on its own, the umbilical cord is cut, and the newborn must now rely on its own organ systems to survive.

During the **placental stage**, the uterus continues to contract to break the placenta free of the endometrium and deliver it out of the body as the **afterbirth**, illustrated in Figure 20.10c. This stage is usually short, and many women deliver the afterbirth within five to ten minutes after the birth of the fetus.

Laboratory Activity   THE SECOND AND THIRD TRIMESTERS AND BIRTH_____

MATERIALS

second and third trimester models
parturition models

PROCEDURES

1. Label Figures 20.9 and 20.10.
2. Examine the models of a second and third trimester fetus. Describe how the fetus is positioned in the uterus. If shown, describe the location of the amnion and the placenta.

   _____

   _____

3. Examine the models of parturition. Describe the contractions that force the fetus out of the uterus.

   _____

   _____

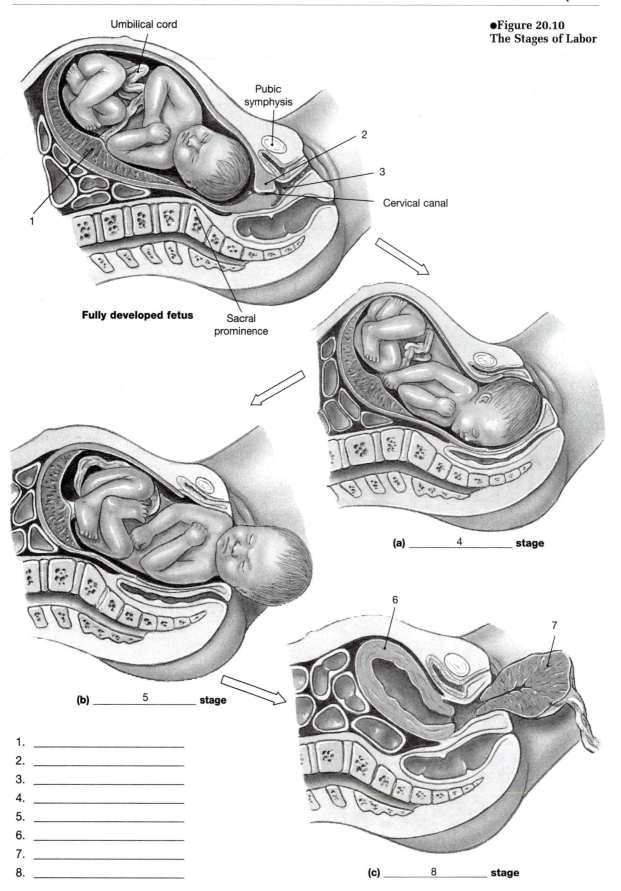

●Figure 20.10
The Stages of Labor

Umbilical cord

Pubic
symphysis

2

3

Cervical canal

1

Fully developed fetus   Sacral
                              prominence

(a) _____4_____ **stage**

(b) _____5_____ **stage**

6

7

(c) _____8_____ **stage**

1. _____
2. _____
3. _____
4. _____
5. _____
6. _____
7. _____
8. _____

HUMAN DEVELOPMENT
Exercise 20 Laboratory Report

A. Matching

Match each structure on the left with the correct description on the right.

1. _____ blastomere		A. has villi
2. _____ amnion		B. differentiates between epiblast and hypoblast
3. _____ allantois		C. cavity of blastocyst
4. _____ morula		D. forms part of urinary bladder
5. _____ blastocoele		E. becomes the ectoderm
6. _____ haploid		F. migration of cells
7. _____ chorion		G. water bag
8. _____ mesoderm		H. birth
9. _____ endoderm		I. cell produced by early cleavage
10. _____ epiblast		J. produces lining of respiratory tract
11. _____ parturition		K. solid ball of cells
12. _____ gastrulation		L. cell with half the number of chromosomes

B. Short Answer Questions

1. Describe the process of fertilization. Where does it occur? Why are so many spermatozoa required for the process?

2. Compare spermatogenesis and oogenesis.

3. How does the amniotic cavity form?

4. List the three stages of labor.

5. Discuss the formation of the three germ layers.

6. Describe the structure of the blastocyst.

7. How does a fetus obtain nutrients and gases from the maternal blood?

8. List the four extraembryonic membranes and their specific functions.

C. Completion

    1. Complete Figure 20.11 by identifying the cells produced during meiosis.

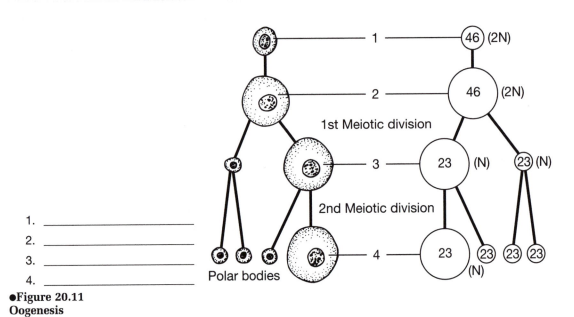

1. _____

2. _____

3. _____

4. _____

●**Figure 20.11**
Oogenesis

APPENDIX A
WEIGHTS AND MEASURES

TABLE 1	The U.S. System of Measurement		
Physical Property	*Unit*	*Relationship to Other U.S. Units*	*Relationship to Household Units*
Length	inch (in.)	1 in. = 0.083 ft	
	foot (ft)	1 ft = 12 in.	
		= 0.33 yd	
	yard (yd)	1 yd = 36 in.	
		= 3 ft	
	mile (mi)	1 mi = 5.280 ft	
		= 1,760 yd	
Volume	fluidram (fl dr)	1 fl dr = 0.125 fl oz	
	fluid ounce (fl oz)	1 fl oz = 8 fl dr	= 6 teaspoons (tsp)
		= 0.0625 pt	= 2 tablespoons (tbsp)
	pint (pt)	1 pt = 128 fl dr	= 32 tbsp
		= 16 fl oz	= 2 cups (c)
		= 0.5 qt	
	quart (qt)	1 qt = 256 fl dr	= 4 c
		= 32 fl oz	
		= 2 pt	
		= 0.25 gal	
	gallon (gal)	1 gal = 128 fl oz	
		= 8 pt	
		= 4 qt	
Mass	grain (gr)	1 gr = 0.002 oz	
	dram (dr)	1 dr = 27.3 gr	
		= 0.063 oz	
	ounce (oz)	1 oz = 437.5 gr	
		= 16 dr	
	pound (lb)	1 lb = 7000 gr	
		= 256 dr	
		= 16 oz	
	ton (t)	1 t = 2000 lb	

TABLE 2	The Metric System of Measurement

Physical Property Unit		Relationship to Standard Metric Units	Conversion to U.S. Units	
Length	nanometer (nm)	$1 \text{ nm} = 0.000000001 \text{ m } (10^{-9})$	$= 4 \times 10\text{-}8 \text{ in.}$	$25{,}000{,}000 \text{ nm} = 1 \text{ in.}$
	micrometer (µm)	$1 \text{ µm} = 0.000001 \text{ m } (10^{-6})$	$= 4 \times 10\text{-}5 \text{ in.}$	$25{,}000 \text{ mm} = 1 \text{ in.}$
	millimeter (mm)	$1 \text{ mm} = 0.001 \text{ m } (10^{-3})$	$= 0.0394 \text{ in.}$	$25.4 \text{ mm} = 1 \text{ in.}$
	centimeter (cm)	$1 \text{ cm} = 0.01 \text{ m } (10^{-2})$	$= 0.394 \text{ in.}$	$2.54 \text{ cm} = 1 \text{ in.}$
	decimeter (dm)	$1 \text{ dm} = 0.1 \text{ m } (10^{-1})$	$= 3.94 \text{ in.}$	$0.25 \text{ dm} = 1 \text{ in.}$
	meter (m)	standard unit of length	$= 39.4 \text{ in.}$	$0.0254 \text{ m} = 1 \text{ in.}$
			$= 3.28 \text{ ft}$	$0.3048 \text{ m} = 1 \text{ ft}$
			$= 1.09 \text{ yd}$	$0.914 \text{ m} = 1 \text{ yd}$
	dekameter (dam)	$1 \text{ dam} = 10 \text{ m}$		
	hectometer (hm)	$1 \text{ hm} = 100 \text{ m}$		
	kilometer (km)	$1 \text{ km} = 1000 \text{ m}$	$= 3280 \text{ ft}$	
			$= 1093 \text{ yd}$	
			$= 0.62 \text{ mi}$	$1.609 \text{ km} = 1 \text{ mi}$
Volume	microliter (µl)	$1 \text{ µl} = 0.000001 \text{ l } (10^{-6})$		
		$= 1 \text{ cubic millimeter } (\text{mm}^3)$		
	milliliter (ml)	$1 \text{ ml} = 0.001 \text{ l } (10^{-3})$	$= 0.03 \text{ fl oz}$	$5 \text{ ml} = 1 \text{ tsp}$
		$= 1 \text{ cubic centimeter } (\text{cm}^3 \text{ or cc})$		$15 \text{ ml} = 1 \text{ tbsp}$
				$30 \text{ ml} = 1 \text{ fl oz}$
	centiliter (cl)	$1 \text{ cl} = 0.01 \text{ l } (10^{-2})$	$= 0.34 \text{ fl oz}$	$3 \text{ cl} = 1 \text{ fl oz}$
	deciliter (dl)	$1 \text{ dl} = 0.1 \text{ l } (10^{-1})$	$= 3.38 \text{ fl oz}$	$0.29 \text{ dl} = 1 \text{ fl oz}$
	liter (l)	standard unit of volume	$= 33.8 \text{ fl oz}$	$0.0295 \text{ l} = 1 \text{ fl oz}$
			$= 2.11 \text{ pt}$	$0.473 \text{ l} = 1 \text{ pt}$
			$= 1.06 \text{ qt}$	$0.946 \text{ l} = 1 \text{ qt}$
Mass	picogram (pg)	$1 \text{ pg} = 0.000000000001 \text{ g } (10^{-12})$		
	nanogram (ng)	$1 \text{ ng} = 0.000000001 \text{ g } (10^{-9})$		
	microgram (µg)	$1 \text{ µg} = 0.000001 \text{ g } (10^{-6})$	$= 0.000015 \text{ gr}$	$66{,}666 \text{ mg} = 1 \text{ gr}$
	milligram (mg)	$1 \text{ mg} = 0.001 \text{ g } (10^{-3})$	$= 0.015 \text{ gr}$	$66.7 \text{ mg} = 1 \text{ gr}$
	centigram (cg)	$1 \text{ cg} = 0.01 \text{ g } (10^{-2})$	$= 0.15 \text{ gr}$	$6.7 \text{ cg} = 1 \text{ gr}$
	decigram (dg)	$1 \text{ dg} = 0.1 \text{ g } (10^{-1})$	$= 1.5 \text{ gr}$	$0.67 \text{ dg} = 1 \text{ gr}$
	gram (g)	standard unit of mass	$= 0.035 \text{ oz}$	$28.35 \text{ g} = 1 \text{ oz}$
			$= 0.0022 \text{ lb}$	$453.6 \text{ g} = 1 \text{ lb}$
	dekagram (dag)	$1 \text{ dag} = 10 \text{ g}$		
	hectogram (hg)	$1 \text{ hg} = 100 \text{ g}$		
	kilogram (kg)	$1 \text{ kg} = 1000 \text{ g}$	$= 2.2 \text{ lb}$	$0.453 \text{ kg} = 1 \text{ lb}$
	metric ton (kt)	$1 \text{ mt} = 1000 \text{ kg}$	$= 1.1 \text{ t}$	
			$= 2205 \text{ lb}$	$0.907 \text{ kt} = 1 \text{ t}$

Temperature	Centigrade	Fahrenheit
Freezing point of pure water	0°	32°
Normal body temperature	36.8°	98.6°
Boiling point of pure water	100°	212°
Conversion	$^\circ\text{C} \to {}^\circ\text{F:} \quad {}^\circ\text{F} = (1.8 \times {}^\circ\text{C}) + 32$	$^\circ\text{F} \to {}^\circ\text{C:} \quad {}^\circ\text{C} = ({}^\circ\text{F} - 32) \times 0.56$